Studies in Computational Intelligence

Volume 1132

The series "Studies in Computational Intelligence" (SCI) publishes new developments and advances in the various areas of computational intelligence—quickly and with a high quality. The intent is to cover the theory, applications, and design methods of computational intelligence, as embedded in the fields of engineering, computer science, physics and life sciences, as well as the methodologies behind them. The series contains monographs, lecture notes and edited volumes in computational intelligence spanning the areas of neural networks, connectionist systems, genetic algorithms, evolutionary computation, artificial intelligence, cellular automata, self-organizing systems, soft computing, fuzzy systems, and hybrid intelligent systems. Of particular value to both the contributors and the readership are the short publication timeframe and the world-wide distribution, which enable both wide and rapid dissemination of research output.

Indexed by SCOPUS, DBLP, WTI Frankfurt eG, zbMATH, SCImago.

All books published in the series are submitted for consideration in Web of Science.

D. P. Acharjya · Kun Ma
Editors

Computational Intelligence in Healthcare Informatics

Springer

Editors
D. P. Acharjya
School of Computer Science
and Engineering
Vellore Institute of Technology University
Vellore, India

Kun Ma
School of Information Science
and Engineering
University of Jinan
Jinan, China

ISSN 1860-949X ISSN 1860-9503 (electronic)
Studies in Computational Intelligence
ISBN 978-981-99-8852-5 ISBN 978-981-99-8853-2 (eBook)
https://doi.org/10.1007/978-981-99-8853-2

This Springer imprint is published by the registered company Springer Nature Singapore Pte Ltd.
The registered company address is: 152 Beach Road, #21-01/04 Gateway East, Singapore 189721, Singapore

Dedicated to
My wife Asima Nanda, and loving children
Aditi and Aditya
D. P. Acharjya

Preface

The world's population is increasing daily, leading to several real-world problems. One such problem is health care. In the present situation, health care is essential for ordinary people as they fall sick often due to various factors. Thus, recovery and cure at an earlier stage of health treatment is crucial. It makes physicians, professionals, integrators, stakeholders, and researchers worldwide look for cost-effective, innovative, and technology-based solutions for the patient community. All these lead to a diagnosis of diseases, illness prevention, management, and patient care. Simultaneously, it leads to extensive data analysis.

A considerable amount of healthcare data is collected worldwide for various diseases. The ability to inspect these data without using automated techniques is very challenging. These clinical data contain many uncertainties, and they also have vagueness. Besides, much information is hidden in the enormous data. Obtaining information from these data without any computational intelligence techniques is tough. So, it is demanded for a new generation of computational intelligence to assist humans in distilling knowledge, self-learning, and rule generation from massive data. It, in turn, helps the physicians to make appropriate decisions. Computational intelligence, soft computing, knowledge discovery in databases, decision-making, cloud computing, parallel and distributed computing, network security, data mining, knowledge representation, deep learning, and machine learning are the fields that have evolved into an essential and active area of healthcare research because of challenges associated with the healthcare problems of discovering intelligent solutions for smart inspect of massive data.

The prime objective of this edited book is to provide the advancements of computational intelligence in perception with healthcare applications. Besides, the concepts, theory, and application in various domains of healthcare systems, including decision-making in healthcare management, disease diagnosis, and electronic health records, will be presented lucidly. To achieve these objectives, theoretical advances and their applications to healthcare problems will be stressed. This has been done to make the edited book more flexible and stimulate further research interest. The book is partitioned into four parts: Theoretical Foundation of Computational Intelligence Techniques, Computational Intelligence in Analyzing Health Data, Computational

Intelligence in Electronic Health Record, and Computational Intelligence in Ethical Issues in Healthcare. This edited volume is expected to provide a launch pad for future research in healthcare informatics.

The first part deals with the Theoretical Foundation of Computational Intelligence Techniques and comprises six chapters. A network-based method is proposed in chapter "Refining Metabolic Network by Fuzzy Matching of Metabolite Names for Improving Metabolites Ranking Toward the Diseases" for identifying and ranking disease-related metabolites. The metabolite's impact on the complex diseases of humans is a very crucial one. The proposed method uses Infer Disease Similarity (InfDisSim) and miRNA Similarity (MISIM) to calculate the disease and metabolite similarities. Further, MISIM is improved by presenting the Subcellular Localization Weight-Based MISIM (SLWBMISM), where subcellular localization of metabolites is considered to find the more similar metabolites. A fuzzy matching algorithm is adopted to identify the identical names from the two models to reduce the computational complexity of SLWBMISM. Further, exact terms of metabolite similarity are calculated using SLWBMISM, and then the process of reconstructing the metabolic network based on disease similarity and metabolite similarity is carried out.

Healthcare and medical-related datasets need to be more balanced. A classification dataset is considered imbalanced if there are significantly fewer instances of one class than others. Because standard classifiers tend to favor the decision classes in many cases, such unbalanced datasets require particular consideration. In medicine, the type with fewer instances might represent a rare or unusual incident that might be interesting. Ignoring the imbalance hurts classifier performance, which hinders the identification of rare cases in areas like disease screening, severity analysis, spotting adverse medication reactions, grading cancer malignancy, and identifying unusual, chronic illnesses in the general population. Chapter "Learning from Imbalanced Data in Healthcare: State-of-the-Art and Research Challenges" discusses the state-of-the-art research challenges and learning from imbalanced data in health care.

Optimization is necessary while solving real-world applications. But, in many real-life problems, the traditional approach is not applicable. Therefore, conventional optimization techniques have been directed toward metaheuristic approaches, and it is standard while solving complex optimization problems in science, engineering, and technology. Keeping the broad view in mind, chapter "A Review on Metaheuristic Approaches for Optimization Problems" presents a comprehensive view of metaheuristic approaches to optimization. The study explores all dimensions of metaheuristic approaches, their representatives, improvement and hybridization, successful application, research gaps, and future direction.

Numerous chronic illnesses have recently harmed human health. The development of technology has shown that early disease treatment is possible. However, while some infections can be prevented, they cannot always be treated entirely. Among them is diabetes. Numerous consequences result from untreated and undiagnosed diabetes. The patient visits a diagnostic facility and sees a doctor at the end of the laborious diagnostic process. However, the development of technical solutions resolves this significant issue. The study presented in chapter "Diabetes Prediction: A Comparison Between Generalized Linear Model and Machine Learning" forecasts whether a patient would develop diabetes. To detect diabetes early, this chapter examines the outcomes of two classification algorithms: a binary logistic regression model and a support vector machine. Further, a comparison of the algorithms is also presented.

Chapter "Prediabetes Prediction Using Response Surface Methodology and Probabilistic Neural Networks Model in an Ethnic South Indian Population" presents a novel method of prediabetes prediction using statistical models and neural network models for early diagnosis. The proposed approach integrates Pearson correlation analysis, response surface method, artificial neural network, and probabilistic neural networks for data analysis. A case study is presented to analyze the model. The subjects' biochemical and anthropometric variables are analyzed, and glycemic levels are predicted. It found that salivary glucose, HbA1C, waist circumference, BMI, and LDL were predictors of prediabetes.

Chapter "Localization and Classification of Brain Tumor Using Multi-layer Perceptron" proposes a multi-layer perceptron technique to detect and classify the brain tumor type. Brain tumors have become a significant cause of death in recent years. The proposed approach can analyze and organize the heterogeneous data into different tumor types. Using MRI images without any surgery scans gives the structure of the brain that helps in further processing in the detection of the brain tumor.

Healthcare data generated by various means increases exponentially. Besides, the healthcare data contains uncertainties. Analyzing these uncertainties present in data is a challenging task. Thus, computational intelligence plays a significant role in handling uncertainties. With this objective, the second part of this book emphasizes the importance of Computational Intelligence in Analyzing Health Data. Chapter "Information Retrieval from Healthcare Information System" provides the advancements of computational intelligence, big data, and the Internet of Things in perception with healthcare applications by retrieving information from healthcare information systems. However, collecting clinical data during an emergency is a critical task. Therefore, a computational intelligence framework is proposed to facilitate ubiquitous content access. The proposed approach delivers rapid and effective care for a variety of patients as compared to non-relational databases.

The data collected from various healthcare domains are only sometimes structured. Analyzing these unstructured data is a critical task. Therefore, it is necessary to find associations between attributes and attribute values. It leads to association rule mining. It is a powerful tool that reveals hidden relationships among characteristics and statistically validates those already known. These relationships can help in understanding diseases and their causes in a better way, which in turn will help to prevent them. Chapter "Association Rule Mining for Healthcare Data Analysis" provides complete information about association rule mining algorithms used in the healthcare information system. The chapter critically analyzes existing association rules relationships among various diseases and discover strong association rule from the healthcare data.

The lethal disorder known as cancer is brought on by the body's aberrant cell multiplication. The use of microarray technology in the diagnosis of such critical diseases has grown. It is crucial to develop a rapid and accurate approach to cancer detection and drug discovery that aids in removing the infection from the body. Precise categorization of the dataset is challenging due to the vast number of variables and small sample size of the raw microarray gene expression data. These microarray genes also contain a noisy, pointless, and redundant gene, which leads to subpar diagnosis and categorization. Researchers extracted the most critical components of gene expression data to accomplish this using several machine learning approaches. Chapter "Feature Selection and Classification of Microarray Cancer Information System: Review and Challenges" presents a comprehensive microarray gene expression data study with feature selection and classification algorithms. It, in turn, helps to diagnose the disease properly.

The illnesses of osteoporosis and osteopenia increasingly strike people more frequently. Earlier disease prediction is essential for the ailments listed above because diet and genetics have been identified as the primary risk factors. Older persons and women who have passed menopause are typically more affected by osteoporosis and osteopenia. The prevention of bone fragility and fractures depends on early detection of osteoporosis and osteopenia. Fragility and fracture of bones are caused by low bone mineral density. A predictive analysis of the illnesses mentioned above uses clinical data such as bone mineral density and radiographic images. When calculating T-Score and Z-Score values, which can be used to classify osteoporosis and osteopenia, bone mineral density values are helpful. Chapter "Early Detection of Osteoporosis and Osteopenia Disease Using Computational Intelligence Techniques" presents a study of various computational approaches to predict osteoporosis and osteopenia diseases and concludes with necessary metrics.

Due to a lack of adequate medical facilities, one of the problematic duties is cancer detection. Patients with cancer must receive treatment and early detection to survive. Most genetic abnormalities that result in cancer tumors are the primary cause of the disease. It takes a lot of time to identify genetic mutations. The molecular pathologist needs to be improved on this. A list of gene variants is chosen by a molecular pathologist for manual analysis. Although there are nine classes in which the clinical evidence strips fall, the classification basis still needs to be determined. Based on clinical data, this implementation suggests a multi-class classifier to categorize genetic variants. The clinical text of gene mutation evidence is analyzed using natural language processing. Chapter "Pathway to Detect Cancer Tumor by Genetic Mutation" mainly focuses on studying nine genetic variations considered multi-class classification problems. Each data point in the information system is classified among the nine classes of gene mutation to assess the risk of cancer and the medications.

Telehealth delivers, administers, and supervises medical services through information and communication technologies. Telemedicine is now relevant in the pandemic since it has great potential to protect patients and medical professionals. Additionally, it restricts patients' social interactions, which lessens the COVID-19 pandemic virus's ability to propagate. Chapter "A Knowledge Perception: Physician and Patient Toward Telehealth in COVID-19" presents a methodology focusing on telemedicine while generalizing how patients and doctors view telehealth. The findings talk about how patients and medical professionals might adapt to telemedicine. Additionally, their perspective and level of pleasure were assessed using various statistical conclusions made with SPSS.

An electronic health record is an electronic version of a patient's medical history maintained by the service provider over time. These data have many uncertainties. Therefore, using computational intelligence techniques while analyzing these data is essential to get meaningful information. The third part of this book highlights using Computational Intelligence in Electronic Health Record. Modern computing methods have made storing and gathering medical data more accessible for precise medical diagnosis. The accuracy of disease diagnosis, the length of the diagnosis process, and the mortality rate are all improved by applying various computational techniques. Using cutting-edge learning techniques to increase efficacy and therapeutic importance is necessary. In healthcare systems, machine learning techniques are frequently utilized for disease screening, risk assessment, prognostication, and decision-making. The results of the classification of cardiovascular disease data using machine learning depend significantly on the sample sizes, characteristics, location of data collection, performance measures, and machine learning techniques used. The effectiveness of several machine learning algorithms concerning cardiovascular disease is covered in chapter "Classification of Cardiovascular Disease Information System Using Machine Learning Approaches".

The most challenging problems in the field of medical image processing are considered to be automatic computer-based diagnosis in medical imaging. The issue of automatically choosing to recognize the subtle edges of medical images predominantly influences pattern recognition, machine learning, and deep learning algorithms. Chapter "Automatic Edge Detection Model of MR Images Based on Deep Learning Approach" presents a deep learning strategy to automatically find fine edges in medical MR images contaminated by noise. Medical image processing uses detection based on a deep learning approach to identify specific organs or tissues and define regions of interest. Metrics such as the peak signal-to-noise ratio (PSNR), structural similarity index measure (SSIM), and edge-keeping index are used to assess how well the strategy suggested performs.

One of the humans' most significant physiological signs is the heart and lungs' sound. When you breathe in and out, your lungs make sounds. A stethoscope can be used to identify these noises. Your breath sounds could be typical or strange. Additionally, a condition might be categorized based on the noise. Chapter "Lung Disease Classification Based on Lung Sounds—A Review" describes the various approaches to organizing lung sounds to identify lung disorders such as chronic obstructive pulmonary disease (COPD) and pneumonia. At first, simple signs are thought to indicate pneumonia. It is also mentioned how computational intelligence is used in health care to address the treatment. Machine learning algorithms are available that can identify lung illnesses from lung sounds. This chapter provides information on different lung disease classifiers.

There have been numerous attempts to analyze and forecast upcoming COVID-19 cases using primary data. Inferential methodology, one of the most popular data science methodologies for studying events like COVID-19 time series analysis, is the foundation of the current work. SIR, ARIMA models, and forecasting are used to analyze and predict COVID-19 cases. Chapter "Analysis of Forecasting Models of Pandemic Outbreak for the Districts of Tamil Nadu" shows the real-time data analysis on data collected from districts in Tamil Nadu.

Physically challenged people can benefit from using a brain-computer interface device to analyze and diagnose various health issues. A significant component of the brain-computer interface system is the signal processing module. The EEG-based brain-computer interface system aims to extract and convert brain activity into command signals that benefit people with physical disabilities. Chapter "Suppression of Artifacts from EEG Recordings Using Computational Intelligence" offers an in-depth analysis of the fundamental concepts underlying various denoising techniques and concisely summarizes some of the pioneers' initiatives. Additionally, the eye blink artifacts are filtered using the EMD, AVMD, SWT, and VME-DWT methods in the comparison analysis. An innovative approach for locating and examining signals' differentiating properties is developed via computational intelligence. Computational intelligence should be used to effectively minimize noise in an EEG-based brain-computer interface system.

Health informatics combines data analytics and health care. Its main goal is to effectively gather, protect, and deliver high-quality health care with the help of information technology, all of which positively influence the relationship between patients and doctors. A few results of information technology's emphasis on health care data analysis include classification, clustering, feature selection, diagnosis, prediction, and decision support systems. Chapter "Rough Computing in Healthcare Informatics" mainly emphasizes rough computing techniques utilized in healthcare informatics. It consequently supports the medical professionals' choices.

Medical information systems are generated rapidly due to various means. Therefore, preserving the privacy of these information systems is necessary, leading to many ethical issues in health care. The ethical problems include care quality balancing and its efficiency, improving care, building the healthcare workforce, improvement of medication, and allocation of donor organs. The fourth part discusses the use of Computational Intelligence in Ethical Issues in Healthcare. Studying a particular clinical problem, obtaining information, and coming to a conclusion using logic are all part of medical ethics. The healthcare industry primarily supports patients, communities, and healthcare practitioners in choosing between different treatment options, medical treatments, and other concerns that crop up. Chronic diseases like cancer are a serious concern today. Using an improper dosage during cancer therapy might have several adverse side effects. Anti-cancer medications are used during cancer chemotherapy, although they can have harmful side effects. Analytical investigations have been carried out in chapter "Ethical Issues on Drug Delivery and Its Impact in Healthcare" to plan pharmaceutical dosage schedules for the required outcomes. As mentioned, it provides a complete elitist genetic algorithm for the problem.

The healthcare industry generates a considerable amount of patient data every second. Advanced deep learning models could be used to analyze medical data, especially patient data, but the private nature of health data restricts utilization. It is necessary to gather enormous amounts of heterogeneous data, frequently only feasible through multi-institutional partnerships. Multi-institutional studies are one technique to build sizable central repositories. This approach is restricted to personal information protection, intellectual property, data identification, standards, and data storage while exchanging data. These difficulties have made cloud data storage more and more practical. Chapter "Privacy-Preserving Deep Learning Models for Analysis of Patient Data in Cloud Environment" discusses the various privacy-protecting cloud approaches for exchanging medical records.

The deep learning approach to machine learning, which has lately gained popularity as a powerful tool for machine learning with the potential to revolutionize artificial intelligence, is built on artificial neural networks. The technology's ability to generate the best high-level features and automatic semantic interpretation from the incoming data, along with rapid advancements in processing power, quick data storage, and parallelization, have all contributed to its early acceptance. Chapter "Computational Intelligence Ethical Issues in Health Care" provides a comprehensive, up-to-date summary of deep learning research in health informatics and critically assesses the approach's relative benefits, potential drawbacks, and future prospects. Deep learning applications in translational bioinformatics, medical imaging, ubiquitous sensing, medical informatics, and public health are discussed in this chapter.

The understanding of how the brain functions has been greatly aided by magnetic resonance imaging (MRI), which is also used to diagnose neurological illnesses like Alzheimer's disease (AD), Parkinson's disease, and schizophrenia. Intelligent algorithms are required to analyze such complicated, high-dimensional data. Chapter "Advances in Deep Learning for the Detection of Alzheimer's Disease Using MRI—A Review" discusses the latest works in the deep learning-assisted MRI identification of AD.

Many researchers in different organizations across the globe have started research in computational intelligence in healthcare informatics. Simultaneously, many data analysis and cloud computing techniques are progressing at the other end. The fusion of computational intelligence and these latest techniques and technologies will take it to a newer dimension. We strove to keep the book reader-friendly to keep abreast with this development in a cohesive manner. The main objective is to bring most of the significant results in the area, as mentioned above, in a precise way so that it can serve as a handbook for many researchers. This edited book will help researchers interested in computational intelligence in healthcare informatics to keep insight into recent advances and applications.

Vellore, India — Dr. D. P. Acharjya
Jinan, China — Dr. Kun Ma

Acknowledgments

Let us rise and be thankful, for if we didn't learn a lot at least we learned a little, and if we didn't learn a little, at least we didn't get sick, and if we got sick, at least we didn't die; so, let us all be thankful.

—Gautama Buddha

It is with great sense of satisfaction that we present our edited book and wish to express our views to all those who helped us both direct and indirect ways to complete this project. First and foremost, we praise and heart fully thank the almighty God, which has been unfailing source of strength, comfort, and inspiration in the completion of this project.

While writing, contributors have referred to several books and journals, and we take this opportunity to thank all those authors and publishers. We are extremely thankful to the reviewers for their constant support during the process of evaluation. Special mention should be made of the timely help given by different persons during the project work, those whose names are not mentioned here. Foremost, we thank the series editor "Janusz Kacprzyk" for giving this opportunity to carry out the project under the series "studies in computational intelligence". We would like to express our sincere thanks to Aninda Bose, and Thomas Ditzinger for providing us an environment and resources for carrying out our research project work. Last but not the least, we would like to thank the production team of *Springer-Verlag, USA* for encouraging us and extending their cooperation and help for a timely completion of this edited book. We trust and hope that it will be appreciated by many readers.

Contents

Editors and Contributors

About the Editors

D. P. Acharjya received his Ph.D. in computer science from Berhampur University, India; M. Tech. degree in computer science from Utkal University, India; and M. Sc. from NIT, Rourkela, India. He has received a Gold Medal from NIT Rourkela. Currently, he is working as a professor in the School of Computer Science and Engineering, at VIT University, Vellore, India. He has authored more than 100 national and international journals, conference papers, and book chapters to his credit. He is acting as an editorial board member of various journals and reviewer of IEEE Fuzzy Sets and Systems, Applied Soft Computing (Elsevier), and Knowledge base Systems (Elsevier). He is associated with many professional bodies CSI, ISTE, IMS, AMTI, ISIAM, OITS, IACSIT, CSTA, IEEE, and IAENG. He was a founder secretary of OITS Rourkela chapter. His current research interests include rough sets, formal concept analysis, knowledge representation, data mining, granular computing, and business intelligence.

Kun Ma received his Ph.D. degree in Computer Software and Theory from Shandong University, Jinan, Shandong, China, in 2011. He is an associate professor in School of Information Science and Engineering, University of Jinan, China, and Shandong Provincial Key Laboratory for Network-Based Intelligent Computing. He has served as the program committee member of various international conferences and reviewer for various international journals. He is the co-editor-in-chief of International Journal of Computer Information Systems and Industrial Management Applications (IJCISIM). He is the managing editor of Journal of Information Assurance and Security (JIAS). He is the editorial board member of International Journal of Grid and Utility Computing and International Journal of Intelligent Systems Design and Computing. His research interests include stream data processing, data intensive computing, and big data management. He has obtained 12 patents for inventions.

Contributors

D. P. Acharjya School of Computer Science and Engineering, VIT, Vellore, India

Rashi Agarwal Department of Computer Science and Engineering, Harcourt Butler Technical University, Kanpur, India

Sandhya Avasthi Department of Computer Science and Engineering, ABES Engineering College, Ghaziabad, India

Shree G B Bakhya Vellore Institute of Technology University, Vellore, Tamilnadu, India

Vishnu Vardhan Battu Annamalai University, Chidamvaram, Tamil Nadu, India

Santosini Bhutia Department of Computer Science and Engineering, SOA University, Bhubaneswar, Odisha, India

Rakesh Chandra Balabantaray IIIT, Bhubaneswar, Odisha, India

Ritu Chauhan Center for Computational Biology and Bioinformatics, Amity University, Noida, India

Ruplal Choudhary Department of Plant, Soil and Agriculture Systems, Southern Illinois University, Carbondale, USA

Raja Das Vellore Institute of Technology University, Vellore, Tamilnadu, India

Swati Das Rourkela Institute of Management Studies, Odisha, India

Rajashree Dash Siksha 'O' Anusandhan, Deemed to be University, Bhubaneswar, Odisha, India

Rasmita Dash Siksha 'O' Anusandhan, Deemed to be University, Bhubaneswar, Odisha, India

Daniel Dasig Graduate Studies College of Science and Computer Studies, De La Salle University, Dasmarinas, Philippines

Bikash Ranjan Debata Kirloskar Institute of Advanced Management Studies, Pune, India

Rema Gopalan CMR Institute of Technology, Bengaluru, India

Bushra Hamid University Institute of Information Technology, University of Arid Agriculture, Rawalpindi, Pakistan

S. Hariharan University of Madras, Chennai, India

Malaya Kumar Hota School of Electronics Engineering, Vellore Institute of Technology, Vellore, Tamil Nadu, India

Mamoona Humayun Department of Information Systems, College of Computer and Information Sciences, Jouf University, Sakakah, Saudi Arabia

P. Iswarya School of Advanced Sciences, Vellore Institute of Technology University, Vellore, India

N. Z. Jhanjhi School of Computer Science and Engineering, Taylor's University, Subang Jaya, Malaysia

M. Kalaiselvi Geetha Annamalai University, Chidamvaram, Tamil Nadu, India

Nimra Khan University Institute of Information Technology, University of Arid Agriculture, Rawalpindi, Pakistan

Shakir Khan Imam Mohammad ibn Saud Islamic University, Riyadh, Saudi Arabia

C. S. Khiran Kumar University of Maryland Baltimore County, Catonsville, Maryland, United States

Ajay Kumar Manipal University Jaipur, Jaipur, India

Yan Ma College of Information, Mechanical and Electrical Engineering, Shanghai Normal University, Shanghai, China

J. Mehena Department of Electronics & Telecommunication Engineering, DRIEMS, Cuttack, Odisha, India

Madhusmita Mishra Dr. Sudhir Chandra Sur Institute of Technology and Sports Complex, Kolkata, India

S. Mishra Department of Electronics and Communication Engineering, Adama Science and Technology University, Adama, Ethiopia

Aniruddha Mohanty Computer Science and Engineering, CHRIST (Deemed to be) University, Bengaluru, India

Anjali Mohapatra Department of Computer Science and Engineering, IIIT, Bhubaneswar, Odisha, India

Prabhujit Mohapatra School of Advanced Sciences, Vellore Institute of Technology University, Vellore, India

Norrima Mokthar Faculty of Engineering, University of Malaya, Kuala Lumpur, Malaysia

Kedar Nath Das NIT Silchar, Silchar, Assam, India

Subham Kumar Padhy Department of Computer Science and Engineering, IIIT, Bhubaneswar, Odisha, India

Shanti Priya Parida Idiap Research Institute, Martigny, Switzerland

Punyaban Patel Department of Computer Science and Engineering, CMR Technical Campus, Hyderabad, India

Riyam Patel Department of Computer Science and Engineering, SRM University, Chennai, India

Bichitrananda Patra Department of Computer Science and Engineering, SOA University, Bhubaneswar, Odisha, India

Sabyasachi Patra Department of Computer Science and Engineering, IIIT, Bhubaneswar, Odisha, India

Alok Ranjan Prusty DGT, RDSDE, NSTI(W), Kolkata, West Bengal, India

Porkodi Rajendran Bharathiar University, Coimbatore, India

T. Ramesh Vellore Institute of Technology, Vellore, Tamilnadu, India

Rasmita Rautray Siksha 'O' Anusandhan, Deemed to be University, Bhubaneswar, Odisha, India

Mitrabinda Ray Department of Computer Science and Engineering, SOA University, Bhubaneswar, Odisha, India

K. Madhusudhan Reddy College of Computing and Information Sciences, University of Technology and Applied Sciences Shinas, Shinas, Sultanate of Oman

Anandarup Roy Sarojini Naidu College for Women, Dum Dum, Kolkata, India

Debashis Roy Department of Computer and System Sciences, Visva Bharati University, Kolkata, India

Utpal Roy Department of Computer and System Sciences, Visva Bharati University, Kolkata, India

Najm Us Sama Faculty of Computer Science and IT, Universiti Malaysia Sarawak, Kota Samarahan, Malaysia

V. Santhi Vellore Institute of Technology, Vellore, Tamilnadu, India

Radha Saraswathy Vellore Institute of Technology University, Vellore, Tamilnadu, India

Aparajita Sengupta Center for Medical Biotechnology, Amity University, Noida, India

H. Sharan Prasad School of Advanced Sciences, Vellore Institute of Technology University, Vellore, India

Bommala Silpa School of Electronics Engineering, Vellore Institute of Technology, Vellore, Tamil Nadu, India

Borra Sivaiah Department of Computer Science and Engineering, CMR College of Engineering & Technology, Hyderabad, India

S Spelmen Vimalraj Bharathiar University, Coimbatore, India

Sreekumar Rourkela Institute of Management Studies, Odisha, India

Sidra Tahir University Institute of Information Technology, University of Arid Agriculture, Rawalpindi, Pakistan

Vijay Viswanathan M V Hospital for Diabetes Prof M Viswanathan Diabetes Research Centre, Royapuram, Chennai, India

Afsana Zannat Ahmed IIT Kharagpur, Kharagpur, West Bengal, India

Kartinah Zen Faculty of Computer Science and IT, Universiti Malaysia Sarawak, Kota Samarahan, Malaysia

Acronyms

AAL	Ambient Assisted Living
AAO	Artificial Algae Optimization
ABC	Artificial Bee Colony
ACACM	Adaptive Crisp Active Contour Models
ACO	Ant Colony Optimization
AD	Alzheimer's Disease
ADA	Americans with Disabilities Act
ADASYN	Adaptive Synthetic
ADNI	Alzheimer's Disease Neuroimaging Initiative
AEGA	Adaptive Elitist population-based Genetic Algorithm
AFNI	Analysis of Functional Neuro Image
AHE	Additively Homomorphic Encryption
AI	Artificial Intelligence
AIBL	Australian Imaging Biomarkers and Lifestyle
AMI	Acute Myocardial Infarction
ANCOVA	Analysis of Covariance
ANFIS	Adaptive Neuro-Fuzzy Inference System
ANN	Artificial Neural Network
ANOVA	Analysis of Variance
API	Application Programming Interface
AR	Auto Regressive
ARM	Association Rule Mining
ARPD	Asbestos-Related Pleural Disease
ASE	Audio Spectral Envelope
ATP	Adult Treatment Panel
AUC	Area Under the Curve
BBO	Biogeography-Based Optimization
BCI	Brain Computer Interface
BERT	Bidirectional Encoder Representations from Transformers
BET	Brain Extraction Tool
BIA	Bio-Inspired Algorithm

BIRCH	Balance Iterative Reducing and Clustering using Hierarchies
BLIMF	Band Limited Intrinsic Mode Functions
BMD	Bone Mineral Density
BMI	Body Mass Index
BN	Bayesian Network
BOLD	Blood Oxygen Level Dependent
BP	Blood Pressure
BPN	Backpropagation Network
BS	Base Station
BSI	Breath Sound Intensity
BSS	Blind Source Separation
CA	Cellular Automata
CAD	Coronary Artery Disease
CAF	Cyclophosphamide, Adriamycin, and Fluorouracil
CAP	Cyclic Alternating Pattern
CBC	Complete Blood Count
CC	Coarse Crackles
CCA	Canonical Correlation Analysis
CEGA	Complete Elitist Genetic Algorithm
CFO	Central Force Optimization
CHF	Congestive Heart Failure
CI	Computing Intelligence
CKD	Chronic Kidney Disease
CMF	Cyclophosphamide, Methotrexate, and Fluorouracil
CNN	Convolutional Neural Network
COBRE	Centre for Biomedical Research Excellence
COPD	Chronic Obstructive Pulmonary Disease
CORSA	Computerized Respiratory Sound Analysis
CPNN	Constructive Probabilistic Neural Network
CRNN	Convolutional Recurrent Neural Network
CS	Cuckoo Search
CSA	Cat Swarm Algorithm
CSF	Cerebrospinal Fluid
CSO	Cuckoo Search Optimization
CT	Computerized Tomography
CURES	Chennai Urban, Rural Epidemiology Study
CVD	Cardiovascular Disease
CWT	Continuous Wavelet Transform
DA	Discrimination Analysis
DBN	Deep Belief Network
DBSCAN	Density-Based Spatial Clustering of Applications with Noise
DE	Differential Evolution
DL	Deep Learning
DMN	Disease-associated Metabolite Network
DNA	Deoxyribonucleic Acid

DNN	Deep Neural Network
DOS	Deterministic Oscillatory Search
DPF	Diabetes Pedigree Function
DPH	Directorate of Public Health
DPN	Diabetic Peripheral Neuropathy
DSS	Decision Support System
DT	Decision Tree
DTCWT	Dual-Tree Complex Wavelet Transform
DWT	Discrete Wavelet Transform
EA	Evolutionary Algorithm
EC	Evolutionary Computing
ECC	Emergency Command Committee
EDA	Exploratory Data Analysis
EDTA	Ethylene Diamine Tetra acetic Acid
EE	Emergency Events
EEMD	Ensemble Empirical Mode Decomposition
EES	Emergency Event Severity
EHR	Electronic Health Records
EID	Explicit Identifiers
EKF	Extended Kalman Filter
EKI	Edge Keeping Index
ELM	Extreme Learning Machine
ELMNN	Extreme Learning Machine Neural Network
EMD	Empirical Mode Decomposition
EMR	Electronic Medical Records
EoR	Entity-oriented Resource
EPI	Echo Planar Image
ERNN	Elman Recurrent Neural Network
ES	Evolutionary Strategy
FBG	Fasting Blood Glucose
FC	Fine Crackles
FCM	Fuzzy C-Means
FD	Fisher Discriminant
FD	Fractal Dimension
FEAT	FMRIB Expert Analysis Tool
FHD	Family History of Diabetes
FILM	FMRIB Linear Model
FIPPS	Fair Information Practices Principles System
FIVD	Fixed Interval Variable Dose
FM	Fuzzy Matching
FMRIB	Functional Magnetic Resonance Imaging of the Brain
FN	Femoral Neck
FP	Frequent Pattern
FRS	Fuzzy Rough Set
FS	Fuzzy Systems

FSL	FMRIB Software Library
GA	Genetic Algorithm
GAN	Generative Adversarial Networks
GAP	Generative Adversarial Privacy
GBCO	Genetic Bee Colony Optimization
GBD	Global Burden of Disease
GBM	Gradient Boosting Machines
GBW	Gaussian Band Width
GCN	Graph Convolutional Network
GCNN	Graph Convolutional Neural Networks
GIP	Gaussian Interaction Profile
GLCM	Gray Level Co-occurrence Matrix
GMM	Gaussian Mixture Models
GP	Genetic Programming
GPAQ	Global Physical Activity Questionnaire
GPS	Global Positioning System
GSHPT	Grid Search-based Hyper Parameter Tuning
GUI	Graphical User Interface
GWO	Grey Wolf Optimizer
HCI	Human-Computer Interaction
HDDT	Hellinger Distance Decision Trees
HDL	High Density Lipoprotein
HDS	High Dimensional Sensing
HE	Homomorphic Encryption
HHS	Hilbert Huang Spectrum
HHT	Hilbert Huang Transform
HIMSS	Health Information and Management System Society
HIPAA	Health Insurance Portability and Accountability Act
HIT	Health Information Technology
HM	Harmony Memory
HMBA	Hybrid Monarch Butterfly Optimization
HMCR	Harmony Memory Consideration Rate
HMDB	Human Metabolome Database
HMM	Hidden Markov Model
HS	Harmony Search
HSA	Harmony Search Algorithm
HSI	High Suitability Index
HTTP	Hyper Text Transmission Protocol
ICA	Independent Component Analysis
ICT	Information Communication Technology
IDE	Integrated Development Environment
IDRS	Indian Diabetes Risk Score
IFG	Impaired Fasting Glucose
IGT	Impaired Glucose Tolerance
IMF	Intrinsic Mode Functions

InfDisSim	Infer Disease Similarity
IoT	Internet of Things
IPF	Interstitial Pulmonary Fibrosis
KNN	K-Nearest Neighbors
LAPO	Lighting Attachment Procedure Optimization
LASSO	Least Absolute Shrinkage and Selection Operator
LDL	Low Density Lipoprotein
LIPC	Local Independent Projection-based Classification
LMS	Least Mean Squares
LOG	Laplacian of Gaussian
LPC	Linear Predictive Coding
LS	Lung Sounds
LSTM	Long Short Term Memory
LVF	Left Ventricular Failure
MAR	Multivariate Autoregressive Models
MBO	Monarch Butterfly Optimization
MCC	Matthew's Correlation Coefficient
MCE	Multiple Classifiers Ensemble
MD	Matching Degree
MDW	Maximal Deflection Width
MFC	Mel Frequency Cepstrum
MFCC	Mel Frequency Cepstral Coefficients
MI	Medical Information
MIRIAD	Minimal Interval Resonance Imaging in Alzheimers Disease
MIRS	Medical Information Retrieval System
MISIM	miRNA Similarity
ML	Machine Learning
MLP	Multilayer Perceptron
MNM	Metabolite Name Matching
MR	Magnetic Resonance
MRI	Magnetic Resonance Imaging
MSKCC	Memorial Sloan Kettering Cancer Centre
NCBO	National Centre for Biomedical Ontolog
NCTRC	National Consortium of Telehealth Resource Centre
NF	Neuro Fuzzy
NIH	National Institutes of Health
NLP	Natural Language Processing
NLTK	Natural Language Toolkit
NMS	Non-Maximum Suppression
OASIS	Open Access Series of Imaging Studies
OGTT	Oral Glucose Tolerance Testing
OPTICS	Ordering Points to Identify the Clustering Structure
PAR	Pitch Adjustment Rate
PBIL	Population-Based Incremental Learning
PCA	Principal Components Analysis

PET	Positron Emission Tomography
PF	Peak Frequency
PHI	Personal Health Information
PHR	Personal Health Records
PNN	Probabilistic Neural Network
PSD	Power Spectral Density
PSM	Platform Specific Model
PSNR	Peak Signal to Noise Ratio
PSO	Particle Swarm Optimization
PTM	Post Translational Modification
QID	Quasi Identifiers
RBF	Radial Basis Function
RBM	Restricted Boltzmann Machine
RCNMF	Relation Completion-based Non-negative Matrix Factorization
RDTF	Remote Diagnostic Testing Facility
REST	Representational State Transfer
RF	Random Forest
RFE	Recursive Feature Elimination
RFID	Radio Frequency Identification
RFT	Random Field Theory
RLMS	Recursive Least Squares Method
RMSE	Root Mean Square Error
RNN	Recurrent Neural Network
ROC	Receiver Operating Characteristic
RS	Relay Station
RSFAS	Rough Set on Fuzzy Approximation Spaces
RSIFAS	Rough Sets on Intuitionistic Fuzzy Approximation Spaces
RSM	Response Surface Methodology
RT	Repetition Time
RWR	Random Walk with Restart
SA	Simulated Annealing
SD	Standard Deviation
SDTENG	Sound Driven Tribo Electric Nano Generator
SH	Safe Harbor
SHFN	Single Hidden-layer Feed-forward Neural network
SI	Swarm Intelligence
SIA	Swarm Intelligence Algorithms
SIMD	Single Instruction Multiple Data
SIV	Suitability Index Variables
SL	Shared Locations
SLWB-MISM	Subcellular Localization Weight-Based MISM
SMC	Secure Multi-party Computation
SMOTE	Synthetic Minority Oversampling Technique
SNN	Supervised Neural Network

SNR	Signal Noise Ratio
SOF	Statistical Overlap Factor
SOM	Self Organizing Map
SPM	Statistical Parametric Mapping
SSIM	Structural Similarity Index Measure
ST	Skin Thickness
SV	Support Vectors
SVD	Single Value Decomposition
SVM	Support Vector Machine
SWT	Stationary Wavelet Transform
TE	Time to Echo
TEWA	Time Expanded Waveform Analysis
TM-SLWBMISM	Threshold Matching SLWBMISM
TRI	Tracheal Sound Index
UAR	Univariate Auto Regressive
UDA	Universal Data Access
UE	User Equipment
URI	Universal Resource Identifier
US	Ubiquitous Sensing
VAERS	Vaccine Adverse Event Reporting System
VAR	Vector Auto Regressive
VIVD	Variable Interval Variable Dose
VMD	Variational Mode Decomposition
VME	Variational Mode Extraction
VQ	Vector Quantization
WHO	World Health Organization
WN	Wavelet Networks
WPD	Wavelet Packet Decomposition
WSI	Whole Slide Images
WT	Wavelet Transform
XGB	Extreme Gradient Boosting

Theoretical Foundation of Computational Intelligence Techniques

Refining Metabolic Network by Fuzzy Matching of Metabolite Names for Improving Metabolites Ranking Toward the Diseases

S Spelmen Vimalraj and Porkodi Rajendran

Abstract The aberrant form of the metabolites inside the human body is known as a disease in other terms. Metabolite's impact on the complex diseases of humans is a very crucial one. Therefore, the depth study of the relationships between the metabolites and diseases is very beneficial in understanding pathogenesis. This chapter proposes a network-based method to identify and rank the disease-related metabolites. In this research, for calculating the disease similarity and the metabolite similarity, infer disease similarity and miRNA similarity have been applied. The miRNA-based disease-related metabolite identification performance was further improved by proposing the Subcellular Localization Weight Based miRNA Similarity (SLWBMISM), where subcellular localization of metabolites was considered to find the more similar metabolites. A fuzzy matching algorithm is adopted to identify the identical names from the two models so that the computational complexity of SLWBMISM can be reduced. After obtaining the identical names of metabolites, two models have been merged without duplication, and metabolite similarity is obtained using SLWBMISM. Then the process of reconstructing the metabolic network based on disease and metabolite similarity was done. At last, a random walk is executed on the reconstructed network to identify and rank disease-related metabolites. For this research work total of 1955 metabolites from network A, 883 metabolites from network B, and 662 diseases were extracted from the experimental datasets. Both networks are merged, and fuzzy matching of metabolite names is applied to avoid the redundant metabolite participating in the metabolite network more than one time. After applying the SLWBMISM method to the merged network, 594521 similarities have been obtained. The proposed method, FM-SLWBMISM, helps find more similar metabolites and enhances the efficiency of SLWBMISM in identifying metabolite prioritization toward complex diseases.

S. Spelmen Vimalraj (✉) · P. Rajendran
Bharathiar University, Coimbatore, India
e-mail: spelmen@gmail.com

P. Rajendran
e-mail: porkodi_r76@buc.edu.in

D. P. Acharjya and K. Ma (eds.), *Computational Intelligence in Healthcare Informatics*, Studies in Computational Intelligence 1132,
https://doi.org/10.1007/978-981-99-8853-2_1

1 Introduction

Metabolism can be referred to as a significant biochemical reaction that travels along the life cycle, and it is particularly vulnerable to infections, which in turn create metabolite defects in urine and blood. The functional modifications of the macromolecules upstream (nucleic acid, protein, etc.) are gradually expressed on the metabolic level, including neurotransmitter changes, hormone modulation, receptor effect, cell signals release, energy transmittance, intercellular communications, etc. Metabolomics [1] is an emerging research field in biological processes, and it not only helps identify the metabolic signatures for the detection of illness but is also helpful in explaining the fundamental pathways that cause a molecular disease-causing mechanism [2–4]. Metabolites are often more studied substances to identify the molecular diseases related to genes, mRNA transcripts, and proteins associated with the disease, with many complex interactions. Metabolism can be a significant consideration in explaining the dynamics of the disease as the final result of cellular regulatory processes. In this post-medical era, advanced technologies allow researchers to analyze diseases at the molecular level [5]. In addition, more researchers have contributed their priceless efforts to the concept of metabolomics to discover more disease knowledge. As a relation between phenotype and genotype, a metabolite is not always correlated with a particular disease, and the influence of illness stretches through a network of functionally associated metabolites. The neighboring metabolites in this network contribute to the same or related diseases. Disease similarities can access the functional relationship of metabolites. Hence, it is aimed to identify more disease-related metabolites by analyzing the diseases and metabolites data. In the next-generation cellular network, cellular networks will be able to include different kinds of relays. Due to the use of relaying technologies into the centrally controlled star network, different kinds of connections as User Equipment (UE) to UE, Relay Station (RS) to RS, and Base Station (BS) to RS can thus be established. The hybrid topology radio network will thus naturally be the future mobile access network, which can help to overcome current and future difficulties and challenges in an efficient manner.

A method [6] was proposed to identify the disease-related metabolites. Initially, infer disease similarity (InfDisSim) was used to compute the similarity of the diseases. Then, the similarity of metabolites was obtained by comparing the diseases. Next, a network of metabolites was constructed using miRNA similarity (MISIM). Finally, a random walk was executed to find more metabolites that are related to the diseases. The performance of this method was improved by proposing subcellular localization weight-based MISIM (SLWBMISM) [7], where subcellular localization of metabolites was used to enhance the finding of similar metabolites. The metabolite network was reconstructed based on disease similarity, and SLWBMISM and the random walk were applied over the reconstructed network to identify disease-related metabolites. The SLWBMISM method's accuracy is further improved by collecting the metabolite names from different models.

The identical names of the metabolites in different models should be determined to reduce the computational complexity of the SLWBMISM method. However, finding metabolite pairs with identical names in different models is computationally trivial. Still, very often, the comparison algorithms lack flexibility, and manual findings of metabolite pairs should do some work to recognize that the difference is caused by symbols like brackets, quotes, apostrophes, spaces, upper/lower case letters, or similar ones. However, it has high time complexity and labor-intensive problem. So, fuzzy matching is introduced in this chapter, which provides an enhanced ability to process the matching names of metabolites for finding the similarity between metabolites. First, the membership function of each element in the metabolite names and a threshold value are used to check whether the metabolite names in the two models are identical. After the extraction of identical names of metabolites for the two models, disease similarity and metabolite similarity are calculated, and based on these similarities; the metabolic network is reconstructed. Finally, a random walk is processed on the reconstructed network for identifying and ranking the disease-related metabolites. This proposed work is named fuzzy matching with SLWBMISM (FM-SLWBMISM).

The rest of the chapter is prepared as follows: Sect. 2 studies the research on identifying disease-related metabolites. Section 3 describes the functioning of FM-SLWBMISM for the identification of disease-related metabolites. Section 4 portrays its performance. Finally, Sect. 5 summarizes this research work and suggests the future scope.

2 Literature Survey

Initially, metabolites prioritization based on the composite network (MetPriCNet) was introduced for identifying and ranking disease-related metabolites. It identified and ranked the candidate metabolites based on their global distance similarity with seed nodes in a composite network that combined multi-omics information from the interactome, genome, metabolome, and phenome. Though MetPriCNet performed well with an incomplete network, its efficiency could be further improved after a more accurate and complete reconstruction of the composite network [8]. Further, a method to identify the potential disease-related metabolites according to the metabolite functional similarity network is proposed. Initially, a modified recommendation strategy was processed to compute the similarity of metabolites. After that, a disease-related metabolic network was constructed with similarities between metabolites and weight. Finally, a random walk was executed on the network to identify disease-related metabolites. However, it is not much effective for the sparse dataset [9].

Likewise, a computational model for identifying disease-related metabolites is also invented. The information about the relationship between diseases and metabolites was collected from human metabolome database (HMDB), and the similarity of metabolites was computed through a modified recommendation strategy. Based on the similarities between metabolites, a disease-associated metabolite network (DMN) was constructed, and a score was refined. A random walk with restart (RWR) was used in this network to identify candidate metabolites related to diseases. However, a threshold value was used in DMN to prune the weak link, and the user-specified threshold value influences the performance of the computational model [10].

Furthermore, a bi-random walks-based method for predicting disease-related metabolites was proposed. First, the metabolite functional similarity network and disease similarity network were reconstructed by combining Gaussian interaction profile (GIP) kernel similarity of metabolites and GIP kernel similarity of diseases correspondingly. Then, the bi-random walks-based method was used in the reconstructed network to predict the disease-related metabolites. However, the GIP kernel similarity of metabolites and disease was over-dependent on known relationships between metabolites and diseases that lead to biased similarity calculations, which is the major drawback of this method [11]. Similarly, a computational model to predict the association between metabolite and disease was also proposed. From the HMDB database, information about metabolite-disease pairs was collected. Using the semantic similarity and enhanced disease GIP kernel similarity, a more consistent disease similarity was obtained. It improved the prediction performance of the computational model. However, due to the sparse data, this model is out-off-balance between the proportion of negative and positive samples [12].

Further, a probability matrix factorization method for combining deep textual features to predict the metabolite-disease association was proposed. Deep Neural Network (DNN) integrating gated recurrent unit network and Convolutional Neural Network (CNN) was used for extracting the equivalent features from text annotations of metabolites and diseases. Finally, matrix factorization was applied to the textual features for predicting the association between metabolites and diseases. This method can be further improved by integrating other previous information on metabolites and diseases [13]. Additionally, a computational method to identify disease-related metabolites using graph deep learning approaches. First, the metabolite and disease networks were constructed by calculating the similarity among the metabolites and diseases, respectively. Then, a graph convolutional network (GCN) was processed on the constructed networks to encode the metabolites and diseases' features. Finally, Principal Components Analysis (PCA) and DNN were applied for feature selection and identification of metabolite-disease pairs. However, the computational complexity of this method depends on the number of nodes in DNN [14].

Another method based on Artificial Bee Colony (ABC) and spy strategy to predict the metabolite-disease association was proposed. A spy strategy was accepted for extracting consistent negative samples from unconfirmed pairs. ABC algorithm was used to fine-tune the effect of parameters. However, this method depends on the quality of similarity matrices [15]. Likewise, a method for predicting the metabolite-disease associations according to the linear neighborhood similarity with an improved bipartite network projection algorithm was also found in the literature. However, in this method, more known confirmed human metabolite-disease associations would enhance the development and performance of computational human metabolite-disease prediction methods [16].

Furthermore, a Relation Completion-based Non-negative Matrix Factorization (RCNMF) to predict the relationship between metabolites and diseases is proposed. Initially, molecular fingerprint similarity and semantic similarity of metabolites were calculated, and then the raw relationship between metabolite and diseases was altered to swap 0 with numbers between 0 and 1. At last, a non-negative matrix factorization was used to predict the potential relationship between the metabolite and disease. The downside of RCNMF is that some of the disease semantic similarities are empty. It will lead to biases in changing the association matrix, further affecting the prediction efficiency [17].

3 Materials and Methods

For the experimental purpose, the metabolites data has been downloaded from HMDB, consisting of three kinds: biochemical data, chemical data, and clinical data. It is collected from thousands of public sources. It consists of 40,000 kinds of metabolites. Here, disease ontology is used to annotate the diseases. Disease ontology was initiated as a part of the NUgene project at North-western University in 2003. National center for Biomedical Ontology (NCBO) is a data-sharing project comprising six core components, including promoting biology projects and external research collaboration, computer science and biomedical informatics research, education, infrastructure, communication, and management. Through NCBO, HMDB data can be easily understood. From the collected data, 1955 metabolites and 662 diseases are used in this experiment. PubChem is the available chemical information database of National Institutes of Health (NIH). The term "open" refers to the fact that you can upload your scientific data to PubChem and allow others to use it. PubChem has grown in popularity among the general public, students, and scientists since its inception in 2004. The PubChem database provides data with the help of the website it owns. PubChem primarily contains little molecules, however, conjointly larger molecules like nucleotides, carbohydrates, lipids, peptides, and chemically modified macromolecules. The user tends to collect data on identifiers, chemical structures, physical and chemical properties, patents, toxicity knowledge, biological activities, health, and safety, as well as plenty of other data.

3.1 Pairwise Disease Similarity

Pairwise disease similarity is calculated by equipping the method named as infer disease similarity (InfDisSim). The InfDisSim method uses the disease ontology and the disease-related genes to find the weight vector for all diseases. Moreover, one of the networks, called a functional gene network, can be formed, and through that network weight vector can be obtained, and all the weight vectors could be cosine to obtain the disease similarity. The disease terms presented in this network are not to be linked directly based on genes associated with the disease they have been connected to in the network.

$$WV_{t1} = \{w_{1,1}, w_{1,2}, w_{1,3}, \ldots, w_{1,i}, \ldots, w_{1,N}\} \tag{1}$$

In Eq. 1, N represents genes in number, $w_{1,i}$ represents the disease weight score, and WV represents the disease weight vector for the given disease set $t1$. Calculating the cosine between the weight vectors is represented as follows in Eq. 2.

$$Inf(t_1, t_2) = \frac{\sum_{i=1}^{N} w_{1,i}.w_{2,i}}{\sqrt{\sum_{i=1}^{N} w_{1,i}^2}\sqrt{\sum_{i=1}^{N} w_{2,i}^2}} \tag{2}$$

The semantic associations of genes and the associations between the gene and disease can be reflected in the disease similarity, and it can be defined as follows in Eq. 3.

$$InfDisSim(t_1, t_2) = Inf(t_1, t_2)\frac{|G_1||G_2|}{|G_{MICA}|^2} \tag{3}$$

In Eq. 3, G_1 and G_2 are the metabolite sets of disease sets t_1 and t_2. G_{MICA} is the metabolite set of t_3 which is a common ancestor for t_1 and t_2. Therefore, based on Eqs. 1 to 3, the disease similarity can be calculated and used for further calculations in this research.

3.2 Threshold Matching Based Metabolite Name Mapping

The threshold matching approach-based metabolite name mapping consisted of three phases to match the metabolite name from different models. In the first phase of the threshold matching approach, a Levenstein edit distance and Levenstein similarity ratio for each metabolite pair (metabolite name in model A and metabolite name in model B) are calculated. Then, the metabolite pairs are removed, which have an edit distance more significant than a user-specified threshold. The second phase of the threshold matching approach discards the pars with a minor similarity. However, many pairs with the same metabolite may have a similarity more significant than the user-specified threshold. So, in the third phase of the threshold matching approach, the pairs with a minor similarity are removed. Next, it removes the multiple connections to the same metabolite in the same model by removing pairs with a similarity ratio less than the highest. Finally, it returns a list that contains all metabolite names from model A and model B with no duplicate metabolite names. A string a is said to be empty if $|a| = 0$. Likewise, string b is empty if $|b| = 0$. In such cases, there is no need of calculating the edit distance. On contrary, the edit distance, $lev(a, b)$ is calculated using Eq. 4.

$$lev(a,b) = \begin{cases} |a| & if \quad |b| = 0 \\ |b| & if \quad |a| = 0 \\ lev(tail(a), tail(b)) & if \quad \text{a[0]=b[0]} \\ 1 + min \begin{cases} lev(tail(a), b) \\ lev(a, tail(b)) \\ lev(tail(a), tail(b)) \end{cases} & Otherwise \end{cases} \quad (4)$$

The task of metabolite name mapping is to map the metabolite name in model A with each metabolite name in model B to find the identical names of the metabolite. Comparing two lists of metabolites can be labor-intensive because it involves screening all possible element combinations. So, fuzzy matching is used to match the metabolite names. For example, a fuzzy matching defining A as a model, maps to the interval [0, 1]; that is $\mu_A(x) \rightarrow [0, 1]$ for all $x \in A$. It is to be noted that $\mu_A(x)$ denotes the membership function of each element x in metabolic name A.

A trapezoidal membership function is used to compute the $\mu_A(x)$. Assume that a, b, c, and d are the x coordinates of the four vertices of $\mu_A(x)$ in a fuzzy set. The degree of membership increases between a and b, with one-degree flatness between c and d. The membership function $\mu_A(x)$ is calculated as in Eq. 5, where $a < b < c < d$.

$$\mu_A(x; a, b, c, d) = \begin{cases} 0 & \text{if} \quad x \leq a \quad or \quad x \leq d \\ \frac{x-a}{b-a} & \text{if} \quad a \leq x \leq b \\ 1 & \text{if} \quad b \leq x \leq c \\ \frac{d-x}{d-c} & \text{if} \quad c \leq x \leq d \end{cases} \quad (5)$$

The matching degree between the metabolite names is calculated with the help of the membership function of each element in the metabolite name $\mu_A(x)$, and the matching degree between the metabolite names is calculated. The similitude degree between two fuzzy sets is the matching degree. In order to compute the matching degree, closing degree and hamming distance are used. Given two metabolite names, A and B, from two models belonging to the same range, the hamming distance between A and B is calculated using Eq. 6.

$$Hamming(a,b) = \frac{1}{n} \sum_{i=1}^{n} |\mu_A(x_i) - \mu_B(y_i)| \quad (6)$$

In Eq. 6, $\mu_A(x_i)$ is the membership function of each element x_i in metabolic name A for U, $\mu_B(y_i)$ is the membership function of each element y_i in metabolic name A for V, and n is the metabolite name number. The matching degree, MD, between metabolite name A and metabolite name B is calculated using Eq. 7.

$$MD = 1 - Hamming(a,b) \quad (7)$$

The less distance between A and B denotes the more similarity and higher matching degree. If $A = B$, the matching degree between them indicates two metabolite names superposition and homology completeness. After calculating the matching degree between A and B, a threshold δ is considered to check whether the matching is successful or a failure. If $MD \geq \delta$, then the matching between metabolite name A and metabolite name B is successful, and it replaces the identical name of the metabolite in models U and V. If $MD < \delta$, then the matching between metabolite name A and metabolite name B is a failure and then this process is continued with the next metabolite name in model V. For each element in the metabolite name, a sub-threshold δ is considered to save the time and efficiency of fuzzy matching. When some element membership function degree is zero, do not check the other elements of metabolite name. The matching results based on threshold δ, MR_{thre}, is defined in Eq. 8. Equation 9 denotes the sub-threshold, where a_i denotes the element membership function degree.

$$MR_{thre} = \gamma \begin{cases} 1, & MD \geq \delta \\ 0, & MD < \delta \end{cases} \tag{8}$$

$$Sub_{thre} = \gamma \begin{cases} 1, & a_i \geq \delta \\ 0, & a_i < \delta \end{cases} \tag{9}$$

3.3 *Fuzzy Matching Based Metabolite Name Matching Algorithm*

This section describes the fuzzy matching-based metabolite name matching (FM-MNM) algorithm.

Algorithm 1 (FM-MNM Algorithm)
Input: Metabolite names from two models
Output: New Metabolite list

For every metabolic name in model U
    Sum=0
    For every element x_i in metabolic name a in model U
        If $\mu_a(x_i) < \delta$
            Break
        Else
            Compute the Hamming (a, b) in models U and V
            $Sum = Sum + \mu_a(x_i) - \mu_b(y_i)$
    End for
    $Hamming(a, b) = Sum/n$
    $MD = 1 - Hamming(a, b)$
    If $MD \geq \delta$
    Put a into the new metabolic list
    Replace the name of b with a in model V
End for

3.4 Pairwise Metabolite Similarity

The cellular compartment is said to be the closed parts presented inside a eukaryotic cell's cytosol, and also it is always incorporated by a solitary, in other terms, two-fold lipid layer called a membrane. Here cellular location plays a significant role in computing the confidence score between all the metabolites. The confidence score provides the reliability of similarity where the M_1 and M_2 are metabolites. For computing the SLWBMISM, all the metabolites are placed inside the range of compartments. The ratio of the number of metabolites present in the compartment SC_x and the compartment with the highest number of metabolites is known as the compartment range. The range of compartment I can be computed using the Eq. 10, where $0 < Score(I) < 1$.

$$Score(I) = \frac{SC_X(I)}{SC_M} \tag{10}$$

According to the range between 0 and 1, all the presented metabolites, represented by N, can be added with some weight. Hence, metabolite (M_1, M_2) can be annotated by the shared locations (SL) or compartments, i.e., $SL(M_1, M_2) = L(M_1) \cap L(M_2)$. Moreover, the range of the metabolites (M_1, M_2) is defined in Eq. 11, where C_{MS} refers to the compartment with lowest number of metabolites.

$$W(M_1, M_2) = \begin{cases} Max(Score(I)) \text{ if } SL(M_1, M_2) \neq \phi \\ Score(C_{MS}) \qquad \text{otherwise} \end{cases} \tag{11}$$

The weight value between a pair of metabolites can be used to calculate the similarity between metabolites. It is defined in Eq. 12.

$$Sim(M_1, M_2) = \frac{W(M_1, M_2)\sum_{i=1}^{num} S(h_i, D_2) + W(M_1, M_2)\sum_{j=1; j\neq i}^{num} S(h_j, D_1)}{m + n} \tag{12}$$

3.5 Identification and Ranking of Disease-Related Metabolites

The identical names of the metabolite are replaced in model U and model V. After that, combining the metabolite names of the two models can be used in the SLWBMISM method to find a better relationship between the metabolites. The metabolic network is reconstructed based on the disease similarity and metabolite similarity. The adjacency matrix of the reconstructed metabolic network is represented as $D = \begin{bmatrix} D_M & D_{MT} \\ D_{TM} & 0 \end{bmatrix}$ where D_M, D_{MT}, and D_{TM} are the adjacency matrix of metabolite-metabolite, metabolite-disease and disease-metabolite, respectively.

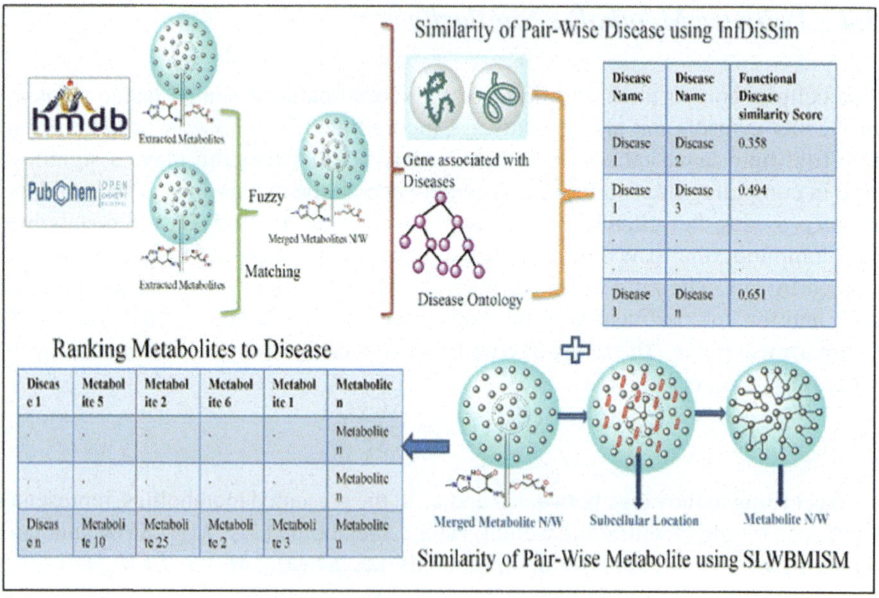

Fig. 1 Identification of disease-related metabolites using FM-SLWBMISM

Similarly, $W = \begin{bmatrix} W_M & W_{MT} \\ W_{TM} & 0 \end{bmatrix}$ refers to the transition matrix of the reconstructed metabolic network, where W_M, W_{MT}, and W_{TM} are the transition matrix of metabolites, metabolite-disease, and disease-metabolite, respectively. A RWR is implemented on the reconstructed metabolic network to identify disease-related metabolites. RWR can be defined as in Eq. 13.

$$P_{t+1} = (1 - \gamma) D P_t + \gamma P_0 \tag{13}$$

In Eq. 13, D represents the adjacency matrix, and P_0 is said to be the initial probability vector element at node i and step t. After some steps, the probability of P_{t+1} will reach a steady state. Then, metabolites are ranked for diseases based on probability. Figure 1 represents the overall methodology for identifying disease-related metabolites using FM-SLWBMISM.

4 Result Analysis

Pharmaceutical Research can be done with a wide range of databases that contain information about chemical compounds. The researchers can use the non-systematic identifiers to retrieve information related to compounds such as structures and mol files from the databases. Moreover, the source-dependent identifiers include generic names, chemical abstracts, Brand names, and research codes that can be assigned while the compound has been registered to that database. The exact matching of non-systematic identifiers is done manually, but it is a complex and lengthy process. Even though it is cumbersome, it is possible to check their name while registering or combining two databases to get more information about the compound. Because some of the information may not be included at the time of registration in one database. The researchers want to get more information about the compound means by combining as many databases as possible.

Text mining methods exist that extract and provide the disease-drug relationships or target-compound relationships from the text file; they can produce insights about relationships. Researchers are thirsty to use different databases to discover the related metabolites and reactions. Due to the mapping inconsistencies, there is a chance of errors arising. While merging two different databases, mapping is not precise with the compound names or IDs. There would be an inconsistency in many mapping if the non-duplicate identifiers were used because they can link to duplicate names. With low-level interoperability, in practice, it is impossible to directly compare models, as metabolites will hardly be cross-mapped, which successively makes it impossible to match reactions in each model.

So here, fuzzy matching-based metabolite name mapping has been introduced to map the metabolite name among the two models, A (HMDB) and B (PubChem). While performing the two metabolite name mapping methods, namely, threshold matching and fuzzy matching on the above-said models, 883 rows of possible duplicate compounds were obtained by both methods among 1955 metabolites. From that prediction, the threshold matching-based metabolite name mapping method produced 535 rows of ambiguous names, and fuzzy matching-based metabolite name mapping produced 428 rows of ambiguous names. While compared with threshold matching-based metabolite name mapping, fuzzy matching has produced more possible combinations among the metabolite names. These results have been cross-checked with chemists, and it is proved that among the 428-ambiguity mapping of metabolites, it is identified that only 48 false predictions among 428 combinations. Whereas in the threshold matching, it is identified that there are 117 wrong predictions out of 535 possible combinations. After eliminating all the duplicated entries, two models have been merged, and the pairwise metabolite similarity has been calculated. Then finally, the random walk was applied to the constructed metabolite network, and the ranking process was done.

Figure 2 depicts the metabolite names presented in two databases, HMDB and PubChem, which are the same but are stated in another name. However, it is not possible to state one compound name in all the databases as it is. Different databases may use different nomenclature, so it is impossible to state the compound name as one in all the databases. The above mention figure is an example of duplicate names identified by the fuzzy matching-based metabolite name mapping, whereas the threshold matching-based metabolite name mapping failed to identify these combinations. It implies that fuzzy matching-based metabolite name mapping can identify more combinations among two databases. If two databases have been merged, there is a possibility of getting a more significant number of instances. So that the method

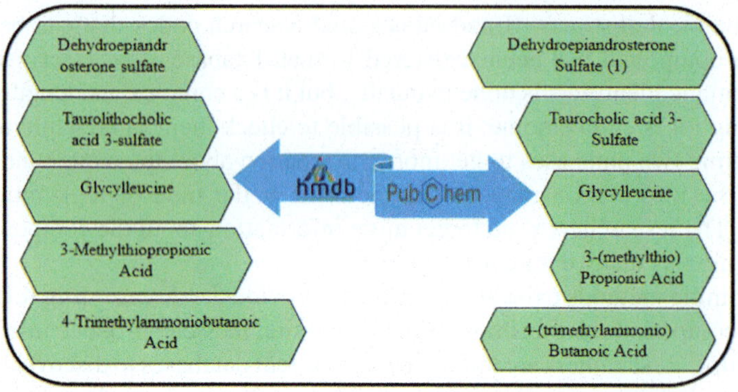

Fig. 2 Pairwise comparison of fuzzy matching-based metabolite name mapping

Table 1 Metabolites similarities for the existing and proposed methods

Similarity range	SLWBMISM with one model	SLWBMISM after threshold matching with 2 model	SLWBMISM after fuzzy matching with 2 models
0.1–0.2	663844	226480	103302
0.2–0.3	178384	242939	188501
0.3–0.4	40836	92690	289027
0.4–0.5	6050	2615	8864
0.5–0.6	294	4014	4825
0.6–0.7	180	170	200
0.7–0.8	79	0	0
0.8–0.9	53	0	0

can identify a more significant number of relationships. It implies that with more instances, the efficiency of SLWBMISM has been improved.

Table 1 describes the improvement of the existing method SLWBMISM with fuzzy matching based on metabolite name mapping. In Table 1, all the similarities have fallen inside the range 0.1 to 0.7. Whereas in the SLWBMISM, the values fell outside 0.8. In the proposed method FM-SLWBMISM, all the possible similarities have gradually fallen on the glance of ranges, whereas in the SLWBMISM, most of them fell within the range 0.1–0.2. The number of similarities has been minimized to get closer to the metabolite-disease associations. Here the values between the range 0.5 and 0.7 have been considered for further processing. Because the threshold for eliminating the relationships among the metabolites must be beyond 0.5. The threshold has been set according to the previous study. More relationships have been identified within the threshold limit because more instances have been added to this study. So, the method could identify more relationships among the metabolites and diseases. Figure 3 compares similarities falling in different ranges from 0.1 to 0.9.

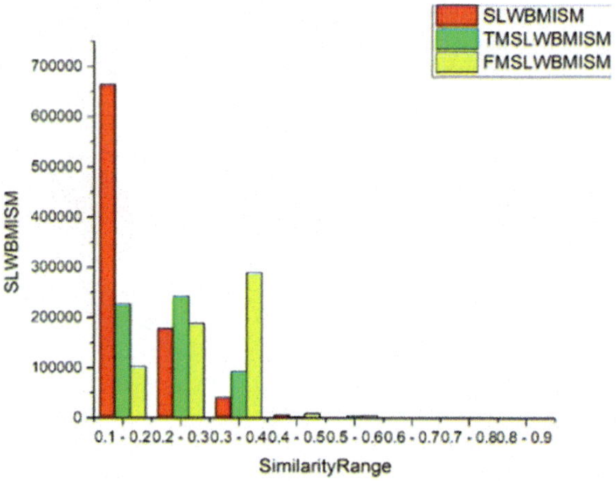

Fig. 3 Comparison of similarities fallen on different ranges of development

4.1 Performance Measures of Threshold and Fuzzy Matching

The precision is defined as the comparison of the existing subcellular localization weight-based miRNA similarity method (SLWBMISM), threshold matching subcellular localization weight-based miRNA method (TM-SLWBMISM), and proposed fuzzy matching with subcellular localization weight-based miRNA similarity method (FM-SLWBMISM) is analyzed from the given dataset. The recall is the ratio of the identified similar metabolites to total metabolites in the given database, which is analyzed from the existing and proposed methods. Table 2 shows the precision and recall values compared with the existing SLWBMISM and proposed TM-SLWBMISM and FM-SLWBMISM for identifying the similarity of metabolites.

Figure 4 represents the comparison of SLWBMISM, TM-SLWBMISM, and FM-SLWBMISM methods regarding precision values. The X-axis denotes the existing and proposed methods, and Y-axis denotes the precision value. The precision of FM-SLWBMISM is 5.76% greater than the SLMBMISM method and 2.9% greater than the TM-SLWBMISM method because the true positive rate is higher than the rest of the two methods. Because of the true positive rate achieved by the FM-SLWBMISM while the rest failed to achieve high. This analysis proves that the proposed FM-SLWBMISM has higher precision and efficiency than SLWBMISM and TM-SLWBMISM methods for identifying the similarity of metabolites.

Table 2 Precision and recall value for existing and proposed methods

Methods	Precision	Recall
SLWBMISM	0.833	0.803
TM-SLWBMISM	0.852	0.821
FM-SLWBMISM	0.881	0.868

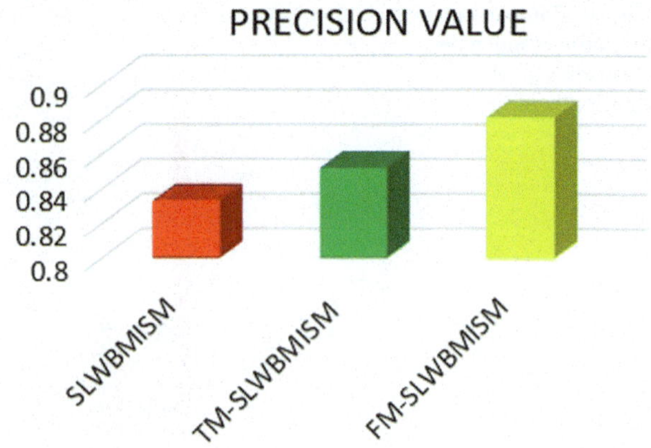

Fig. 4 Comparison of precision values for existing and proposed methods

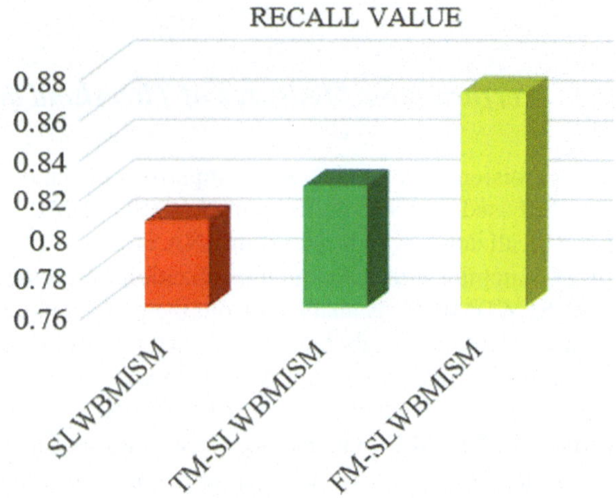

Fig. 5 Comparison of recall values for existing and proposed methods

Figure 5 represents the comparison SLWBMISM method and TM-SLWBMISM, and FM-SLWBMISM for recall. The X-axis denotes the existing and proposed methods, and Y-axis denotes the recall value. The recall of FM-SLWBMISM is 6.5% greater than the SLMBMISM method and 4.7% greater than TM-SLWBMISM because of the highest number of true positive rates and less in numbers of false negatives. This analysis proved that the proposed FM-SLWBMISM has higher recall than the TM-SLWBMISM and SLWBMISM method for identifying the similarity of metabolites relation.

The ROC curve is the relation between the false positive and true positive rates. The AUC ranges in value from 0 to 1. Figure 6 represented the ROC values compared with the existing SLWBMISM and proposed FM-SLWBMISM method for identifying the similarity of metabolites. Figure 6 compares the ROC curve for SLWBMISM and FM-SLWBMISM methods on the HMDB online portal dataset. Hence, it is analyzed

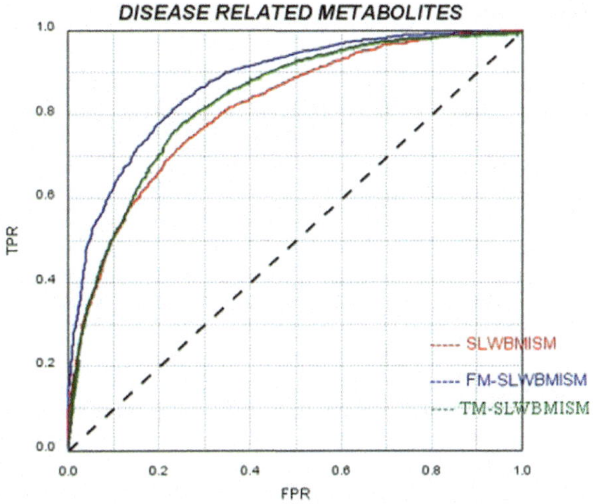

Fig. 6 ROC measure for existing and proposed methods

that the ROC value is more significant for the proposed FM-SLWBMISM method than for the existing SLWBMISM method for identifying the similarity of metabolic relation.

5 Conclusion

This chapter, FM-SLWBMISM and TM-SLWBMISM, is proposed to improve the efficiency of SLWBMISM-based identification and ranking of disease-related metabolites. The metabolites from the two models are used to find the more similar metabolites effectively. The two models may have similar metabolite names with different notations, so fuzzy matching and threshold matching are introduced to find the identical names of metabolites. The identical names of the metabolites are replaced in the model and then form a new metabolite list. The metabolites in the new metabolite list are used in SLWBMISM. Then, the metabolite network is reconstructed based on disease and metabolite similarity. Finally, an RWR is implemented in the reconstructed network to identify and rank disease-related metabolites. The experimental results show that the FM-SLWBMISM returns more related metabolites for diseases.

References

1. Qu, X., Gao, H., Sun, J., Tao, L., Zhang, Y., Zhai, J., Song, Y., Hu, T., Li, Z.: Identification of key metabolites during cisplatin-induced acute kidney injury using an HPLC-TOF/MS-based non-targeted urine and kidney metabolomics approach in rats. Toxicology **431**, 152366 (2020)
2. Nordström, A., Lewensohn, R.: Metabolomics: moving to the clinic. J. Neuroimmun. Pharmacol. **5**(1), 4–17 (2010)
3. Dexter, D.T., Jenner, P.: Parkinson disease: from pathology to molecular disease mechanisms. Free Radical Biol. Med. **62**, 132–144 (2013)
4. Stirling, P.C., Hieter, P.: Canonical DNA repair pathways influence R-loop-driven genome instability. J. Mol. Biol. **429**(21), 3132–3138 (2017)
5. Zhao, Z.J., Shen, J.: Circular RNA participates in the carcinogenesis and the malignant behavior of cancer. RNA Biol. **14**(5), 514–521 (2017)
6. Hu, Y., Zhao, T., Zhang, N., Zang, T., Zhang, J., Cheng, L.: Identifying diseases-related metabolites using random walk. BMC Bioinform. **19**(5), 37–46 (2018)
7. Vimalraj, S.S., Rajendran, P.: Convalescing the process of ranking metabolites for diseases using subcellular localization. Arabian J. Sci. Eng. **47**(2), 1619–1629 (2022)
8. Yao, Q., Xu, Y., Yang, H., Shang, D., Zhang, C., Zhang, Y., Sun, Z., Shi, X., Feng, L., Han, J., Su, F., Li, C., Li, X.: Global prioritization of disease candidate metabolites based on a multi-omics composite network. Sci. Rep. **5**(1), 1–14 (2015)
9. Wang, Y., Juan, L., Liu, C., Zang, T., Wang, Y.: Identifying candidate diseases-related metabolites based on disease similarity. In: Proceedings of IEEE International Conference on Bioinformatics and Biomedicine, pp. 1281–1285 (2018)
10. Wang, Y., Juan, L., Peng, J., Zang, T., Wang, Y.: Prioritizing candidate diseases-related metabolites based on literature and functional similarity. BMC Bioinform. **20**(18), 1–11 (2019)
11. Lei, X., Tie, J.: Prediction of disease-related metabolites using bi-random walks. PloS One **14**(11), e0225380 (2019)
12. Lei, X., Zhang, C.: Predicting metabolite-disease associations based on KATZ model. BioData Min. **12**(1), 1–14 (2019)
13. Zhu, Q., Han, C., Zhu, Q., He, T., Jiang, X.: Integrating deep textual features to probability matrix factorization for metabolite-disease association prediction. In: Proceedings of IEEE International Conference on Bioinformatics and Biomedicine, pp. 628–633 (2019)
14. Zhao, T., Hu, Y., Cheng, L.: Deep-DRM: a computational method for identifying disease-related metabolites based on graph deep learning approaches. Brief. Bioinform. **22**(4), bbaa212 (2021)
15. Lei, X., Zhang, C., Wang, Y.: Predicting metabolite-disease associations based on spy strategy and ABC algorithm. Front. Mol. Biosci. **7**, 603121 (2020)
16. Lei, X., Zhang, C.: Predicting metabolite-disease associations based on linear neighborhood similarity with improved bipartite network projection algorithm. Complexity **2020** (2020)
17. Lei, X., Tie, J., Fujita, H.: Relational completion based non-negative matrix factorization for predicting metabolite-disease associations. Knowl. Based Syst. **204**, 106238 (2020)

Learning from Imbalanced Data in Healthcare: State-of-the-Art and Research Challenges

Debashis Roy, Anandarup Roy, and Utpal Roy

Abstract Datasets associated with medical and healthcare domains are imbalanced in nature. An imbalanced dataset refers to a classification dataset where the number of instances of a given class is much lower than for other classes. Such imbalanced datasets require special attention because traditional classifiers tend to favor the classes with many instances. On the other hand, in healthcare, the class with fewer instances may correspond to a rare and uncommon event of potential interest. Ignoring the imbalance affects the performance of classifiers which hampers the detection of rare cases such as disease screening, severity analysis, detecting adverse drug reactions, cancer malignancy grading, and the identification of uncommon, chronic illnesses in the population. Therefore, the design of classifiers should aim at successfully classifying such classes as compared to classifying other classes. Many significant research works accept the intrinsic imbalance in healthcare data and propose methodologies to handle this. Considering the importance and identifying research challenges in imbalance, a compilation of such works is the aim of this chapter.

1 Introduction

Class imbalance refers to classification problems where more instances are available for some classes than others. Notably, in a two-class scenario, one class contains the majority of instances rather than the minority class. To name a few, imbalanced datasets may originate from many real-life problems, including biomedical diagnosis

D. Roy (✉) · U. Roy
Department of Computer and System Sciences, Visva Bharati University, Kolkata, India
e-mail: debashisjcc@gmail.com

U. Roy
e-mail: utpal.roy@visva-bharati.ac.in

A. Roy
Sarojini Naidu College for Women, Dum Dum, Kolkata, India
e-mail: roy.anandarup@sncwgs.ac.in

D. P. Acharjya and K. Ma (eds.), *Computational Intelligence in Healthcare Informatics*, Studies in Computational Intelligence 1132,
https://doi.org/10.1007/978-981-99-8853-2_2

[1] and text classification [2]. When the dataset is imbalanced, conventional classifiers typically favor the majority class and thus fail to correctly classify the minority instances, which results in performance loss [3]. Upon realizing the importance of class imbalance in classification, various techniques have been developed to address this problem. These approaches can be categorized into four categories [4] such as algorithm base, data level, cost-sensitive, and multiple classifiers ensemble.

Algorithm-level approaches adapt existing classifier learning algorithms to tune them for class imbalance. Examples include the modifications for k-nearest neighbors [5], support vector machine (SVM) [6] and the Hellinger distance decision trees (HDDT) [7]. Data-level approaches include preprocessing algorithms that rebalance the class distribution by resampling the data space, thus avoiding modifications of the learning algorithm [8]. For example, the synthetic minority oversampling technique (SMOTE) is used to increase data instances in a balanced way [9]. A recent survey on handling imbalanced data was carried out in the literature [10]. Cost-sensitive learning frameworks lie in between the data and algorithm-level approaches. Such methods assign a different cost to instances and modify the learning algorithm to accept the cost. Well-known methods in this category are cost-sensitive SVM [11] and the adacost family [12]. Multiple classifiers ensemble (MCE) approaches are designed to increase the accuracy of a single classifier by training several different classifiers and combining their decisions to output a single class label [13]. Such classifiers are, however, focused on improving the accuracy and hence not directly applicable to solving the imbalance problem. Most often, ensemble classifiers are combined with one of the techniques, usually a preprocessing technique.

Among the above categories, algorithm level and cost-sensitive level approaches depend on the choice of classifiers. Since these approaches attempt to modify the underlying classifier, they may not be suitable for all types of available classifiers. On the other hand, the data-level approaches are independent of any classifier and the data at hand. Furthermore, such approaches work with the principle of balancing the majority and minority classes. It can be done by increasing the minority class size or decreasing the majority class size. Due to their independent nature, data-level approaches are most widely used for handling imbalances.

Hospitals, diagnostic labs, and insurance companies store large datasets about the patients according to their medical history, disease-wise. Analyzing medical data is crucial because it may lead to the diagnosis of certain diseases as well as their severity [14]. Consequently, treatment and drug administration depend on proper biomedical data analysis. However, it is challenging to differentiate between unhealthy and healthy patients during disease prediction. This is because, most of the time, medical datasets have very few cases of the target disease compared to healthy patients. Therefore, misclassification in an imbalanced medical dataset is extremely risky in medical science, which may cause death by wrong prediction [10]. Researchers have been working on imbalanced datasets for the last few years to realize the importance of this imbalance. As a result, several significant articles have been available so far. However, no significant compilation and review of these articles have been available. Therefore, this chapter aims to compile the current state of the art for handling imbalances in healthcare information systems.

The organization of this chapter is as follows. First, in Sect. 2, a detailed review of available literature concerning imbalance in healthcare data is prepared. Next, Sect. 3 presents the various methodologies for handling imbalanced data in the healthcare consortium. Further, Sect. 4 enlists publicly available imbalanced healthcare data and points to the repositories where such datasets are found. Finally, Sect. 5 depicts the current state of the art and outlines some future research aspects in this field.

2 Literature Review of Imbalance Healthcare Data

As pointed out in an earlier section, class imbalance is a well-taken problem in machine learning. However, the articles focus on healthcare-related problems in the medical domain. Hence imbalance is often considered only a subproblem. Most research works adopt an already developed technique to handle the imbalance issue. Now, let us understand that the problem at hand in medical science is of utmost importance due to its connection to human life. For this reason, let us focus on the problem domain to categorize available literature. Accordingly, six such important domains can be identified, where imbalance learning has been observed. Before describing them, it is worthwhile to mention that the following domains are based on previous works published recently, and there is enough room to add more such domains with the advancement in the study.

2.1 Imbalanced Cancer Data Diagnosis

Cancer is highly life-threatening and is a crucial application of data analysis in healthcare. Related works often aim toward the early detection of cancer and thus assist doctors and medical practitioners in preparing a treatment protocol. Unfortunately, cancer data is generally imbalanced; hence, misclassification is a significant problem in this domain. Researchers proposed different oversampling, undersampling, and algorithm-level techniques to balance and classify the dataset efficiently. In an early work, a boosting algorithm with evolutionary undersampling for classifying breast cancer in an imbalanced dataset is used [15]. Concerning undersampling, a clustering-based undersampling method to predict breast cancer with an imbalanced dataset using classifiers K+ boosted C5.0, SMOTEBoost, RUSBoost, SMOTE-Boosted C5.0 is proposed [16].

Apart from undersampling, oversampling was also considered for rebalancing. One notable work in this line predicted breast cancer with a convolutional neural network and different oversampling methods [17]. It also used undersampling and concluded that synthetic oversampling is an effective way of rebalancing. This fact, however, is also a conclusion in the comprehensive analysis of oversampling and undersampling methods in cancer diagnosis. It is observed that in 90% of the cases, classifier performance is significantly improved after rebalancing. It was also found that each kind of cancer dataset responded differently to different balancing techniques and classifiers. Considering more recent works, the SMOTE oversampling technique is applied to handle imbalanced leukemia data. Similarly, SMOTE is also used to rebalance a dataset of liver cancer [18].

Further, rebalancing is combined with a stacking ensemble of deep learning models to classify cancer types [19]. In this work, both undersampling and oversampling are employed. Interestingly, in this case, after a detailed analysis, the authors observed that the performance of ensemble classifiers is improved if undersampling is applied compared to oversampling. However, this observation seems to contradict the conclusion of previous research work [14]. Moreover, the datasets were also different.

2.2 *Prediction of Imbalanced Covid-19 Data*

After the onset of the Covid-19 pandemic, scientists devote themselves to predicting the spreading of Covid-19 and its diagnosis, using different types of data like blood tests, chest X-rays, and CT thorax. However, in this domain, data collection is somewhat limited; therefore, the data became heavily imbalanced. Most of the works in this domain are during the years 2020–2021. In the literature, ensemble learning is used for diagnosing COVID-19 from specific blood test results [20]. Nevertheless, the major limitation of the analysis is that it uses only 9.9% SARS-CoV-2 positive patients against 90.1% healthy persons. Due to this imbalance, SMOTE oversampling is used. It increases classification results after using SMOTE. On the other hand, chest X-ray data is used to predict Covid-19. It is also observed that there is an imbalance in the data. However, using the convolution neural network model claimed to overcome the imbalance issue [21].

Complete blood count (CBC) tests are among the most widely used medical tests. In fact, due to its broad applicability, CBC datasets are highly imbalanced, particularly when considering Covid-19. Besides, comparing five different resampling techniques and eight classifiers is carried out for diagnosing Covid-19. Nevertheless, these experiments do not suggest any sampling technique to be a clear winner. The data and metrics condition the performance of each sampling technique and classifier at hand [22]. Additionally, a detailed investigation of various decision tree ensembles to classify Covid-19 positive cases is presented [23]. It uses random undersampling and SMOTE oversampling methods and ensembles to handle imbalance. The results suggested that the resampling methods do not affect all the ensembles similarly. Random undersampling with the XGBoost ensemble method yielded the best result.

Several works have been done on predicting the spread and severity of Covid-19. Another model predicts the severity of the Covid-19 pandemic in a country into low, moderate, or high. It helps policymakers to control the vaccination process. It categorizes the dataset into three categories; namely, low, moderate, and high [24]. This categorization makes the dataset imbalanced. Hence, to deal with imbalance, an undersampling approach is used. For classification, the random forest classifier outperformed other competitors. Likewise, forecasting the infections of Covid-19 is also carried out [25]. Such a prediction requires a continuous flow of related information. However, such datasets are highly imbalanced mainly due to concept drift, i.e., the nonstationary nature of gathered information. To overcome the limitation, a Continuous-SMOTE oversampling is used to tackle the imbalance issue.

2.3 *Drug Prediction with Imbalanced Data*

Arguably the most complex and essential domain of medical science is discovering a particular drug and predicting its risk. Both these tasks are expensive and time-consuming if performed manually. However, publicly available resources enable machine learning researchers to develop classification models to detect active compounds for drug discovery. However, an imbalance in such data is usually unavoidable. After the preparation of a drug, clinical trials are performed for the risk prediction of a particular drug. Data obtained from such trials are often imbalanced because adverse effects of a drug may only be noticeable in a small group of subjects during the trial. Machine learning researchers trying to predict and mitigate the risk of a drug over the short or long term have to deal with this imbalance carefully.

One noticeable recent work on detecting adverse drug reactions was carried out in 2019 [26]. It is found that adverse drug reactions are indeed a rare event and lead to an imbalance ratio. In many cases, 1 out of 100 has adverse reactions. Further, experiments are conducted with resampling, cost-sensitive learning, ensemble learning, and one-class classification to treat this high imbalance. The aforementioned individual approaches are combined to deal with such a high imbalance. However, the experimental results show that the individual techniques, except resampling, do not meet the expectations. Therefore, it is worth combining resampling with cost-sensitive learning during analysis.

Likewise, SMOTE oversampling, along with four classifiers, are applied for drug risk prediction [27]. Apart from the logistic regression classifier, all the others are ensemble classifiers used for drug risk prediction. The result shows significant improvement in classifier performances after combining them with SMOTE. Considering drug discovery, the classification performance of deep neural networks on imbalanced datasets after applying various rebalancing methods is also explored [28]. It is found that imbalance hurts the network's performance, and the effect can be reduced using rebalancing methods.

Virtual screening technology is a simulation of the drug development process on a computer. A framework that virtually screens drug proteins to aid the drug discovery process is also proposed [29]. However, it is pointed out that there is a significant difference between the number of active and inactive compounds. Therefore it is essential to apply SMOTE to handle this imbalance. Similarly, a genetic algorithm is combined with SMOTE to generate new samples. It showed improvement in dealing with imbalance.

2.4 Imbalance Classification of Diabetes

Diabetes is one of the most common metabolic diseases that cause high blood sugar. It is a global health problem that increases many complications, including cardiovascular, kidney failure, and eye damage, to name a few. To prevent silent killer disease, early diagnosis and treatment are very important. First, however, a predictive model for diabetes has to overcome the imbalance issue.

In 2019, it was pointed out that diabetes mellitus classification is challenging due to missing values and the class imbalance seen in the diabetes data [30]. Therefore, further $DMPMI$ is proposed. It is a framework that uses oversampling by the adaptive synthetic (ADASYN) method along with the random forest classifier. The said framework outperformed several other candidate machine learning approaches for diabetes classification. Diabetic peripheral neuropathy (DPN) is a type of nerve damage that can occur with diabetes. Magnetic resonance (MR) neuroimaging can effectively classify painful DPN. It, in turn, aids a satisfying pain treatment for a patient. However, the data obtained from MR images are also found imbalanced. Furthermore, oversampling methods combined with classifiers are imposed to overcome the imbalance. It is observed that oversampling, in general, improves the classifier performance.

In recent work, a predictive model that can achieve a high classification accuracy of type 2 diabetes has been developed. According to it, assessing symptoms of diabetes is a challenging task whereby 50% of patients with type 1 and type 2 symptoms remain underdiagnosed due to no noticeable signs of disease. Therefore, their predictive model could assist clinical staff in diagnosing type 2 diabetes more accurately. However, it was found after analyzing the imbalance in the number of samples that were diabetic compared to those that did not have diabetes. Therefore, the SMOTE oversampling method is applied to treat this imbalance [31].

2.5 Rare Disease Prediction with Imbalanced Data

Accurate diagnosis of rare diseases is essential in patient risk mitigation and targeted therapies. However, due to the very nature of rare diseases, collected data become incredibly imbalanced.

In 2016, a cost-sensitive boosted probabilistic classifier was proposed for identifying people with rare diseases using behavioral data. The proposed framework is compared with SMOTE to assess its effectiveness [32]. In 2017, a framework termed hyper-ensemble of SMOTE undersampled random forests was developed for predicting rare diseases. This framework applies SMOTE oversampling on different partitions of the majority class. Afterward, an ensemble of random forest classifiers solves the prediction problem [33]. According to the experiments, rebalancing improved the classifier performance. SMOTE is also combined with logistic regression in a later stage to improve rare event classification [34].

Recently, the observations and opinions of medical domain experts in the form of knowledge graphs have been considered a supplement to the classifier [35]. The classifier, on the other hand, analyzes clinical documents with the help of knowledge graphs. Therefore, this work does not have imbalance as a prime issue. However, oversampling is considered an alternative method. Notably, the performance of oversampling is very unstable; hence, resampling is unsuitable for highly imbalanced text classification tasks.

2.6 *Depression and Suicidal Detection with Imbalanced Data*

This section discusses some more works on imbalance learning relating to depression and suicidal ideation. These works aimed at topics like depression and suicidal detection. These topics should ideally form new categories as above. However, all research works are put forth in the same category because of the lack of many publications. An early attempt to detect suicide attempters among suicide ideators was carried out in 2019 [36]. It uses SMOTE oversampling along with random forest to predict suicide attempters. In this direction, combined SMOTE with different machine learning algorithms is used to predict future suicidal behavior based on data of psychological measures [37]. Online communication channels, and social media, are becoming a novel way for people to express their feelings, suffering, and suicidal tendencies. Hence, data collected from social media may be analyzed to detect future suicidal attempts. In 1921, exciting work was carried out in this direction by collecting data from Twitter and social media [38]. Since the number of suicide intention tweets is much less than all tweets, this leads to an imbalanced classification problem. However, the researchers did not use data-level methods to deal with the imbalance. Instead, new attractive measures are introduced to form association rules for the classification task. Therefore, it may be considered an algorithm-level approach to handle imbalance.

Similarly, depression prediction analysis is also carried out in the literature [39]. A model is introduced for predicting university student depression using SMOTE and borderline-SMOTE oversampling with the random forest classifier. Likewise, smartphone sensor data for identifying behavioral markers indicative of depression is also studied [40]. It uses SMOTE along with several different classifiers for the prediction task. Additionally, SMOTE is also used in another study where a person who is depressed can be found by analyzing various socio-demographic and psychosocial information.

3 Methodologies for Handling Imbalance Data in Healthcare

In Sect. 2, the research works that deal with imbalance are compiled according to the particular healthcare domain. It is justifiable since most articles do not belong to any single category. However, readers with a machine intelligence background may be required to categorize the articles according to the four categories mentioned in Sect. 1. Realizing it, this section discusses the review methodologies adopted to handle an imbalance of healthcare data.

3.1 Algorithm Approach

A good amount of work is available that adopts and tunes existing classification algorithms to make them suitable for imbalanced data in healthcare. One such work is presented as the integration of a novel multilayer classification model and dynamic generative adversarial net [41]. Further, a combination of K-means and C5.0 classifiers is introduced to handle imbalanced data. This approach uses K-means to select certain representatives from the majority and minority classes. Similarly, the K-nearest neighbor and SVM classifier are hybridized to handle imbalances [42].

An improvement of the existing random forest classifier is also proposed, which successfully learns imbalance data [43]. Improvements to the neural network classifier are also proposed [44] to train the classifier to identify hypertensive patients in a highly imbalanced scenario. Research is also done on improving deep learning classifiers. In this direction, researchers studied a semi-supervised deep learning architecture and proposed a weighting scheme for the observations to modify the loss function of this network [45]. In effect, the neural network handled imbalanced data effectively. Moreover, a modified graph convolutional network is proposed to handle imbalanced data with a re-weighting scheme. The resultant classifier determines the boundary between classes with more attention to the actual samples.

3.2 Data-Level Approach

Data-level approaches consider preprocessing to reduce the amount of imbalance either by oversampling the minority class or by undersampling the majority class. These approaches have been the subject of research for a long time, and considerable work is available. Let us further categorize such works into two categories: oversampling and undersampling approaches.

Oversampling Approaches: Most of the oversampling researchers consider SMOTE and its modifications as preprocessing algorithms. For example, in [46], the authors' used naive SMOTE and its significant modifications and evaluated four classifiers to measure performance enhancement after preprocessing. Other recent articles that employed SMOTE are [27, 47]. SMOTE is, in fact, a widely adopted preprocessing technique. Apart from SMOTE, another widely used preprocessing technique is ADASYN [22, 30]. Although this technique is adopted less often than SMOTE, readers may find comparisons of ADASYN with other preprocessing approaches in [22].

Undersampling Approaches: Although oversampling is more popular, significant works have also been done with random undersampling. For example, random undersampling is considered for detecting health reports on social media [48]. Similarly, overlap-based undersampling is used to design a predictive framework for healthcare data [49]. Performance comparison of oversampling and undersampling is also performed considering hepatitis diagnosis [50]. According to the comparison, oversampling outperforms the undersampling technique. Undersampling can also be combined with oversampling. Toward this direction, a framework is proposed that combines undersampling with oversampling and clustering [51]. Such a framework is used for the early prediction of a heart attack. Likewise, undersampling is combined with an ensemble classifier and studied over disease diagnosis [52].

3.3 Cost-Sensitive Approach

As noted in Sect. 1, cost-sensitive approaches lie between algorithm and data-level techniques. Therefore, pure cost-sensitive methods are not very common in healthcare. Among the significant works, GentleBoost ensemble strategy for classification is noteworthy as a cost-sensitive classifier [53]. Further, cost-sensitive classifiers by modifying the objective functions of well-known classification algorithms are developed [54]. Interestingly, the strategy does not alter the original data distribution and hence does not employ any preprocessing.

3.4 Multiple Classifier Ensemble Approach

A significant attempt has been carried out on combining bagging ensemble classifier with random subspace generation strategy to analyze imbalance data [55]. In addition, it uses SVM as base classifiers for the ensemble due to its high accuracy. Another work performs an active example selection for training the ensemble classifier. It adopted the data level and ensemble approaches and outperformed the ensemble with undersampling methods [56].

In recent days, specific significant works have been done on Covid-19 datasets. Among them, an extreme gradient boosting ensemble classifier, along with SMOTE, is applied for disease diagnosis [20]. Further, a decision tree is employed with undersampling and oversampling preprocessing techniques for the disease diagnosis [23]. Likewise, ensemble classier is also combined with clustering-based undersampling strategy [52] for modeling ICU healthcare-associated infections. Additionally, a framework is developed using an ensemble of random forest classifiers and SMOTE to study imbalances [33]. It identifies imbalance-aware machine learning as a critical aspect of designing robust and accurate prediction algorithms.

4 Source of Imbalanced Healthcare Data

With the development of machine learning techniques, it became equally important to conduct experiments on widely and publicly available datasets. Such datasets are well characterized, and the research challenges are well-known. On the other hand, comparison and demonstration of different algorithms become more straightforward and cost-effective with such datasets. Thus, many public repositories consisting of machine learning datasets have been built with a growing need. Here, let us focus on available medical datasets which are imbalanced. This section provides the source repositories of these data to the readers instead of analyzing them. Furthermore, these imbalanced medical data are organized according to their focusing sub-domains.

1. Breast Cancer:
   - Breast Cancer Wisconsin Diagnostic: This database is available in the Kaggle data repository.
   - Breast Cancer Histopathological: This image database is available in the UFPR database.
   - Breast Cancer Image: This database was created and maintained by Andrew Janowczyk.
   - Breast Cancer expression data: This dataset is available at the tissue bank of the breast cancer organization.
2. Liver and Lung Cancer:
   - ILPD data: This dataset is included in UC Irvine (UCI) machine learning repository.
   - India Liver Patient records: This database is available in the Kaggle data repository.
   - Liver Disorders Data: This dataset is included in the UCI machine learning repository.
   - Pathological lung cancer: The lung cancer explorer database created by Quantitative Biomedical Research Center.
   - Lung diagnostic data: Available through Cancer Data Access System of National Cancer Institute.
3. Cerebral Stroke and Diabetes:
   - Cerebral Stroke Prediction: This database is available in the Kaggle data repository.

- ATLAS Dataset: Human brain atlases and databases generated at UNC-Chapel Hill.
- Diabetes Dataset: This dataset is included in the UCI machine learning repository.
- Early prediction diabetes dataset: This dataset is included in the UCI machine learning repository.

4. Covid-19 and Drug Discovery:

- COVID-19 Surveillance Dataset: This dataset is included in the UCI machine learning repository.
- Covid-19 imaging dataset: Available at European Institute For Biomedical Imaging Research.
- Covid chest x-ray dataset: This database is available in the GitHub database.
- Covid-19 CT images: This database is available in the GitHub database.
- Chemical compounds and their biological activities: This database is available in PubChem, an open chemistry database at the National Institutes of Health.
- Drug-like compounds: This database is maintained by EMBL's European Bioinformatics Institute.

5. Depression and Suicide Prediction:

- The Depresjon Dataset: The dataset was collected by Simula research laboratory and SimulaMet to study motor activity in schizophrenia and major depression.
- Kaggle Depression dataset: This database is available in the Kaggle data repository.
- Prediction of suicidal rate: This database is available in the Kaggle data repository.
- Prediction of suicidal attempt: This database is available in the Kaggle data repository.
- Mental health datasets: This database is available in the GitHub database.

5 Conclusions and Future Aspects

This chapter compiles significant recent studies regarding the imbalance classification of healthcare or medical data. Most of these studies consider reducing the degree of imbalance as a remedy, while only a few consider algorithm-level approaches. However, many available studies employed hybrid approaches that do not fall under any particular category. Again, most researchers focused on combining data level with algorithm-level approaches. Also observed that data-level approaches have generally been combined with ensemble methods, although the combination of cost-sensitive approaches with ensemble classifiers has also been observed. Multiple classifier systems produce satisfactory results in most cases. Recent works focus on classifier ensembles, and it may expect significant future applications of this in healthcare data.

Now, considering the research challenges of imbalance classification, let us point out important future directions, especially in the healthcare domain. First, regarding class imbalance, an important issue is the overlap among various classes. Due to a lack of confidence, a classifier often leads to type-I or type-II errors in the overlapped region. Such errors might be crucial, especially when dealing with health-related data. However, this issue is not focused on by researchers so far. Therefore, performance enhancement in the overlapped region might be an essential future aspect. Regarding classifier design, one popular approach toward imbalance is a one-class classifier. Such classifiers can separate the minority class by treating it as an outlier. This approach may be practical for classifying rare diseases, in particular. Besides, class imbalance itself is a well-explored topic in machine learning. Various sophisticated methods have been developed so far. Employing such methods for healthcare data might be another interesting direction for future research.

References

1. He, H., Garcia, E.A.: Learning from imbalanced data. IEEE Trans. Knowl. Data Eng. **21**(9), 1263–1284 (2009)
2. Zheng, Z., Wu, X., Srihari, R.: Feature selection for text categorization on imbalanced data. ACM Sigkdd Explor. Newslett. **6**(1), 80–89 (2004)
3. Prati, R.C., Batista, G.E., Silva, D.F.: Class imbalance revisited: a new experimental setup to assess the performance of treatment methods. Knowl. Inf. Syst. **45**(1), 247–270 (2015)
4. Galar, M., Fernandez, A., Barrenechea, E., Bustince, H., Herrera, F.: A review on ensembles for the class imbalance problem: bagging-, boosting-, and hybrid-based approaches. IEEE Trans. Syst. Man Cybern. **42**(4), 463–484 (2011)
5. Garcia, V., Mollineda, R.A., Sanchez, J.S.: On the k-NN performance in a challenging scenario of imbalance and overlapping. Pattern Anal. Appl. **11**(3), 269–280 (2008)
6. Batuwita, R., Palade, V.: FSVM-CIL: fuzzy support vector machines for class imbalance learning. IEEE Trans. Fuzzy Syst. **18**(3), 558–571 (2010)
7. Cieslak, D.A., Hoens, T.R., Chawla, N.V., Kegelmeyer, W.P.: Hellinger distance decision trees are robust and skew-insensitive. Data Min. Knowl. Discov. **24**(1), 136–158 (2012)
8. García, V., Sánchez, J.S., Mollineda, R.A.: On the effectiveness of preprocessing methods when dealing with different levels of class imbalance. Knowl. Based Syst. **25**(1), 13–21 (2012)
9. Chawla, N.V., Bowyer, K.W., Hall, L.O., Kegelmeyer, W.P.: SMOTE: synthetic minority oversampling technique. J. Artif. Intell. Res. **16**, 321–357 (2002)
10. Branco, P., Torgo, L., Ribeiro, R.P.: A survey of predictive modeling on imbalanced domains. ACM Comput. Surv. (CSUR) **49**(2), 1–50 (2016)
11. Akbani, R., Kwek, S., Japkowicz, N.: Applying support vector machines to imbalanced datasets. In: Proceedings of European Conference on Machine Learning, pp. 39–50. Springer, Berlin (2004)
12. Sun, Y., Kamel, M.S., Wong, A.K., Wang, Y.: Cost-sensitive boosting for classification of imbalanced data. Pattern Recogn. **40**(12), 3358–3378 (2007)
13. Kuncheva, L.I.: Combining Pattern Classifiers: Methods and Algorithms. John Wiley & Sons (2014)
14. Fotouhi, S., Asadi, S., Kattan, M.W.: A comprehensive data level analysis for cancer diagnosis on imbalanced data. J. Biomed. Inform. **90**, 103089 (2019)

15. Ameta, D.: Ensemble classifier approach in breast cancer detection and malignancy grading-a review. arXiv:1704.03801 (2017)
16. Zhang, J., Chen, L., Abid, F.: Prediction of breast cancer from imbalance respect using cluster-based undersampling method. J. Healthcare Eng. **2019** (2019)
17. Reza, M.S., Ma, J.: Imbalanced histopathological breast cancer image classification with convolutional neural network. In: Proceedings of 14th IEEE International Conference on Signal Processing, pp. 619–624. IEEE (2018)
18. Patsadu, O., Tangchitwilaikun, P., Lowsuwankul, S.: Liver cancer patient classification on a multiple-stage using hybrid classification methods. Walailak J. Sci. Technol. **18**(10), 9169–14 (2021)
19. Mohammed, M., Mwambi, H., Mboya, I.B., Elbashir, M.K., Omolo, B.: A stacking ensemble deep learning approach to cancer type classification based on TCGA data. Sci. Rep. **11**(1), 1–22 (2021)
20. AlJame, M., Ahmad, I., Imtiaz, A., Mohammed, A.: Ensemble learning model for diagnosing COVID-19 from routine blood tests. Inf. Med. Unlocked **21**, 100449 (2020)
21. Mursalim, M.K.N., Kurniawan, A.: Multi-kernel CNN block-based detection for COVID-19 with imbalance dataset. Int. J. Electr. Comput. Eng. **11**(3), 2467 (2021)
22. Dorn, M., Grisci, B.I., Narloch, P.H., Feltes, B.C., Avila, E., Kahmann, A., Alho, C.S.: Comparison of machine learning techniques to handle imbalanced COVID-19 CBC datasets. PeerJ Comput. Sci. **7**, e670 (2021)
23. Ahmad, A., Safi, O., Malebary, S., Alesawi, S., Alkayal, E.: Decision tree ensembles to predict coronavirus disease 2019 infection: a comparative study. Complexity **2021**, (2021)
24. Oladunni, T., Tossou, S., Haile, Y., Kidane, A.: COVID-19 County Level Severity Classification with Imbalanced Dataset: A NearMiss Under-sampling Approach. medRxiv (2021)
25. Bernardo, A., Della Valle, E.: Predict COVID-19 Spreading With C-SMOTE. In: Business Information Systems, pp. 27–38 (2021)
26. Santiso, S., Casillas, A., Pérez, A.: The class imbalance problem detecting adverse drug reactions in electronic health records. Health Inform. J. **25**(4), 1768–1778 (2019)
27. Wei, J., Lu, Z., Qiu, K., Li, P., Sun, H.: Predicting drug risk level from adverse drug reactions using SMOTE and machine learning approaches. IEEE Access **8**, 185761–185775 (2020)
28. Korkmaz, S.: Deep learning-based imbalanced data classification for drug discovery. J. Chem. Inform. Model. **60**(9), 4180–4190 (2020)
29. Li, P., Yin, L., Zhao, B., Sun, Y.: Virtual screening of drug proteins based on imbalance data mining. Math. Probl, Eng (2021)
30. Wang, Q., Cao, W., Guo, J., Ren, J., Cheng, Y., Davis, D.N.: DMPMI: an effective diabetes mellitus classification algorithm on imbalanced data with missing values. IEEE Access **7**, 102232–102238 (2019)
31. Roy, K., Ahmad, M., Waqar, K., Priyaah, K., Nebhen, J., Alshamrani, S.S., Ali, I.: An enhanced machine learning framework for type 2 diabetes classification using imbalanced data with missing values. Complexity **2021**, 1–21 (2021)
32. MacLeod, H., Yang, S., Oakes, K., Connelly, K., Natarajan, S.: Identifying rare diseases from behavioural data: a machine learning approach. In: Proceedings of First International Conference on Connected Health: Applications, Systems and Engineering Technologies, pp. 130–139. IEEE (2016)
33. Schubach, M., Re, M., Robinson, P.N., Valentini, G.: Imbalance-aware machine learning for predicting rare and common disease-associated non-coding variants. Sci. Rep. **7**(1), 1–12 (2017)
34. Zhao, Y., Wong, Z.S.Y., Tsui, K.L.: A framework of rebalancing imbalanced healthcare data for rare events' classification: a case of look-alike sound-alike mix-up incident detection. J. Healthcare Eng. **2018** (2018)
35. Li, X., Wang, Y., Wang, D., Yuan, W., Peng, D., Mei, Q.: Improving rare disease classification using imperfect knowledge graph. BMC Med. Inform. Decis. Making **19**(5), 1–10 (2019)
36. Ryu, S., Lee, H., Lee, D.K., Kim, S.W., Kim, C.E.: Detection of suicide attempters among suicide ideators using machine learning. Psychiatry Invest. **16**(8), 588–593 (2019)

37. van Mens, K., de Schepper, C.W.M., Wijnen, B., Koldijk, S.J., Schnack, H., de Looff, P., De Beurs, D.: Predicting future suicidal behaviour in young adults, with different machine learning techniques: A population-based longitudinal study. J. Affect. Disord. **271**, 169–177 (2020)
38. Ben Hassine, M.A., Abdellatif, S., Ben Yahia, S.: A novel imbalanced data classification approach for suicidal ideation detection on social media. Computing **104**(4), 741–765 (2022)
39. Sawangarreerak, S., Thanathamathee, P.: Random forest with sampling techniques for handling imbalanced prediction of university student depression. Information **11**(11), 519 (2020)
40. Asare, K.O., Terhorst, Y., Vega, J., Peltonen, E., Lagerspetz, E., Ferreira, D.: Predicting depression from smartphone behavioral markers using machine learning methods, hyperparameter optimization, and feature importance analysis: exploratory study. JMIR mHealth uHealth **9**(7), e26540 (2021)
41. Zhang, L., Yang, H., Jiang, Z.: Imbalanced biomedical data classification using self-adaptive multilayer ELM combined with dynamic GAN. Biomed. Eng. Online **17**(1), 1–21 (2018)
42. Majid, A., Ali, S., Iqbal, M., Kausar, N.: Prediction of human breast and colon cancers from imbalanced data using nearest neighbor and support vector machines. Comput. Methods Program. Biomed. **113**(3), 792–808 (2014)
43. Paing, M.P., Choomchuay, S.: Improved random forest (RF) classifier for imbalanced classification of lung nodules. In: Proceedings of International Conference on Engineering, Applied Sciences, and Technology, pp. 1–4. IEEE (2018)
44. López-Martínez, F., Núñez-Valdez, E.R., Crespo, R.G., García-Díaz, V.: An artificial neural network approach for predicting hypertension using NHANES data. Sci. Rep. **10**(1), 1–14 (2020)
45. Calderon-Ramirez, S., Yang, S., Moemeni, A., Elizondo, D., Colreavy-Donnelly, S., Chavarría-Estrada, L.F., Molina-Cabello, M.A.: Correcting data imbalance for semi-supervised covid-19 detection using x-ray chest images. Appl. Soft Comput. **111**, 107692 (2021)
46. Teh, K., Armitage, P., Tesfaye, S., Selvarajah, D., Wilkinson, I.D.: Imbalanced learning: Improving classification of diabetic neuropathy from magnetic resonance imaging. Plos One **15**(12), e0243907 (2020)
47. Richardson, A.M., Lidbury, B.A.: Enhancement of hepatitis virus immunoassay outcome predictions in imbalanced routine pathology data by data balancing and feature selection before the application of support vector machines. BMC Med. Inform. Decis. Making **17**(1), 1–11 (2017)
48. Li, X., Wang, Y., Wang, D., Yuan, W., Peng, D., Mei, Q.: Improving rare disease classification using imperfect knowledge graph. BMC Med. Inform. Decis. Making **19**(5), 1–10 (2019)
49. Vuttipittayamongkol, P., Elyan, E.: Overlap-based undersampling method for classification of imbalanced medical datasets. In: Proceedings of IFIP International Conference on Artificial Intelligence Applications and Innovations, pp. 358–369. Springer (2020)
50. Orooji, A., Kermani, F.: Machine learning based methods for handling imbalanced data in hepatitis diagnosis. Front. Health Inform. **10**(1), 57 (2021)
51. Wang, M., Yao, X., Chen, Y.: An imbalanced-data processing algorithm for the prediction of heart attack in stroke patients. IEEE Access **9**, 25394–25404 (2021)
52. Sánchez-Hernández, F., Ballesteros-Herráez, J.C., Kraiem, M.S., Sánchez-Barba, M., Moreno-García, M.N.: Predictive modeling of ICU healthcare-associated infections from imbalanced data. Using ensembles and a clustering-based undersampling approach. Appl. Sci. **9**(24), 5287 (2019)
53. Ali, S., Majid, A., Javed, S.G., Sattar, M.: Can-CSC-GBE: Developing Cost-sensitive Classifier with Gentleboost Ensemble for breast cancer classification using protein amino acids and imbalanced data. Comput. Biol. Med. **73**, 38–46 (2016)
54. Mienye, I.D., Sun, Y.: Performance analysis of cost-sensitive learning methods with application to imbalanced medical data. Inform. Med. Unlocked **25**, 100690 (2021)
55. Yu, H., Ni, J.: An improved ensemble learning method for classifying high-dimensional and imbalanced biomedicine data. IEEE Trans. Comput. Biol. Bioinform. **11**(4), 657–666 (2014)
56. Oh, S., Lee, M.S., Zhang, B.T.: Ensemble learning with active example selection for imbalanced biomedical data classification. IEEE Trans. Comput. Biol. Bioinform. **8**(2), 316–325 (2010)

A Review on Metaheuristic Approaches for Optimization Problems

Rasmita Rautray, Rasmita Dash, Rajashree Dash, Rakesh Chandra Balabantaray, and Shanti Priya Parida

Abstract Based on various sources of inspiration, there are different metaheuristic algorithms, such as swarm intelligence, evolutionary computation, and bio-inspired algorithms, available in the literature. Among these algorithms, few approaches have been proven representative of each category for their excellence in solving a critical real-world problem in various application areas. However, some approaches are inadequately interpreted. A comprehensive general survey of these algorithms is rarely seen in the literature. Therefore, this study aims at a comprehensive view of metaheuristic approaches, performing a survey on existing literature. It explores all the different dimensions of metaheuristic approaches, their representations, improvement and hybridization, successful application, research gaps, and future direction.

1 Introduction

Optimization is indispensable in every real-world application. Optimization techniques are intelligent approaches, basically applied for extracting the best solution for a given assignment with certain constraints out of all available solutions. However, the complexities of optimization problems are very high because it has many locally optimal solutions. The characteristic of this problem again belongs to differ-

R. Rautray (✉) · R. Dash · R. Dash
Siksha 'O' Anusandhan, Deemed to be University, Bhubaneswar, Odisha, India
e-mail: rashmitaroutray@soa.ac.in

R. Dash
e-mail: rashmitadash@soa.ac.in

R. Dash
e-mail: rajashreedash@soa.ac.in

R. Chandra Balabantaray
IIIT, Bhubaneswar, Odisha, India

S. P. Parida
Idiap Research Institute, Martigny, Switzerland

D. P. Acharjya and K. Ma (eds.), *Computational Intelligence in Healthcare Informatics*, Studies in Computational Intelligence 1132,
https://doi.org/10.1007/978-981-99-8853-2_3

ent groups such as constrained or un-constrained, discrete or continuous, static or dynamic, single-objective or multi-objective.

Broadly, the optimization techniques are categorized into two types: the traditional approach and the metaheuristic approach [1]. However, many critical real-life problems exist in engineering, economics, transportation, business, and social science, which cannot be solved in polynomial time using traditional approaches [2]. Thus to improve the findings and accuracy of these problems, researchers are directed toward metaheuristic approaches. Since that, metaheuristic techniques have earned much more popularity in solving complex optimization problems for their simplicity, method of implementation, and robustness.

Metaheuristic algorithms are classified into five basic categories such as swarm intelligence algorithms (SIA), evolutionary algorithms (EA), bio-inspired algorithms (BIA), physics and chemistry-based algorithms (PCA), and other algorithms (OA). Swarm intelligence (SI) based optimization approaches are inspired by creatures' social behavior, which means the collective behavior of intelligent individuals such as birds, ants, bees, and bats. It emphasizes collective intelligence, which shows single intelligence may not be good enough. However, intelligence within a group of individuals can lead to the optimal solution. Some algorithms show the successful implementation of SI-based optimization techniques in various application areas, such as particle swarm optimization (PSO), ant colony optimization (ACO), artificial bee colony algorithms (ABC), cuckoo search optimization (CSO), and cat swarm algorithm (CSA). Various categories of metaheuristic algorithms are presented in Fig. 1.

The second categories of metaheuristic algorithms are evolutionary algorithms. These algorithms are inspired by biological evolution depending on the survival of fitness, implementing the Darwinian theory of evolution. EA implements reproduction,

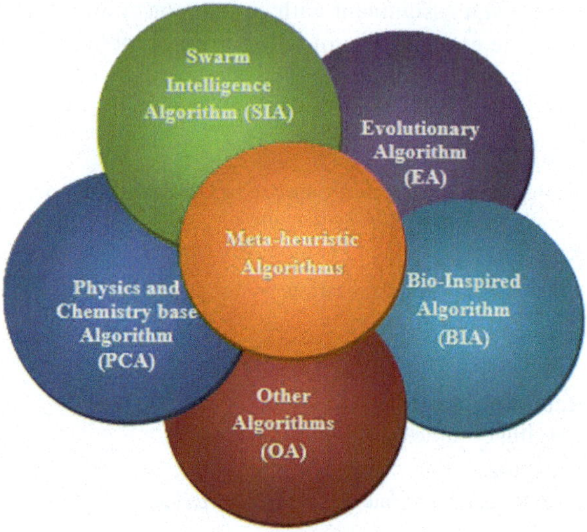

Fig. 1 Categories of metaheuristic algorithms

mutation, and selection operations to obtain a new solution. A few representatives of EA are genetic algorithm (GA), differential evolution (DE), genetic programming (GP), and evolutionary strategy (ES). The third categories of metaheuristic optimization approaches are bio-inspired algorithms. As per these algorithms, computational optimization analysis is adopted through organisms' biological activities. Here the optimization process takes advantage of both individual animal behavior and the collective behavior of all the animals. It mimics the characteristics and movement of intelligent animals such as dolphins, chimpanzees, bats, and fireflies.

There exist a few other categories of metaheuristic algorithms where inspiration comes from physics and chemistry rather than swarm or biology. It follows some principles or laws of physics and chemistry. However, breaking them into either physics or chemistry is difficult because some rules are standard or applicable to both streams. A few successful metaheuristic algorithms of these categories are simulated annealing, gravitational search, and harmony search. Few other algorithms cannot be categorized into any of the abovementioned groups. These algorithms are developed from different sources such as emotion, wind, water behavior, and teaching-the learning process. This chapter presents an overview of these techniques, along with their applications.

In the rest of this chapter, the following points are emphasized. Section 2 presents a brief and comprehensive list of all groups of algorithms and their real-life application areas. Section 3 highlights representatives of each category of metaheuristic approaches with their importance, basic concepts, and pseudo code. Finally, Sect. 4 concludes with some suggestions. This survey cannot claim that categorization is a unique type. However, this survey is a good attempt to cover maximum metaheuristic algorithms.

2 Metaheuristic Approach and Types

In the introductory part, the authors have tried to highlight different types of metaheuristic approaches. This section covers a detailed description of each category of metaheuristic approaches, along with their class and origin.

2.1 Swarm Intelligence Algorithms

The SI is one of the branches of computational intelligence that inspires the collective behavior of groups or self-organizing societies of animals. In literature, many naturalists have been studying the behavior of animals because of their efficiency in solving complex problems. This problem includes finding the shortest path between their food source and nest and organizing nests. However, the behavior of swarms is also used to find prey or mating as numerical optimization techniques. Due to robustness and simplicity, SI-based algorithms have been used in many application areas. The various SI algorithms from the scratch are highlighted in Tables 1 and 2. Besides, one widely used algorithm considered representative of SI is explained subsequently.

In SIA, the colony responds to internal disturbances and external challenges for any number of agents. There is no central control in the colony. The SIA-based algorithms are robust, adaptive, and speedy. The major limitation is that it is challenging to predict the behavior from the individual rules. However, a slight change in the rules results in different group-level behavior. Also, the functions of the ant colony could not be understood without the knowledge of an agent's functioning. Simultaneously, its behavior looks like noise as the action of choice is stochastic.

2.2 Evolutionary Algorithm

Evolutionary algorithms (EA) are randomized search heuristics based on the evolutionary theory of Charles Darwin. The basis of the approach is a natural selection from an individual population exposed to environmental pressure, which is called survival of the fittest. The population's fitness is measured regarding the degree of adaption of an organism to the environment. The higher fitness value shows the organism is more fit and accommodated to the environment. However, EAs are mainly focused on mechanisms characterized by the evolutionary process of biology. The various types of EAs are illustrated in Table 3.

The evolutionary computation algorithm is conceptually simple. It supports parallelism and robustness for dynamic change. These algorithms are hybridized with other methods and used in various application areas due to their simplicity. The major limitation is that an algorithm's outcome depends on many self-adaptive parameters. Furthermore, tuning parameters and use of functions meta-model make this algorithm computationally expensive. Thus, there is no guarantee of obtaining an optimal solution within a finite time.

2.3 Bio Inspired Algorithms

The biological evolution of nature inspires the BIA. It refers to the biological behaviors of animals or birds, which are used to design novel and robust competing techniques. BIA's key feature is maintaining equilibrium between diversity and solution speed of techniques. Besides that, it also finds the global best solution with less computing effort. Few BIAs are highlighted in Table 4.

The main advantage is that these algorithms are proven their efficiency and robustness by the natural selection process. However, it deals with a too complex problem that is not solved by a human-derived solution. Besides, it has some limitations. The major disadvantage is that sometimes BIAs may lead to slow and premature convergence in local minima. Therefore, it gives less accurate results. Moreover, the

Table 1 Swarm intelligence algorithms

Sl. No.	Algorithm name	Inspired by	Year
1	Particle swarm optimization	Social behavior of bird flock	1995
2	Ant colony system	Foraging behavior of ants	1997
3	Bee colony optimization	Foraging habits of honeybees	2005
4	Ant colony optimization	Behavior of real ants	2006
5	Firefly algorithm	Social behavior of fireflies	2008
6	Fast bacterial swarming algorithm	Swarming behaviors of bacteria	2008
7	Beehive	Foraging behavior of honey bee swarm	2009
8	Honey bees mating optimization	Mating behavior of the bumble bees	2009
9	Bat algorithm	Echolocation behavior of bats	2010
10	Firework algorithm	Process of fireworks explosion	2010
11	Social emotional optimization algorithm	Human behavior guided by emotion	2010
12	Bees swarm optimization	Foraging behavior of honey bees	2012
13	Fruit fly optimization	Foraging behavior of fruit flies	2012
14	Green herons optimization algorithm	Fishing skills of the bird	2013
15	Bumble bees mating optimization	Mating behavior of the bumble bees	2014
16	Foraging agent swarm optimization	–	2014
17	Radial movement optimization	–	2014
18	Pigeon optimization algorithm	Swarm behavior of passenger pigeons	2014
19	Simplified swarm optimization	–	2015
20	Monarch butterfly optimization	Migration of monarch butterflies	2015
21	Cat swarm optimization algorithm	Behavior of cats	2015
22	Jaguar algorithm	Hunting behavior of jaguar	2015
23	Locust swarm algorithm	Hunting mechanism of jaguar	2015
24	African buffalo optimization	Behavior of African buffalos	2016
25	Galactic swarm optimization algorithm	Galactic motion	2016
26	Spider monkey optimization	Behavior of spider monkeys	2016
27	AntStar	Integrating the AS and A* algorithm	2016
28	Crow search algorithm	Food hiding behavior of crows	2016
29	Raven roosting optimization algorithm	Foraging activities of various organisms	2016
30	Bacterial foraging optimization	Foraging behavior of escherichia	2016
31	Selfish herd optimizer	Hamilton's selfish herd theory	2017
32	Mouth brooding fish algorithm	Life cycle of mouthbrooding fish	2017
33	Weighted superposition attraction	Superposition principle	2017
34	Butterfly-inspired algorithm	Mate searching mechanism of butterfly	2017
35	Beetle antennae search	Searching behavior of longhorn beetles	2017
36	Grasshopper optimization algorithm	Swarming behavior of grasshoppers	2017

Table 2 Swarm intelligence algorithms (continued)

Sl. No.	Algorithm name	Inspired by	Year
37	Artificial bee colony	Foraging and waggle dance behaviors of real bee	2005
38	Glowworm swarm optimization	Animal searching behavior and group living theory	2006
39	Intelligent water drops	Observing natural water drops that flow in rivers	2008
40	Monkey algorithm	Monkey climbing process on trees while looking for food	2008
41	Consultant-guided search	Decisions of real people based on advice received from consultants	2010
42	Cyber swarm algorithm	Social behavior of bird flock with the strategies of scatter search and path relinking	2010
43	Marriage in honey bees optimization	Mating and insemination process of honey bees	2010
44	Stochastic diffusion search	Direct communication between the agents similar to the tandem calling mechanism employed by one species of ants	2011
45	Social spider optimization	Foraging strategy of social spiders by utilizing the vibrations on the spider web to determine the positions of preys	2013
46	Seven-spot ladybird optimization	Foraging behavior of a seven-spot ladybird	2013
47	Wolf pack algorithm	Hunting behaviors and distributive mode for prey	2014
48	Cuckoo search	The obligate brood parasitism of some cuckoo species	2014
49	Smart dispatching and metaheuristic swarm flow	Natural systems and processes	2014
50	Shark smell optimization	Ability of shark in finding its prey by smell sense	2014
51	Seeker optimization algorithm	Simulate the act of humans' intelligent search with their memory	2014
52	Artificial fish swarm algorithm	Collective movement and social behaviors of the fish	2015
53	Grey wolf optimizer	Leadership hierarchy and hunting mechanism of Grey wolves in nature	2015
54	Social network-based swarm optimization	Neighborhood strategy and individual learning behavior	2015
55	Weighted superposition attraction	Superposition principle in combination with the attracted movement of agents	2015
56	Krill herd algorithm	Herding behavior of krill individuals in nature	2015
57	Dragonfly algorithm	Static and dynamic swarming behaviors of dragonflies	2016
58	Across neighborhood search	Assumptions and issues of population-based search algorithms	2016
59	Dolphin swarm optimization algorithm	Mechanism of dolphins in detecting, chasing and preying on swarms of sardines	2016
60	Salp swarm algorithm	Swarming behavior of salps during navigating and foraging in oceans	2017

Table 3 Evolutionary algorithms

Sl. No.	Algorithm name	Inspired by	Year
1	Genetic algorithm	Charles Darwin's natural evolution	1992
2	Shuffled complex evolution	Synthesis of global optimization	1993
3	Evolutionary algorithm	Biological evolution	1994
4	Differential evaluation	Natural evolution	1997
5	Grammatical evolution algorithm	Biological process	1998
6	Evolutionary Programming	Evolution by means of natural selection	1999
7	Evolution strategies	Natural evolution	2002
8	Weed optimization algorithm	Colonizing weeds	2006
9	Contour gradient optimization	Local cooperation behavior in real-world	2013
10	Backtracking search optimization	Natural evolution	2013
11	Differential search	Migration of super organisms	2014
12	Hyper-spherical search	Hyper-sphere center and its particle	2014
13	Cultural algorithm	Principle of cultural evolution	2016
14	Imperialistic competitive	Socio-political behaviors	2016
15	Fish electrolocation optimization	Electrolocation principle of elephant nose fish and shark	2017

exploitation rate and exploration are unreasonable in search space and unsuitable for continuous constrained optimization problems.

2.4 *Physics and Chemistry Based Algorithms*

The physics and chemistry-based algorithms are inspired by the law and concepts of physics and chemistry. The various laws and concepts followed to design optimization techniques are gravitational law and Newton's law of motion, concepts of thermodynamics, gravity, and many more. Few PCA algorithms are highlighted in Table 5.

2.5 *Other Algorithms*

The algorithms inspired by nature but cannot be social or emotional are presented here. The names of such algorithms are listed in Table 6.

Table 4 Bio-inspired algorithms

Sl. No.	Algorithm name	Inspired by	Year
1	Biogeography-based optimization	Species distribution through time and space	2008
2	Roach infestation algorithm	Social behavior of cockroaches	2008
3	Paddy field algorithm	Growth process of paddy field	2009
4	Group search optimizer algorithm	Animal searching behavior	2009
5	Artificial algae algorithm	Living behaviors of microalgae	2010
6	Invasive weed optimization	Colonizing weeds	2010
7	Brain storm optimization algorithm	Collective behavior of insects like ants, bee	2011
8	Eco-inspired evolutionary algorithm	Habitats, and successive relations	2011
9	Bacterial colony optimization	Behaviors of escherichia lifecycle	2012
10	Japanese tree frogs calling	Male Japanese tree frogs	2012
11	Great salmon run	Annual natural events in the North America	2012
12	Egyptian vulture	Egyptian vultures for acquiring food	2013
13	Atmosphere clouds model	Cloud in the natural world	2013
14	Coral reefs optimization algorithm	Corals' biology and reefs formation	2014
15	Flower pollination algorithm	Pollination process of flowers	2014
16	Chicken swarm optimization	Behaviors of chicken, roosters, and hens	2014
17	Lifecycle-based swarm optimization	Biology life cycle	2014
18	Artificial root foraging algorithm	Root foraging behaviors	2014
19	Bottlenose dolphin optimization	Foraging behavior of bottlenose dolphin	2015
20	Elephant search algorithm	Behavioral aspects of elephant herds	2015
21	Lion optimization algorithm	Cooperation characteristics of lions	2016
22	Dolphin echolocation optimization	Search space due to echolocation	2016
23	Whale optimization algorithm	Hunting behavior of humpback whales	2016
24	Spotted hyena optimizer	Social behavior of spotted hyenas	2017
25	Squirrel search	Foraging behavior of flying squirrels	2018

3 Category Wise Representatives of Metaheuristic Approaches

According to applications of different metaheuristic approaches, few techniques have been considered representative of each category of these intelligent approaches. Therefore, a detailed discussion of category-wise representative algorithms is highlighted in the following sub-sections.

Table 5 Physics and chemistry-based algorithms

Sl. No.	Algorithm name	Inspired by	Year
1	Simulated annealing	Annealing in metallurgy	1983
2	Harmony search	Improvisation process of jazz musicians	2001
3	Big bang big crunch	Ordering of randomly distributed particles	2006
4	Gravitational search algorithm	Law of gravity and mass interactions	2009
5	Charged system search	Interaction between charged particles in electric field	2010
6	Galaxy-based search algorithm	Spiral arm of spiral galaxies	2011
7	Ray optimization	Snell's light refraction law	2012
8	Gases Brownian motion	Brownian, turbulent rotation motion of gas	2013
9	Magnetic charged system search	Interaction between charged particles in magnetic field	2013
10	Central force optimization	Metaphor of gravitational kinematics	2015
11	Chemical reaction optimization	Nature of chemical reactions	2015
12	Gradient gravitational search	Gravity force	2015
13	Gradient evolution algorithm	Gradient-based search method	2015
14	Electromagnetic field optimization	Electromagnets behavior with different polarities and golden ratio	2016
15	Electro search algorithm	Electrons orbital movement of atom nucleus	2017
16	Thermal exchange optimization	Newton's law of cooling	2017
17	Lightning procedure optimization	Lightning attachment process	2017

3.1 Cuckoo Search Algorithm

Cuckoo search (CS) is one of the latest and most widely used SI-based metaheuristic algorithms. CS algorithm is inspired by the kind of bird species called the cuckoo. Cuckoos are fascinating birds because of their aggressive reproduction strategy and the beautiful sounds; they can make. The adult cuckoos lay their eggs in the nests of other host birds, or species [3]. The egg in the nest represents a solution, and one egg can be laid by each cuckoo, which means a new and possibly better solution. Three idealized rules can characterize the standard CS algorithm. Initially, the cuckoo lays one egg in a random nest, representing a solution. Then, the nests containing the best eggs will take over to the next generation. Further, a host bird with a probability (P_a) can discover an alien egg for the fixed number of available nests. If this condition satisfies, either the egg can be discarded or abandoned the nest by the host and built a new nest elsewhere.

Table 6 Other algorithms

Sl. No.	Algorithm name	Inspired by	Year
1	Greedy randomized adaptive search	–	1989
2	Tabu search	–	1989
3	Variable neighborhood search	Local search	1997
4	Termite colony optimization	Behaviors of termites	2006
5	Water flow-like algorithm	Water flow from higher to lower	2007
6	Warping search	Movements of celestial bodies	2008
7	Key cutting algorithm	Work of locksmiths to defeat the lock	2009
8	Mosquito host-seeking algorithm	Host-seeking behavior of mosquitoes	2009
9	Human inspired algorithm	Search strategies of mountain climbers	2009
10	League championship algorithm	Competition of sport teams in a sport league	2009
11	Coalition-based metaheuristic	–	2010
12	Fireworks algorithm	Fireworks explosion	2010
13	Hunting search	Group hunting of animals	2010
14	Wind driven optimization	–	2010
15	Anarchic society optimization	Anarchical behavior of social group	2011
16	Teaching-learning optimization	Classroom teaching-learning	2011
17	Spiral dynamics algorithm	Spiral phenomena	2011
18	Migrating birds optimization	V flight formation of the migrating birds	2012
19	Mine blast algorithm	Mine bomb explosion	2013
20	Colliding bodies optimization	Collision between objects	2014
21	Golden ball algorithm	Soccer concepts	2014
22	Interior search algorithm	Interior design and decoration	2014
23	Adaptive dimensional search	–	2015
24	Cloud particles differential evolution	Matter state transition and cloud formation	2015
25	Enhanced best performance	Local search	2015
26	Leaders and followers	–	2015
27	Moth-flame optimization	Navigation method of moths in nature	2015
28	Runner-root algorithm	Plants such as strawberry and spider plant	2015
29	Stochastic fractal search	Natural phenomenon of growth	2015
30	Search group algorithm	–	2015
31	Vortex search algorithm	Vortex pattern of vortical flow	2015
32	Water cycle algorithm	Water cycle process	2015
33	Water wave optimization	Shallow water wave theory	2015
34	Lightning search algorithm	Natural phenomenon of lightning	2015
36	Joint operations algorithm	Joint operations strategy of military units	2016
38	Open source development model	Open source development model	2016
39	Passing vehicle search	Overtaking of vehicles on highway	2016
40	Sine Cosine algorithm	–	2016
41	Symbiotic organism search	Relationship among the living beings	2016
42	Virus colony search	Diffusion and infection strategies of virus	2016
43	Water evaporation optimization	Evaporation of small quantity water particles	2016
44	Multi-verse optimizer	Multi-verse theory	2016

For easy implementation, the CS algorithm can be used for each nest with a single egg. There is no difference between a nest, egg, or cuckoo in such a situation. The nest corresponds to an egg that represents one cuckoo. In some complicated cases, the algorithm can be extended to each nest with multiple eggs that represent a set of solutions. For new solutions, x_i^{t+1} generation, a switching parameter, P_a, is used to control the balanced combination of a local random walk and the global explorative random walk. The local random walk is defined in Eq. 1, where x_j^t and x_k^t are two different solutions selected randomly by random permutation, $H(u)$ is a Heaviside function, ε is a random number drawn from a uniform distribution and S is the step size.

$$x_i^{t+1} = x_i^t + \alpha \times S \otimes H(P_a - \varepsilon) \otimes (x_j^t - x_k^t) \tag{1}$$

On the other hand, the global random walk is carried out by using levy flights. A levy flight contains successive random steps [4]. A sequence of rapid jumps characterizes it. The rapid jumps are represented using the Eq. 2, where α is step size and is proportional to scale of optimization problem, i.e., $\alpha > 0$. The notion $\otimes$ is entry wise move during multiplication, and $Levy(\lambda)$ is random numbers drawn from the Levy distribution. The workflow of the cuckoo search algorithm is depicted in Fig. 2.

$$x_i^{t+1} = x_i^t + \alpha \otimes Levy(\lambda) \tag{2}$$

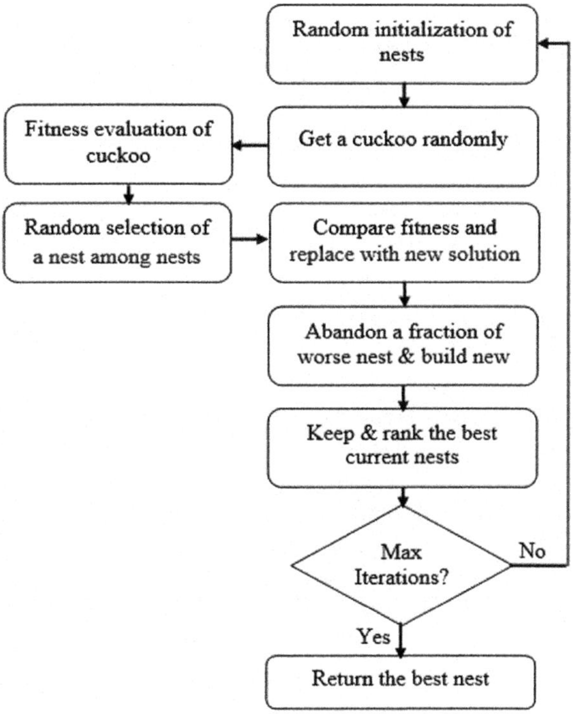

Fig. 2 Workflow of cuckoo search algorithm

Algorithm 1 (Cuckoo Search Algorithm)

1. *Random initialization of n host nests*
2. *While (!Termination Condition) do*
3. *Random generation of a cuckoo*
4. *Fitness evaluation of cuckoo, F*
5. *Random selection of a nest among n nests, j (say)*
6. *If* $F_i > F_j$
7. *Replace the new solution with j*
8. *Abandon a fraction* (P_a) *of the worse nest and build new ones*
9. *Keep and rank the best current nests*
10. *Return the best nest*

Thermal system, heat transfer problems, clustering, object tracker and Kalman filter, traffic signal controller, engine optimization, hydrothermal scheduling, network configuration, power system, computer security, heat exchangers, energy conservation, solar radiation, sheet nesting problem, phase equilibrium problem, phase stability calculation, fuzzy system, Software engineering, satellite image segmentation, FIR Differentiator design, speech signal, information retrieval are various applications of the cuckoo search algorithm. Detail description is presented in Table 7.

Table 7 Description of few variants and application of CS algorithm

Sl. No.	Variation of CS algorithm	Application
1	Multi-objective cuckoo search	Electrical machines and transformers [5]
2	Improved binary cuckoo search	Electric power generation system [6]
3	Extended cuckoo search	Motion tracking [7]
4	Traditional	Magnetic flux leakage signal [8]
5	Hybrid cuckoo search	Internet of things [9]
6	Efficient cuckoo search	Fault diagnosis [10]
7	Improved cuckoo search	Monopulse antennas [11]
8	Adaptive cuckoo search	Satellite images [12]
9	Hybrid cuckoo search	Chaotic system [13]
10	Hierarchical cuckoo search	Wireless communication systems [14]
11	Traditional	Auto industry [15]
12	Traditional	Multi-machine power systems [16]
13	Adaptive cuckoo search	Wiener filter [17]
14	Modified cuckoo search	Power dispatch [18]
15	Hybrid cuckoo search	Concentric circular antenna array [19]

3.2 Genetic Algorithm

The genetic Algorithm (GA) is an evolutionary-based optimization algorithm. In the early 1970s, John Holland invented GA. Later, it became widely popular by Goldberg [20, 21]. It is a stochastic search method based on the concept of Darwin's theory of evolution. GA provides the best solution for the evaluation function of an optimization problem. Simultaneously, it deals with many solutions and one evaluation function. Therefore, each solution is coded as a "string". The "string" is called an individual chromosome for solving an optimization problem. A collection of such chromosomes is called a population. Initially, the finite size of the randomly generated population is processed [22]. Then, algorithm processing evolves three bio-inspired operators, selection, crossover, and mutation, to develop a new population from the current population. The selection procedure examines the survival of the fittest, whereas crossover represents mating between chromosomes, and mutation introduces the random modifications of chromosomes. The overview of the algorithm of GA is presented below. The workflow of GA is shown in Fig. 3. The GA has been extended in many directions. A short overview of various extensions and applications is presented below.

Fig. 3 Workflow of genetic algorithm

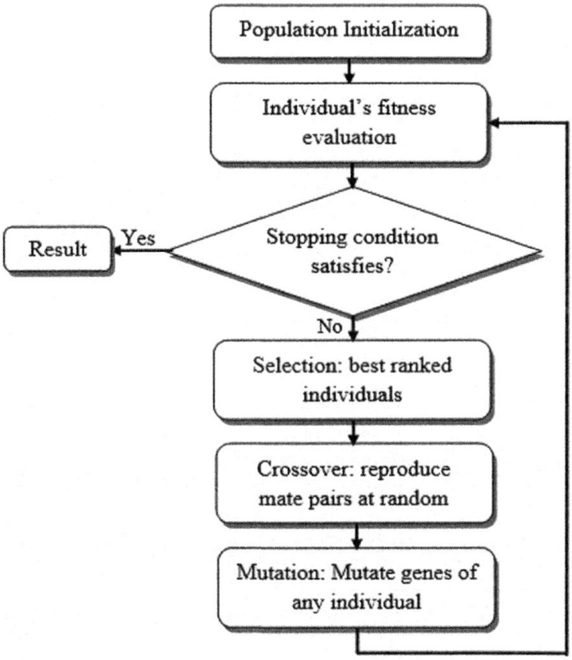

Algorithm 2 (Genetic Algorithm)

1. *Initialization of the population*
2. *Evaluate each individual's fitness*
3. *While (! Termination condition) do*
4. *Select best ranked individuals*
5. *To reproduce mate pairs at random, apply the crossover operator*
6. *Apply mutation operator*
7. *Evaluate each individual's fitness*
8. *Return the best individual*

The GA is generally used in various kinds of optimization problems such as data mining, website optimizations, text classification, computer-aided design, path planning of mobile robots, transportation problems, scheduling problems, assignment problems, flight control system design, pattern recognition, reactive power dispatch, sensor-based path planning of robot, training of different types of the neural network, information retrieval, vehicle routing problem, wireless sensor networks, software engineering problems, pollutant emission reduction problem in the manufacturing industry, power system optimization problems, portfolio optimization, web page classification system, closest string problem in bioinformatics. Besides, GA is also used in other application areas, as discussed in Table 8.

Table 8 Description of few variants and application of genetic algorithm

Sl. No.	Variation of GA	Application
1	Traditional GA	Augmented reality [23]
2	Real coded GA	Nonlinear least square regression [24]
3	Robust GA	Electromagnetic machines and devices [25]
4	Multi-objective GA	Booster stations in a drinking water distribution system [26]
5	Traditional GA	Helical antenna [27]
6	Traditional GA	Integrated starter alternator [28]
7	Traditional GA	Cancer prevention [29]
8	Traditional GA	Radar application [30]
9	Traditional GA	Vehicle-to-grid [31]
10	Hybrid GA	Induction motor [32]
11	Niche GA	Robot manufacturing [33]
12	Traditional GA	Real-time traffic surveillance system [34]
13	Traditional GA	Scientific applications [35]
14	Hybrid GA	Antenna design [36]
15	Taguchi GA	Marine vessel propulsion system [37]

3.3 Harmony Search

In 2001, Harmony Search (HS) algorithm was introduced [38]. It is a stochastic metaheuristic algorithm. This algorithm is motivated by the different improvisation processes of music to achieve optimal output or perfect harmony. It can also be obtained by pitch adjustment, unlike exploration of perfect harmony state by improvising instrument pitches [39–41]. The HS algorithm comprises three basic steps: initialization of population and parameters, improvisation of harmony vector, and harmony memory update. The illustrations of the steps are presented below. Further, the algorithm is presented.

Step 1: The parameters involved in HS algorithm are size of harmony memory ($HMS \geq 1$), consideration rate of harmony memory ($0 \leq HMCR \leq 1$), the pitch adjusting rate ($0 \leq PAR \leq 1$), the distance bandwidth ($BW \in R^N > 0$).

Step 2: Create a new harmony vector $X^{new} = (x_1^{new}, x_2^{new}, \cdots, x_n^{new})$, which uses three improvisation rules such as memory consideration, pitch adjustment and random selection is given by Eq. 3.

$$x_j^{new} = x_j^{new} + r_3 \times BW \tag{3}$$

Step 3: Update the HM as $X^{worst} = X^{new}$, if $f(X^{new} < X^{worst})$, where X^{new} is new survival vector, X^{worst} is the worst harmony vector and f is fitness value comparison of survival vector and worst harmony vector.

Based on the above concept, the HS algorithm involves the following steps. The workflow diagram of the HS algorithm is presented in Fig. 4. Further, various applications of the HS algorithm are presented.

Algorithm 3 (Harmony Search)

1. *Initialization of population and parameters*
2. *Initialization of Harmony memory*
3. *While (! Termination condition) do*
4.   *Improvisation new harmony*
5.   *Updation of harmony memory*
6. *Write Result*

The various applications of the HS algorithm include two-bar truss problem, pressure vessel design, optimum design of R.C. concrete beam, steel engineering problems, shell, and tube heat exchangers, wireless sensor network, construction, and engineering structures, water and groundwater system management, robotics, application in medical issues, power, and energy, feature selection, clustering and scheduling, information retrieval, chemical engineering, visual tracking, bioinformatics, mathematics, logistics, image processing, neural network training, puzzle solving, commercial consulting services, stock market prediction. In brief, it is presented in Table 9.

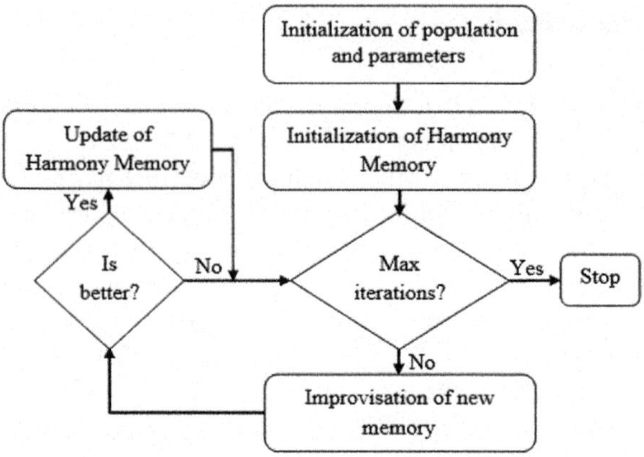

Fig. 4 Workflow of HS algorithm

Table 9 Description of few variants and application of HS algorithm

Sl. No.	Variation of HS	Application
1	Gaussian HS algorithm	Loney's solenoid problem [42]
2	Improved HS algorithm	Thrust allocation problem [43]
3	Traditional HS algorithm	Electromagnetic railgun [44]
4	Traditional HS algorithm	Transformer design [45]
5	Multi-population HS algorithm	Underwater acoustic sensor networks [46]
6	Traditional HS algorithm	Sparse linear arrays [47]
7	New harmony search	Transmission line [48]
8	New harmony search	Semiconductor manufacturing [49]
9	Improved harmony search	Distributed generation problem [50]
10	Quasi-oppositional HS algorithm	Power system [51]
11	Improved HS algorithm	Constrained numerical optimization [52]
12	Levy-harmony algorithm	Microgrid [53]
13	Improved harmony search	Traffic flow forecasting [54]
14	Modified HS algorithm	Magnetic flux leakage [55]
15	Adaptive HS algorithm	Linear quadratic regulator design [56]

3.4 *Biogeography Based Optimization*

Biogeography-based optimization (BBO) is a metaheuristic optimization approach. It is based on the equilibrium theory of biogeography. This theory illustrates how species migrate from one habitat to another, how new species arise, and how species become extinct depending on different factors. These factors include rainfall, diversity of topographic features, temperature, and land area. An island is the habitat of a species if it is isolated from other habitats and its high suitability index (HSI) is high. It means the habitat is geographically well-suitable for species living. There is the possibility of changing species' habitat into their neighboring habitat if more species reside in a habitat with high HSI or the habitat has low HSI. This process is named emigration. However, the species changing the habitat with high HSI and having few species is called immigration. The emigration and immigration of species in a habitat are known as migration [57].

From an implementation point of view, each candidate solution to an optimization problem is considered a habitat with HSI. The HSI is the same as the fitness value in other evolutionary computations. The different features of solutions are called suitability index variables (SIV). New candidate solutions can be generated using two operators: migration and mutation. Migration is the process of changing the existing SIVs of habitat. For improvisation of the solution, features of habitat are mixed based on immigration rate λ and emigration rate μ. Each individual has its immigration rate λ and emigration rate μ. The immigration rate is defined in Eq. 4 whereas emigration rate is defined in Eq. 5, where IM and EM are the maximum immigration rate and emigration rate respectively. p is the number of species in the pth habitat. N is the maximum number of species or population size.

$$\lambda_p = IM\left(1 - \frac{p}{N}\right) \tag{4}$$

$$\mu_p = EM\left(\frac{p}{N}\right) \tag{5}$$

Change habitat or generate a solution to an optimization problem is modified based on its probability, PR, that a single habitat contains p species. PR_p can be the change time T to $(T + \Delta T)$. The probability at time $(T + \Delta T)$ is defined in Eq. 6.

$$PR_p(T + \Delta T) = \begin{array}{l} PR_p(T)(1 - \lambda_p \Delta T - \mu_p \Delta T) + \\ PR_{p-1}\lambda_{p-1}\Delta T + PR_{p+1}\mu_{p+1}\Delta T \end{array} \tag{6}$$

In the above Eq. 6, it holds the condition to evaluate p species at the time. In the first part, no immigration or emigration occurred between T and $T + \Delta T$ for p species. In the second and third parts, one species emigrated from other habitats at time T for species $(p - 1)$ and $(p + 1)$ respectively. Assume that ΔT is very small, so the probability for one or more immigration and emigration is ignored. On taking the limit $\Delta T \to 0$, PR_p is defined as in Eq. 7.

$$PR_p = \begin{cases} -\lambda_0 PR_0 + \mu_1 PR_1 & \text{if } p = 0 \\ -(\lambda_p + \mu_p)PR_p + \lambda_{p-1}PR_{p-1} + \mu_{p+1}PR_{p+1} & \text{if } 1 \le p \le (N-1) \\ -\mu_N PR_N + \lambda_{N-1}PR_{N-1} & \text{if } p = N \end{cases} \tag{7}$$

When $\mu_p = 0$ and $\lambda_p = 0$, Eq. 4 is valid for $p = 0, \cdots, N$. The detail discussion of probability score PR and PR^* is presented in [58]. As the HSI of the habitat

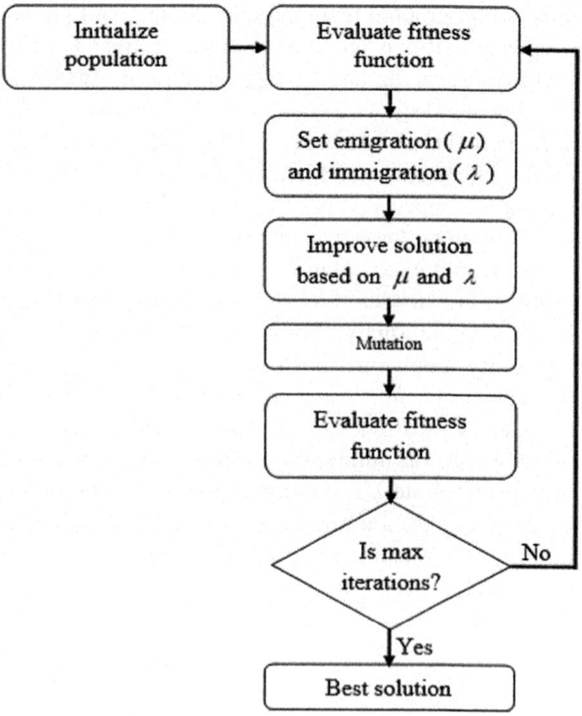

Fig. 5 Workflow of biogeography-based optimization algorithm

changes drastically, a mutation operator, M_p, is employed to modify the solutions. The mutation operator is expressed in Eq. 8, where M_{max} is the maximum probability of mutation, PR_p is the probability of a habitat with p species. Additionally, $PR_{max} = \max\left(PR_p\right) \forall p = 1, 2, \cdots, N$.

$$M_p = M_{max} \cdot \left(\frac{1 - PR_p}{PR_{max}}\right) \tag{8}$$

The algorithm of biogeography-based optimization is given below. An overview of the biogeography-based optimization algorithm is presented in Fig. 5.

Algorithm 4 (Biogeography-based Optimization)

1. *Initialize population*
2. *While (!Termination condition) do*
3. *Set emigration and immigration probability*
4. *Calculate HSI*
5. *Perform mutation*
6. *Recalculate HSI*
7. *Result*

Table 10 Description of few variants and application of BBO algorithm

Sl. No.	Variation of BBO	Application
1	Traditional BBO	Load dispatch problem [59]
2	Traditional BBO	Antenna design [60]
3	Hybrid BBO	Wireless sensor network [61]
4	Multi-objective BBO	Electromagnetics field [62]
5	Binary BBO	Microarray gene expression experiment [63]
6	Differential BBO	Classification [64]
7	Multi-objective BBO	Task scheduling [65]
8	Traditional BBO	Transmission systems [66]
9	Traditional BBO	MEMS design [67]
10	Hybrid chaotic BBO	Flowshop scheduling problem [68]
11	Hybrid BBO	Induction generator [69]
12	Chaotic BBO	Spectrum utilization [70]
13	Efficient BBO	reliability optimization problem [71]
14	Modified BBO	AC–AC converter [72]
15	Hybrid BBO	power flow controller [73]

Statistical mechanics, constrained optimization, equilibrium analysis, PID controller, motion estimation, power planning problem, positioning system, consolidation of soft soil, manufacturing system scheduling, voltage stability assessment, design of silica fume concrete, climbing rate of slip formwork system, renewable energy utilization are few applications of biogeography-based optimization. The applications of biogeography-based optimization are depicted in Table 10

4 Conclusion and Future Scope

This chapter presents a systematic review of various metaheuristic approaches. It highlights the classification of metaheuristic approaches such as swarm intelligence, evolutionary computation, bio-inspired algorithms, physics and chemistry-based algorithms, and other nature-inspired algorithms. Besides that, it also includes representatives of each category of metaheuristic approaches. Also, some specific problems where it has been implemented are highlighted. It is found that the characteristics of metaheuristic approaches and their mathematical computation vary a lot from method to method. These methods must be adaptive in terms of different random parameters to meet the local, global, and convergence optima. Besides, the scalability of designed metaheuristic approaches is required to be suitable for the problem statement of an application area. The process should be self-adaptive, self-tuned, and self-evolved for high-dimensional, complex decision-making problems.

This research does not focus on the successful metaheuristic technique for a real-world application and its reason. Thus, it is considered one of the future research directions for the authors.

References

1. Adam, S., Halina, K.: Nature inspired methods and their industry applications-Swarm intelligence algorithms. IEEE Trans. Ind. Inform. **14**(3), 1004–1015 (2018)
2. Jain, M., Singh, V., Rani. A.: A novel nature-inspired algorithm for optimization: Squirrel search algorithm. Swarm Evol. Comput. **44**, 148–175 (2018)
3. Yang, X.-S., Deb, S.: Cuckoo search: recent advances and applications. Neural Comput. Appl. **24**(1), 169–174 (2014)
4. Rautray, R., Balabantaray, R.C.: An evolutionary framework for multi document summarization using cuckoo search approach: MDSCSA. Appl. Comput. Inform. **14**(2), 134–144 (2018)
5. Coelho, L.S., Guerra, F., Batistela, N.J., Leite, J.V.: Multiobjective cuckoo search algorithm based on Duffing's oscillator applied to jiles-atherton vector hysteresis parameters estimation. IEEE Trans. Mag. **49**(5), 1745–1748 (2013)
6. Zhao, J., Liu, S., Zhou, M., Guo, X., Qi, L.: An improved binary cuckoo search algorithm for solving unit commitment problems: Methodological description. IEEE Access **6**, 43535–43545 (2018)
7. Zhang, H., Zhang, X., Wang, Y., Qian, X., Wang, Y.: Extended cuckoo search-based kernel correlation filter for abrupt motion tracking. IET Comput. Vis. **12**(6), 763–769 (2018)
8. Han, W., Xu, J., Zhou, M., Tian, G., Wang, P., Shen, X., Hou, E.: Cuckoo search and particle filter-based inversing approach to estimating defects via magnetic flux leakage signals. IEEE Trans. Mag. **52**(4), 1–11 (2016)
9. Jiang, M., Luo, J., Jiang, D., Xiong, J., Song, H., Shen, J.: A cuckoo search-support vector machine model for predicting dynamic measurement errors of sensors. IEEE Access **4**, 5030–5037 (2016)
10. Xuan, H., Zhang, R., Shi, S.: An efficient cuckoo search algorithm for system-level fault diagnosis. Chin. J. Electron. **25**(6), 999–1004 (2016)
11. Li, X., Ma, S., Yang, G.: Synthesis of difference patterns for monopulse antennas by an improved cuckoo search algorithm. IEEE Antennas Wirel. Propag. Lett. **16**, 141–144 (2017)
12. Suresh, S., Lal, S., Reddy, C.S., Kiran, M.S.: A novel adaptive cuckoo search algorithm for contrast enhancement of satellite images. IEEE J. Sel. Top. Appl. Earth Obs. Remote Sens. **10**(8), 3665–3676 (2017)
13. Wei, J., Yu, Y.: An effective hybrid cuckoo search algorithm for unknown parameters and time delays estimation of chaotic systems. IEEE Access **6**, 6560–6571 (2018)
14. Sun, G., Liu, Y., Li, J., Zhang, Y., Wang, A.: Sidelobe reduction of large-scale antenna array for 5G beamforming via hierarchical cuckoo search. Electr. Lett. **53**(16), 1158–1160 (2017)
15. Osman, H., Baki, M.F.: A cuckoo search algorithm to solve transfer line balancing problems with different cutting conditions. IEEE Trans. Eng. Manag. **65**, 505–518 (2018)
16. Chitara, D., Niazi, K.R., Swarnkar, A., Gupta, N.: Cuckoo search optimization algorithm for designing of a multimachine power system stabilizer. IEEE Trans. Ind. Appl. **54**(4), 3056–3065 (2018)
17. Suresh, S., Lal, S., Chen, C., Celik, T.: Multispectral satellite image denoising via adaptive cuckoo search-based Wiener filter. IEEE Trans. Geosci. Remote Sens. **56**(8), 4334–4345 (2018)
18. Zhao, J., Liu, S., Zhou, M., Guo, X., Qi, L.: Modified cuckoo search algorithm to solve economic power dispatch optimization problems. IEEE/CAA J. Automatica Sinica **5**(4), 794–806 (2018)
19. Sun, G., Liu, Y., Chen, Z., Liang, S., Wang, A., Zhang, Y.: Radiation beam pattern synthesis of concentric circular antenna arrays using hybrid approach based on cuckoo search. IEEE Trans. Antennas Propag. **66**(9), 4563–4576 (2018)

20. Holland, J.H.: Adaptation in Natural and Artificial Systems. MIT Press, USA (1975)
21. Goldberg, D.E.: Genetic Algorithms in Search. Optimization and Machine Learning. Addison Wesley, Reading, MA (1992)
22. Rautray, R., Balabantaray, R.C.: Bio-inspired Algorithms for Text Summarization: A Review. Bio-Inspired Computing for Information Retrieval Applications, pp. 71–92. IGI Global, USA (2017)
23. Yu, Y.K., Wong, K.H., Chang, M.M.Y.: Pose estimation for augmented reality applications using genetic algorithm. IEEE Trans. Syst. Man Cybern. Part B (Cybern) **35**(6), 1295–1301 (2005)
24. Tomioka, S., Nisiyama, S., Enoto, T.: Nonlinear least square regression by adaptive domain method with multiple genetic algorithms. IEEE Trans. Evol. Comput. **11**(1), 1–16 (2007)
25. Maruyama, T., Igarashi, H.: An effective robust optimization based on genetic algorithm. IEEE Trans. Mag. **44**(6), 990–993 (2008)
26. Ewald, G., Kurek, W., Brdys, M.A.: Grid implementation of a parallel multiobjective genetic algorithm for optimized allocation of chlorination stations in drinking water distribution systems: Chojnice case study. IEEE Trans. Syst. Man Cybern. Part C (Appl. Rev.) **38**(4), 497–509 (2008)
27. Moghaddam, E.S.: Design of a printed quadrifilar-helical antenna on a dielectric cylinder by means of a genetic algorithm. IEEE Antennas Propag. Mag. **53**(4), 262–268 (2011)
28. Mirahki, H., Moallem, M., Rahimi, S.A.: Design optimization of IPMSM for 42 V integrated starter alternator using lumped parameter model and genetic algorithms. IEEE Trans. Mag. **50**(3), 114–119 (2014)
29. Nguyen, D.C., Azadivar, F.: Application of computer simulation and genetic algorithms to gene interactive rules for early detection and prevention of cancer. IEEE Syst. J. **8**(3), 1005–1013 (2013)
30. Euziere, J., Guinvarch, R., Uguen, B., Gillard, R.: Optimization of sparse time-modulated array by genetic algorithm for radar applications. IEEE Antennas Wirel. Propag. Lett. **13**, 161–164 (2014)
31. Thirugnanam, K., TP, E.R.J., Singh, M., Kumar, P.: Mathematical modeling of Li-ion battery using genetic algorithm approach for V2G applications. IEEE Trans. Energy Convers. **29**(2), 332–343 (2014)
32. Kawecki, L., Niewierowicz, T.: Hybrid genetic algorithm to solve the two point boundary value problem in the optimal control of induction motors. IEEE Latin Am. Trans. **12**(2), 176–181 (2014)
33. Chen, C.H., Liu, T.K., Chou, J.H.: A novel crowding genetic algorithm and its applications to manufacturing robots. IEEE Trans. Ind. Inform. **10**(3), 1705–1716 (2014)
34. Lee, G., Mallipeddi, R., Jang, G.J., Lee, M.: A genetic algorithm-based moving object detection for real-time traffic surveillance. IEEE Sig. Process. Lett. **22**(10), 1619–1622 (2015)
35. Cui, L., Zhang, J., Yue, L., Shi, Y., Li, H., Yuan, D.: A genetic algorithm based data replica placement strategy for scientific applications in clouds. IEEE Trans. Services Comput. **11**(4), 727–739 (2015)
36. Choi, K., Jang, D.H., Kang, S.I., Lee, J.H., Chung, T.K., Kim, H.S.: Hybrid algorithm combing genetic algorithm with evolution strategy for antenna design. IEEE Trans. Mag. **52**(3), 1–4 (2015)
37. Ho, W.H., Tsai, J.T., Chou, J.H., Yue, J.B.: Intelligent hybrid taguchi-genetic algorithm for multi-criteria optimization of shaft alignment in marine vessels. IEEE Access **4**, 2304–2313 (2016)
38. Geem, Z.W., Kim, J.H., Loganathan, G.V.: A new heuristic optimization algorithm: harmony search. Simulation **76**(2), 60–68 (2001)
39. Rautray, R., Balabantaray, R.C.: Bio-inspired approaches for extractive document summarization: A comparative study. Karbala Int. J. Mod. Sci. **3**(3), 119–130 (2017)
40. Dash, R.: An adaptive harmony search approach for gene selection and classification of high dimensional medical data. J. King Saud University-Comput. Inform. Sci. **33**(2), 195–207 (2021)

41. Dash, R., Dash, P.K., Bisoi, R.: A differential harmony search based hybrid interval type2 fuzzy EGARCH model for stock market volatility prediction. Int. J. Approximate Reasoning **59**, 81–104 (2015)
42. Duan, H., Li, J.: Gaussian harmony search algorithm: A novel method for loney's solenoid problem. IEEE Trans. Mag. **50**(3), 83–87 (2014)
43. Yadav, P., Kumar, R., Panda, S.K., Chang, C.S.: Optimal thrust allocation for semisubmersible oil rig platforms using improved harmony search algorithm. IEEE J. Oceanic Eng. **39**(3), 526–539 (2014)
44. Chao, T., Yan, Y., Ma, P., Yang, M., Hu, Y.W.: Optimization of electromagnetic railgun based on orthogonal design method and harmony search algorithm. IEEE Trans. Plasma Sci. **43**(5), 1546–1554 (2015)
45. Ayala, H.V.H., Dos Santos Coelho, L., Mariani, V.C., Da Luz, M.V.F., Leite, J.V.: Harmony search approach based on ricker map for multi-objective transformer design optimization. IEEE Trans. Mag. **51**(3), 1–4 (2015)
46. Lin, C.C., Deng, D.J., Wang, S.B.: Extending the lifetime of dynamic underwater acoustic sensor networks using multi-population harmony search algorithm. IEEE Sens. J. **16**(11), 4034–4042 (2016)
47. Yang, S.H., Kiang, J.F.: Optimization of sparse linear arrays using harmony search algorithms. IEEE Trans. Antennas Propag. **63**(11), 4732–4738 (2015)
48. Abdelgayed, T.S., Morsi, W.G., Sidhu, T.S.: A new harmony search approach for optimal wavelets applied to fault classification. IEEE Trans. Smart Grid **9**(2), 521–529 (2018)
49. Zhang, L., Liu, M., Hao, J., Wang, X., Dong, J.: Scheduling semiconductor wafer fabrication using a new harmony search algorithm based on receipt priority interval. Chin. J. Electron. **25**(5), 866–872 (2016)
50. Sheng, W., Liu, K.Y., Liu, Y., Ye, X., He, K.: Reactive power coordinated optimisation method with renewable distributed generation based on improved harmony search. IET Gener. Transm. Distrib. **10**(13), 3152–3162 (2016)
51. Mahto, T., Mukherjee, V.: Fractional order fuzzy PID controller for wind energy-based hybrid power system using quasi-oppositional harmony search algorithm. IET Gener. Transm. Distrib. **11**(13), 3299–3309 (2017)
52. Portilla-Flores, E.A., Sanchez-Marquez, A., Flores-Pulido, L., Vega-Alvarado, E., Yanez, M.B.C., Aponte-Rodrguez, J.A., Nino-Suarez, P.A.: Enhancing the harmony search algorithm performance on constrained numerical optimization. IEEE Access **5**, 25759–25780 (2017)
53. Li, P., Li, R.X., Cao, Y., Li, D.Y., Xie, G.: Multiobjective sizing optimization for island microgrids using a triangular aggregation model and the levy-harmony algorithm. IEEE Trans. Ind. Inform. **14**(8), 3495–3505 (2018)
54. Chen, X., Cai, X., Liang, J., Liu, Q.: Ensemble learning multiple LSSVR with improved harmony search algorithm for short-term traffic flow forecasting. IEEE Access **6**, 9347–9357 (2018)
55. Li, F., Feng, J., Zhang, H., Liu, J., Lu, S., Ma, D.: Quick reconstruction of arbitrary pipeline defect profiles from MFL measurements employing modified harmony search algorithm. IEEE Trans. Instrum. Meas. **67**(9), 2200–2213 (2018)
56. Nascimento, L.B.P., Pinto, V., P., Amora, M.A.B.: Harmony search algorithm with adaptive parameters to optimize the linear quadratic regulator design. IEEE Latin Am. Trans. **16**(7), 1862–1869 (2018)
57. Simon, D.: Biogeography based optimization. IEEE Trans. Evol. Comput. **12**(6), 702–713 (2008)
58. Zhao, F., Qin, S., Zhang, Y., Ma, W., Zhang, C., Song, H.: A two-stage differential biogeography-based optimization algorithm and its performance analysis. Expert Syst. Appl. **115**, 329–345 (2019)
59. Bhattacharya, A., Chattopadhyay, P.K.: Biogeography-based optimization for different economic load dispatch problems. IEEE Trans. Power Syst. **25**(2), 1064–1077 (2010)
60. Singh, U., Kumar, H., Kamal, T.S.: Design of yagi-uda antenna using biogeography based optimization. IEEE Trans. Antennas Propag. **58**(10), 3375–3379 (2010)

61. Boussad, I., Chatterjee, A., Siarry, P., Ahmed-Nacer, M.: Hybridizing biogeography-based optimization with differential evolution for optimal power allocation in wireless sensor networks. IEEE Trans. Veh. Technol. **60**(5), 2347–2353 (2011)
62. Silva, M.D.A.C.E., Coelho, L.D.S., Lebensztajn, L.: Multiobjective biogeography-based optimization based on predator-prey approach. IEEE Trans. Mag. **48**(2), 951–954 (2012)
63. Li, X., Yin, M.: Multiobjective binary biogeography based optimization for feature selection using gene expression data. IEEE Trans. NanoBiosci. **12**(4), 343–353 (2013)
64. Zheng, Y.J., Ling, H., Chen, S., Xue, J.Y.: A hybrid neuro-fuzzy network based on differential biogeography-based optimization for online population classification in earthquakes. IEEE Trans. Fuzzy Syst. **23**(4), 1070–1083 (2015)
65. Zheng, Y.J., Ling, H.F., Xue, J.Y.: Disaster rescue task scheduling: An evolutionary multiobjective optimization approach. IEEE Trans. Emer. Top. Comput. **6**(2), 288–300 (2018)
66. Albasri, F.A., Alroomi, A.R., Talaq, J.H.: Optimal coordination of directional overcurrent relays using biogeography-based optimization algorithms. IEEE Trans. Power Delivery **30**(4), 1810–1820 (2015)
67. Di Barba, P., Dughiero, F., Mognaschi, M.E., Savini, A., Wiak, S.: Biogeography-inspired multiobjective optimization and MEMS design. IEEE Trans. Mag. **52**(3), 1–4 (2016)
68. Wang, Y., Li, X.: A hybrid chaotic biogeography based optimization for the sequence dependent setup times flowshop scheduling problem with weighted tardiness objective. IEEE Access **5**, 26046–26062 (2017)
69. Sarker, K., Chatterjee, D., Goswami, S.K.: Modified harmonic minimisation technique for doubly fed induction generators with solar-wind hybrid system using biogeography-based optimisation. IET Power Electron. **11**(10), 1640–1651 (2018)
70. Tegou, T.I., Tsiflikiotis, A., Vergados, D.D., Siakavara, K., Nikolaidis, S., Goudos, S.K., Obaidat, M.: Spectrum allocation in cognitive radio networks using chaotic biogeography-based optimisation. IET Networks **7**(5), 328–335 (2018)
71. Garg, H.: An efficient biogeography based optimization algorithm for solving reliability optimization problems. Swarm Evol. Comput. **24**, 1–10 (2015)
72. Hassanzadeh, M.E., Hasanvand, S., Nayeripour, M.: Improved optimal harmonic reduction method in PWM AC-AC converter using modified biogeography-based optimization algorithm. Appl. Soft Comput. **73**, 460–470 (2018)
73. Rahman, A., Saikia, L.C., Sinha, N.: Maiden application of hybrid pattern search-biogeography based optimisation technique in automatic generation control of a multi-area system incorporating interline power flow controller. IET Gener. Transm. Distrib. **10**(7), 1654–1662 (2016)

Diabetes Prediction: A Comparison Between Generalized Linear Model and Machine Learning

Sreekumar, Swati Das, Bikash Ranjan Debata, Rema Gopalan, and Shakir Khan

Abstract Recently, many chronic diseases have directly impacted human health. Many diseases are rampant and cause significant damage to humankind. Technological advancement has proven that diseases can be cured early. However, some diseases cannot be cured entirely but can be prevented. One of them being diabetes. If diabetes remains untreated and undiagnosed, many complications arise. The tedious diagnostic process leads the patient to visit a diagnostic center and see a doctor. The growth of technological solutions is solving this vital problem. The study's goal is to forecast whether a patient would get diabetes based on eight input variables: pregnancy, glucose, skin thickness (ST), blood pressure (BP), insulin, BMI, diabetic pedigree function (DPF), and age are all factors to consider. The output variable of the study is named "outcome," which is a binary variable taking values 1 or 0. The number 1 represents the existence of diabetes, whereas the value 0 represents the absence of diabetes. This chapter compares the results of classification algorithms, viz., binary logistic regression model and a support vector machine to detect diabetes early. The models are evaluated using the accuracy score metric to find the best model. The data used for the study is taken from GitHub.

Sreekumar (✉) · S. Das
Rourkela Institute of Management Studies, Odisha, India
e-mail: Sreekumar42003@yahoo.com

S. Das
e-mail: swatidasengg@gmail.com

B. R. Debata
Kirloskar Institute of Management, Pune, India
e-mail: bikash95@gmail.com

R. Gopalan
CMR Institute of Technology, Bengaluru, India
e-mail: gopalan.rema@gmail.com

S. Khan
Imam Mohammad ibn Saud Islamic University, Riyadh, Saudi Arabia
e-mail: sgkhan@imamu.edu.sa

D. P. Acharjya and K. Ma (eds.), *Computational Intelligence in Healthcare Informatics*, Studies in Computational Intelligence 1132,
https://doi.org/10.1007/978-981-99-8853-2_4

1 Introduction

Healthcare is one of the modern areas and generates a large amount of data. This data is for technical use and can cause significant research results and engineering [1]. Patient information records are converted to use data mining data [2]. The separation method is widely used in medicine to categorize data into other categories. According to the World Health Organization (WHO), diabetes is a chronic disorder that occurs when the pancreas does not make enough insulin or the body does not use the insulin produced effectively. The presence of glucose in human blood governs the metabolism of the organism. A low blood sugar level occurs when the glucose level in the blood is deficient; a high blood sugar level occurs when the glucose level in the blood is extremely high. A low blood sugar level occurs when the glucose level in the blood is deficient; a high blood sugar level occurs when the glucose level in the blood is extremely high. In diabetes, people are often suffering from elevated glucose levels. A high blood sugar level is called diabetes mellitus, or simply diabetes. People with diabetes may show a different range of symptoms. Due to increased blood sugar levels, they may experience extreme thirst, hunger, and frequent urination. If diabetes is not addressed, it can lead to a slew of problems. Ketoacidosis and non-ketotic hyperosmolar coma are two more serious issues [3].

Diabetes is one of the most common chronic diseases. There are two forms of diabetes: type 1 and type 2. In type 1 diabetes, the body cannot produce insulin [4]. As a result, people with type 1 diabetes must take insulin daily as directed. Type 2 diabetes is characterized by high blood sugar levels, which can be controlled with regular exercise and a healthy diet. However, ignoring both types can lead to chronic health concerns such as heart disease, stroke, and retinopathy. As a result, preventing diabetes is more vital than curing it. Several risk factors associated with diabetes include race, family history of diabetes, age, being overweight, an unhealthy diet, lack of physical activity, and smoking. Furthermore, the failure to recognize diabetes early is believed to lead to the development of other chronic illnesses, such as renal disease. Furthermore, current non-communicable disease outbreaks represent a great danger to patients since they are readily spread and vulnerable to infectious diseases like COVID-19 [5].

Data mining is finding essential patterns in vast sets of information. Therefore, data mining requests are broad. Other applications include ethical, social network analysis, prediction, and medical mining. According to a recent PubMed survey, data mining is growing increasingly important. According to the study, data mining in healthcare has proven effective in various disciplines, including predictive medicine, customer relationship management, fraud and abuse detection, healthcare management, and assessing the efficacy of certain medications. Predictive modeling is perhaps one of data mining's most prevalent and fundamental uses [6]. The present study focuses on two classification algorithms, viz. binary logistic regression model and a support vector machine, for analyzing historical data of patients with diabetes. The study's objective

1. is to predict diabetes using the sample data set of health records.
2. is to study the comparison of the accuracies of the two classifications as mentioned above algorithms.
3. is to present a superior prediction method based on the comparison.

The rest of this chapter is structured as follows. Section 2 describes the data mining procedure and related work in Sect. 3. The research approach is discussed in Sect. 4. Section 5 presents the experimental findings and comments. Section 6 contains the chapter's conclusion.

2 Data Mining Process

Data mining evaluates data from several perspectives and synthesizes it into helpful information. Data analysis may be divided into two types: extracting models that characterize important classes and anticipating future data trends. Data mining responsibilities include data collection, preparation, research, and reporting [7]. Data mining aims to extract information from a data source and convert it into a usable structure. Data mining is a rapidly growing field that focuses on developing tools to aid managers and decision-makers in making sense of massive volumes of data. There are two parts to the data mining process: data preparation and data mining. Data preparation activities include cleansing, integrating, reducing, and transforming data. Finally, data mining is the study of data, the discovery of patterns, the development of mining algorithms, the evaluation of data, and the expression of knowledge.

2.1 Classification

In data mining, classification algorithms are often used to categorize data. These algorithms predict categorical class labels and organize data based on the training set and class labels. It is also used to classify newly available data. There are two phases to classification. The first stage is to create a model using training data using a classification algorithm (supervised learning), and the second step is to extract the model. The extracted model is compared to predefined test data to determine the trained model's performance and accuracy. Several classification methods are employed in data mining, and some of the most popular classification algorithms include decision tree classifiers, neural networks, support vector machines, and Nave Bayes classifiers [8]. Other strategies for categorization include K-nearest neighbor, case-based reasoning, evolutionary algorithms, rough set theory, fuzzy logic, and other hybrid methods [9, 10]. These many categorization methods each have their functions and uses. These algorithms are all used to extract information from a dataset. The use of the appropriate way is determined by the task's purpose and the data type required.

2.2 Types of Classification Techniques

There are two types of classification algorithms: generative and discriminative. These are briefly defined below.

Generative: A generative classification method approximates the distribution of individual classes. It tries to understand the model that generates the data by estimating the distributions and assumptions of the model. We can use generative algorithms to anticipate unknown data. The Naive Bayes classifier is a popular generating approach.

Discriminative: Discriminative classification is a straightforward classification approach for determining a class for a given row of data. It models with observable data and is predicated on the quality of the data rather than its distributions. Both support vector machines and logistic regression are effective discriminative classifiers.

2.3 Major Classification Algorithms

Several algorithms have been developed for classifying real-world problems [10]. It is difficult to state all those algorithms. This chapter discusses some of the prime classification algorithms briefly.

Decision Trees: A decision tree is a categorization system that creates a tree and a set of rules from provided data to reflect the model of distinct classes. A training set and a test set are often separated from the collection of information used to build classification techniques. The classifier is derived from the former. Similarly, the latter is used to assess the classifier's accuracy. The percentage of successfully categorized test instances determines the classifier's accuracy [11–13].

Logistic Regression: Logistic regression is a statistical technique for producing a binomial outcome by employing one or more explanatory variables. This method attempts to identify whether a variable instance belongs to a category. A regression may be utilized in applications such as forecasting revenue by product and marketing campaign success rate.

Naive Bayes: Naive Bayes is a primary classification method that predicts the categorization of incoming data based on past data. It calculates the likelihood of an event occurring if another event has already happened. The circumstances of the occurrence in question are known. Filtering emails as spam and analyzing news items about technology, politics, or sports are real-world examples of Naive Bayes classification. They are also utilized in software for facial recognition.

K-Nearest Neighbors: This is a typical classification procedure that requires the selection of a classification metric. To begin, the algorithm is trained using a collection of data. The distance between the education and the new data is then computed to categorize the latest data. This method can operate intensely as computations depending on the size of the educational environment. On the complete data set, the KNN algorithm is utilized to make predictions. In addition, the KNN algorithm predicts that such objects are nearby. That is, related products are clustered together. As a result, they receive unidentified data and calculate the distance between new and previously categorized data. It is now employed for simplicity of usage, ease of research and use, and ease of analysis.

3 Related Research Work

Many researchers employed machine learning algorithms to extract insights from existing medical data in their diabetes study. Some authors, for example, have used support vector machines to develop predictive analytics models [14]. Similarly, ten-fold cross-validation has been used as an estimation method in three different algorithms: logistic regression, naive Bayesian network, and SVM [4]. According to the study, SVM surpassed other algorithms' performance and accuracy by 84%. Random forest, KNN, SVM, Naive Bayes, decision trees, and logistic regression predict diabetes early on [15].

Furthermore, the literature focuses on heart disease prediction utilizing closest neighbors, decision trees, and Naive Bayes [16, 17]. The study suggested that Naive Bayes predicted the illness more correctly than kNNs and decision trees. Furthermore, SVM and Naive Bayes are used to predict renal disease [18]. It discovered that SVM outperforms Naive Bayes. Several authors have investigated early-stage diabetes detection using classification algorithms such as decision trees, SVM, and Naive Bayes. Consequently, with 76.30% accuracy, Naive Bayes was the best [19]. Similarly, to discover the best method for categorizing medical data, authors employed WEKA to evaluate diabetes data using J48, SMO, Nave Bayes, random forest, and kNN classification algorithms [20]. As a result, SMO showed the highest performance with an accuracy of 78%.

A diabetes prediction system is also being developed to predict the kind of diabetes a candidate will have at a given age [21]. The proposed solution is based on machine learning and employs decision trees. The findings are promising, as the created approach functioned well in accurately predicting diabetes patients of a particular age [22]. Likewise, genetic programming (GP) is used to train and test a diabetes predictive database using diabetes data obtained from the GitHub data source [23]. The results obtained using genetic programming [24] provide optimal accuracy compared to other methods implemented. Reducing the time it takes to create a classifier can significantly improve accuracy. It is proving helpful in predicting diabetes at a low cost. Nevertheless, an algorithm to classify the risk of developing diabetes is also applied to diabetes data [25]. The technique employs four well-known machine learning classification methods to achieve this goal: decision trees, artificial neural networks, logistic regression, and the naive Bayes method are all examples of machine learning techniques. Bagging and boosting processes are also used to increase the model's reliability. The trials indicated that the random forest algorithm delivers the best results of any algorithms studied.

Another study uses a revolutionary training technique, SVM, to data having two or more classes, such as diabetes, heart, satellite, and shuttle data. SVM is a powerful machine-based method developed in statistical learning and has had considerable success in various applications [26]. Similarly, diabetic diseases from the machine learning lab at the University of California, Irvine, are also analyzed [27]. All patient data are trained using SVM. The SVM approach is proposed to successfully detect common diseases with simple clinical measurements without laboratory testing. Likewise, performance analysis using a decision tree and Naive Bayes is carried out for diabetes prediction using WEKA. The study revealed that Naive Bayes showed the highest accuracy of 98.43% [28].

The use of classification methods, including bagging, SVM, multilayer perceptron, simple logistics, and decision trees to effectively predict type 2 diabetes, is also found in the literature [29]. It effectively predicts type 2 diabetes with an accuracy of 94%. Specific authors employed WEKA to predict type 2 diabetes using the Bayesian classifier, J48, Naive Bayes, multilayer perceptron, SVM, and random forest algorithms. As a result, the Bayesian classifier J48, multilayer perceptron, and random forest all reached 100% accuracy [30]. The performance of the lib SVM classifier with the deep learning architecture is compared over the Pima dataset. The classification accuracy obtained is 75%, and the time required to build the model is 0.13 s [31]. PCA enhanced the accuracy of the k-means clustering method and logistic regression classifier in another experimental investigation compared to prior published studies [32]. Furthermore, k-means produces 25 more precisely categorized data and 1.98% greater logistic regression accuracy. Therefore, this model is helpful for the automatic prediction of diabetes using data from the patient's electronic health records. To predict type 2 diabetes in Pima Indian women, Joshi et al. employed a machine learning algorithm, a logistic regression model, and a decision tree. Their chosen specification achieves 78.26% prediction accuracy and a 21.74% cross-validation error rate. The scientists stated that the model might be used to forecast type 2 diabetes intelligently and as an addition to traditional preventative strategies to lessen the onset of diabetes and its related expenses [33].

4 Computational Methodology

The research methodology consists of several steps to achieve the specified research objectives: collecting a diabetes dataset with relevant patient attributes, preprocessing the values of numerical facts, applying various machine learning classification methods, and performing appropriate predictive analysis using these data. Figure 1 depicts these processes in short.

Diabetes data are gathered from GitHub for this project. Diabetes mellitus characteristics or risk factors from 768 participants are included in the dataset. Table 1 outlines the attributes, their categories, and the values they correspond to.

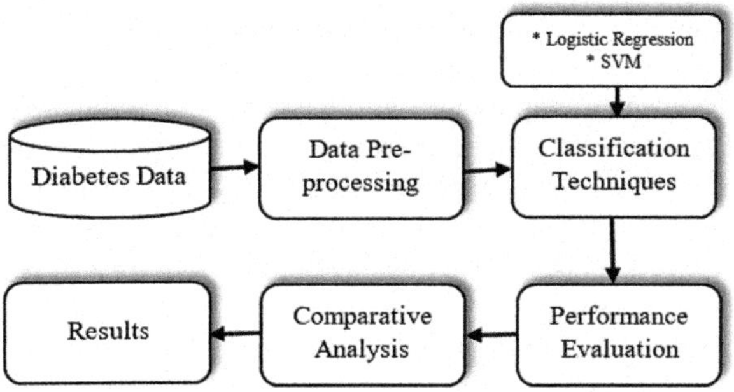

Fig. 1 Overall process of evaluation

Table 1 Descriptions of various parameters under study

Sl. No.	Parameter	Type	Values
1	Pregnancies	Numeric	{0–17}
2	Glucose	Numeric	{0–199}
3	BP	Numeric	{0–122}
4	ST	Numeric	{0–99}
5	Insulin	Numeric	{0–846}
6	BMI	Numeric	{0.00–67.10}
7	DPF	Numeric	{0.08–2.42}
8	Age	Numeric	{21–81}
9	Outcome	Numeric	{0, 1}

4.1 Data Pre-processing

Some data preprocesses have been carried out on the diabetes dataset to achieve the research objectives. For instance, the missing values of the attributes are corrected. Then, data is cleaned and prepared for the application of classification algorithms. All the data inconsistencies are removed to avoid errors. After the cleaning and preprocessing, the data is applied to the models for training and testing.

4.2 Application of Classification Techniques

When the data is available for modeling, two widely used categorization algorithms predict diabetes mellitus. An overview of these procedures is provided before proceeding to the results and analysis.

1. *Support Vector Machines:* Support vector machines (SVMs) are a family of generalized linear classifiers that collect related supervised learning algorithms for

classification and regression. SVMs were created to tackle classification difficulties, but they have lately been expanded to handle regression problems [34]. SVMs are systems that employ the hypothesis space of a linear function in a high dimensional feature space and are trained with an optimization theory learning algorithm that applies a statistical learning theory learning bias. SVMs begin as binary classifiers, with the learned function's output being positive or negative. The basic SVM solves two-class issues with data split by a hyperplane specified by many support vectors.
The SVM may be used to draw a line or hyperplane between two data sets to classify them. The SVM tries to draw a straight line between the two classes and orient it to maximize the margin. The SVM attempts to position the edge so that each class's distance between the border and the nearest data point is as short as possible. The frame is between two locations in the middle of this margin. The margins are defined by the closest data points, known as support vectors (SVs). Once the SVs are chosen, the other feature sets are eliminated. It is because the SVs include all the information required for the classifier. In general, SVM uses nonlinear machines to discover a hyperplane that minimizes errors for the training set when the data is not linearly segregated. Because, in many circumstances, no separating hyperplane exists, or it may be advantageous to misclassify a few observations to enhance the classification of the remaining ones, the job of margin maximization must be modified.
2. *Logistic Regression:* The likelihood of a specific occurrence or class is modeled using logistic regression to model binary dependent variables. It provides for a single trial. Because logistic regression was designed for categorization, it aids in understanding the impact of several explanatory factors on a single outcome variable. Logistic regression takes the output of a linear function and compresses data in the range [0, 1] using a sigmoid function. If the compressed value exceeds the threshold, 0.5, label 1; otherwise, label 0. The limitation of logistic regression is that it works only when the predictors are binary, and all factors are present and independent. It also assumes that the data has all the values, which is a serious concern.

Data is preprocessed to build the prediction model after collecting data from the GitHub source. Next, the two classification techniques, i.e., support vector machine and logistic regression, are used on the training dataset. Finally, the test dataset is used to measure the performance of the techniques to choose the best classifier for predicting diabetes mellitus. Finally, a comparative analysis is done based on the performance measures for selecting the best classifier among the two.

5 Experimental Results and Discussion

The data description with standard statistics is presented in Table 2 and is used after data cleaning and preprocessing. After that, the logistic regression technique is applied to perform classification. All the steps involved in the computation and the results are explained below.

Table 2 Descriptions of various parameters under study

Parameter	N	Min.	Max.	Mean	Std. Deviation
Pregnancies	768	0	17	3.85	3.370
Glucose	768	0	199	120.89	31.973
BP	768	0	122	69.11	19.356
ST	768	0	99	20.54	15.952
Insulin	768	0	846	79.80	115.244
BMI	768	0.00	67.10	31.99	7.88416
DPF	768	0.08	2.42	0.4719	0.33133
Age	768	21	81	33.24	11.760
Outcome	768	0	1	Count(0) = 500	Count(1) = 268

The above table reveals that of the 768 responses, 268 respondents have diabetes, and 500 are non-diabetic. The average diabetes pedigree function is 0.4719, with a standard deviation 0.331.

5.1 Binary Logistics Regression

There are many cases when the dependent variable may be a categorical variable, like yes or no in a set of variables. These are presented numerically as 0 or 1. The binary logistic regression is an extension of regular regression, which can handle the two categories of the dependent variable. In binary logistic regression, this chapter also attempts to find the relationship between the independent variables "X'' and the dependent variable "Y''. The goal is to find a simple model establishing the relationship between "X'' and "Y''. The approach is appropriate for the classification problem, like the case of predicting whether a person is "diabetic" or "non-diabetic" based on the set of input variables. The method is suitable for predicting the categorical input variable, testing the goodness of fit, and evaluating the accuracy of the classification of cases. The binary logistics has the following assumptions for its implementations.

1. The dependent variable should be measured on a dichotomous scale.
2. The number of independent variables can be one or more, either continuous (i.e., an interval or ratio variable) or categorical (i.e., an ordinal or nominal variable).
3. There shall be independent observations, and the dependent variable should have mutually exclusive and exhaustive categories.
4. There needs to be a linear relationship between any continuous independent variables and the logit transformation of the dependent variable.

Table 3 Initial classification

Step	Observed outcome	Predicted outcome		Percentage correct
		0	1	
0	0	500	0	100.0
	1	268	0	0.00
Overall percentage				65.1

Table 4 Omnibus tests of model coefficients

		Chi-Square	Degree of freedom	Significance
Step 1	Step	270.039	8	0.0
	Block	270.039	8	0.0
	Model	270.039	8	0.0

To introduce the logit model, let us define "odd." The odd of any event is the ratio of the probability that an event will occur to the probability that it will not occur. It is noted that p is the probability that the event will happen, and (1_p) is the probability that the event will not occur.

$$Odd(Event) = \frac{p}{1-p} \tag{1}$$

With the logistic regression model, the natural log odds are a linear function of the explanatory variable [35]. It is to be noted that $x = (x_1, x_2, \ldots, x_n)$ and $\beta = (\beta_1, \beta_2, \ldots, \beta_n)$. The binary logistic regression is implemented using SPSS 25.0 software. The analysis result is presented in Table 3.

$$logit(y) = ln(Odds) = ln\left(\frac{p}{1-p}\right) = \beta_0 + \beta x \tag{2}$$

Table 3 shows the initial classification and describes the null model. It means the model has no predictors and just the intercept. The cut-off value for inclusion in a category is fixed at 0.50. The result shows that all the input variables not in the equations, viz. pregnancies, glucose, BP, ST, insulin, BMI, DPF, and age, are significant at a 10% significance level. The method used to run the algorithm is "Enter." Further, Table 4 presents the Omnibus tests of model coefficients. It is one of the essential outputs and is the overall test of the model and the coefficients and odds ratios. Table 4 shows that $\chi^2(8) = 270.039$ and $p < 0.05$. It shows that the model build is statistically significant.

Table 5 Descriptions of various parameters under study

Parameter	B	Standard error	Wald	DF	Significance	Exp(B)	95% CI for Exp(B)	
							Lower	Upper
Pregnancies	0.123	0.032	14.747	1	0.000	1.131	1.062	1.204
Glucose	0.035	0.004	89.897	1	0.000	1.036	1.028	1.043
BP	−0.013	0.005	6.454	1	0.011	0.987	0.977	0.997
ST	0.001	0.007	0.008	1	0.929	1.001	0.987	1.014
Insulin	−0.001	0.001	1.749	1	0.186	0.999	0.997	1.001
BMI	0.090	0.015	35.347	1	0.000	1.094	1.062	1.127
DPF	0.945	0.299	9.983	1	0.002	2.573	1.432	4.625
Age	0.015	0.009	2.537	1	0.111	1.015	0.997	1.034
Constant	−8.405	0.717	137.546	1	0.000	0.000		

In addition, the model's explained variation is calculated. It is computed using Cox & Snell R^2, and Nagelkerke R^2 values. These numbers are known as pseudo R^2 and can be interpreted similarly. These values will generally be smaller than those obtained using multiple regression. In our model, for example, the Cox & Snell R^2 and Nagelkerke R^2 values are 0.296 and 0.408. As a result, the model-based explained variance in the dependent variable ranges from 29.6% to 40.8%, depending on whether the Cox & Snell R^2 or Nagelkerke R^2 approaches are used.

Similarly, the Hosmer and Lemeshow Test is performed. It examines the null hypothesis that the model's predictions fit exactly with observed group memberships. The observed frequencies are compared to those predicted by the linear model, yielding a chi-square statistic. The chi-square value was 8.323 with an 8 degree of freedom. A chi-square that is not significant suggests that the data match the model well. The model's p-value is 0.403, more effective than 0.05, indicating that the data fits the model.

Furthermore, the Wald test determines the relevance of the decision-making factors. It is shown in Table 5. The Wald test calculates the statistical significance of each independent variable. The significance column contains the test's statistical significance. Pregnancies ($p = 0.000$), glucose ($p = 0.000$), blood pressure ($p = 0.011$), BMI ($p = 0.000$), and DPF ($p = 0.000$) all contributed substantially to the model prediction, while skin thickness, insulin, and age did not. Equation 3 defines the forecasting model. In Table 5, DF stands for 'degrees of freedom,' The letters CI and B stand for confidence interval and regression coefficient, respectively.

$$\begin{aligned} logit(outcome) = & -8.405 + 0.123 \times Pregnancies + 0.035 \times Glucose \\ & -0.013 \times BP + 0.001 \times ST - 0.001 \times Insulin \\ & +0.09 \times BMI + 0.945 \times DPF + 0.015 \times Age \end{aligned} \quad (3)$$

The likelihood of diabetes arising is estimated using logistic regression. SPSS Statistics identifies an event as happening if the estimated chance of occurrence is

Table 6 Binary logistics regression classification

Step	Observed outcome	Predicted outcome		Percentage correct
		0	1	
1	0	445	55	89.0
	1	112	1560	58.2
Overall percentage				78.3

greater than or equal to 0.5. If the likelihood is less than 0.5, the model classifies the event as not occurring. Binomial logistic regression is commonly used to forecast whether instances may be appropriately categorized based on the independent variables. As a result, a mechanism for comparing the efficacy of the predicted categorization to the actual category is required. Table 6 displays the logistic regression computed results.

It is observed from Table 6 that with independent variables added, overall, the model correctly classifies 78.3% of cases. The sensitivity obtained is 58.2%. It indicates that the model also predicted 58.2% of the respondents with diabetes disease to have diabetes. Likewise, the specificity obtained is 89%. It suggests that the model correctly predicted 89% of participants who did not have diabetes to be non-diabetic.

5.2 *Support Vector Machine*

The categorization of the support vector machine (SVM) is highlighted in this section. After cleaning and preparing the data, the SVM is applied. The SVM method is then utilized for categorization. The steps, as well as the outcomes, are detailed below. To begin, a training dataset instructs how to construct a model. It is then utilized to train an algorithm to obtain the desired outcome. An SVM is used to train a model on the training data set. The model is trained using machine learning methods and training data. The model describes the outcomes of a machine learning algorithm applied to a training data set. Following training, test data is utilized as input to calculate accuracy. Test inputs usually determine if a particular set yields the predicted output. The model trained on the test input determines the correctness of the findings. The result of a prediction algorithm indicates whether the input details have diabetes or not. The data set considered for analysis contains 768 re. The experiments are performed using the R tool. R is a free software tool available online for data mining and analysis. R studio is an integrated development environment (IDE) designed specifically for the R language and is one of the leading data mining tools with a well-equipped built-in library for data mining.

The SVM algorithm uses a set of mathematical functions defined by the kernel. The kernel's process is to take data as input and transform it into the required format.

Different SVM algorithms use different types of kernel features. These features can be of various kinds, for example, linear, nonlinear, polynomial, radial basis function (RBF), and sigmoid. The "C" parameter penalizes each incorrectly classified data point. If C is minor, the punishment for incorrectly classifying is low. It means the boundaries of a solution with a large margin due to the wrong classification are selected. If C is large, the SVM is trying to minimize the number of examples of the incorrectly classified due to the high punishment, leading to a few solutions. Penalties are not the same for all incorrect models. It is directly proportional to the distance to the boundary of the solution.

Accuracy is one of the metrics for evaluating a classification model. Informally, accuracy is the percentage of accurate predictions achieved by the model. Analytically, accuracy is defined using Eq. 4.

$$Accuracy = \frac{\text{Number of correct predictions}}{\text{Total number of predictions}} \tag{4}$$

The kappa statistic measures how closely a machine learning classifier classifies the instances. Then, the machine learning classifier matches the data using a random classifier, and the accuracy is measured. These kappa statistics reveal how the classifier itself works. Simultaneously, two models can be compared using the kappa statistics directly in the same classification problem.

SVM with linear model: The analysis has been carried out using SVM with the linear model over 576 samples with eight predictors. The total number of decision classes is 2, i.e., 0 and 1. The ten-fold cross-validation is repeated thrice, considering the tuning parameter C as a constant value 1. The results obtained are presented in Table 7.

Table 7 Various measures of SVM with linear model

Various measures	Values
Accuracy and Kappa	0.7865 and 0.5108
95% CI	(0.7217, 0.8422)
No information rate	0.6354
P-Value [Acc > NIR]	$4.486e - 06$
Mcnemar's Test P-Value	0.004937
Sensitivity	0.9098
Specificity	0.5714
Pos Pred Value	0.7872
Neg Pred Value	0.7843
Prevalence	0.6354
Detection Rate	0.5781
Detection Prevalence	0.7344
Balanced Accuracy	0.7406

Table 8 Re-sampling results across tuning parameters

C	Accuracy	Kappa
0.25	0.7559297	0.4106483
0.50	0.7623329	0.4355184
1.00	0.7611831	0.4343439
2.00	0.7473299	0.4066858
4.00	0.7350985	0.3813173
8.00	0.7206598	0.3523619
16.00	0.7102451	0.3364620
32.00	0.6987303	0.3180239
64.00	0.6986388	0.3232771
128.00	0.6922363	0.3119654

Table 9 Various measures of SVM with radial basis functions model

Various measures	Values
Accuracy and kappa	0.7708 and 0.4664
95% CI	(0.7048, 0.8283)
No information rate	0.6354
P-Value [Acc > NIR]	$3.901e - 05$
Mcnemar's Test P-Value	0.0005256
Sensitivity	0.9180
Specificity	0.5143
Pos Pred Value	0.7671
Neg Pred Value	0.7826
Prevalence	0.6354
Detection Rate	0.5833
Detection Prevalence	0.7604
Balanced Accuracy	0.7162

SVM with radial basis function kernel: The analysis has been carried out using SVM with radial basis function kernel over 576 samples with eight predictors. The total number of decision classes is 2, i.e., 0 and 1. The ten-fold cross-validation is carried out considering the different values of tuning parameter C. The accuracy and Kappa values results are presented in Table 8. From Table 8, it is clear that the maximum accuracy is achieved at $C = 0.5$. Table 9 shows the results obtained for various measures. Hence, according to the above results, the accuracy using SVM with the linear kernel is 78.6%, and the accuracy using SVM with radial basis function is 77.08%. Thus, it is concluded that a linear kernel gives better accuracy. Figure 2 depicts the comparison efficiency of the two classification techniques, i.e., logistic

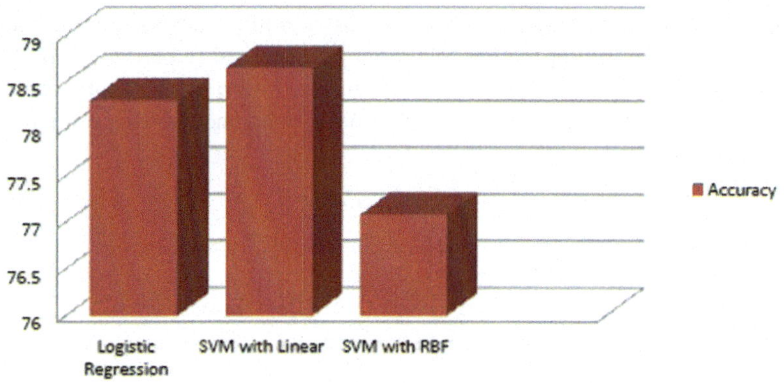

Fig. 2 Accuracy of classification techniques

regression and SVM. The x-axis represents the machine learning algorithm, whereas the y-axis represents the prediction percentage. Hence, SVM with linear function performs better than radial basis function and logistic regression.

6 Conclusion

A literature review suggests an enormous potential for data mining applications in healthcare. The chapter categorizes diabetes patients as anticipated diabetic or non-diabetic using SPSS 25.0 and R-project, a free statistical computing software environment. A comparison of SVM and binary logistics regression is presented in this paper. Diabetes data is gathered from GitHub for the analysis. The dataset included nine diabetes mellitus characteristics or risk variables from 768 people. The data is cleansed and translated into a suitable and legible format for generation by the data mining program. After preprocessing, the test data is used to compare and assess the performance of the two classification approaches outlined above. These training datasets are sent into the classification algorithms. Diabetes was predicted by the prediction method for the given sample datasets and error rate. Finally, the prediction accuracy of the two classification methods is compared. For sample datasets, the prediction effectiveness of logistic regression is 78.3%, while the prediction efficiency of SVM with linear function is 78.65%. For sample datasets, the prediction effectiveness of SVM with radial basis function is 77.08%. As a result, the SVM technique with linear function is shown to be more efficient than the other two in predicting the likelihood of patients with diabetes across sample datasets. In the traditional healthcare system, doctors used to rely on symptoms and multiple tests to detect diabetes and pre-diabetes. Still, these tests could not see the type of diabetes the patient suffered from. In such situations, one can use data mining techniques such as classification algorithms and, specifically, SVM to get results with improved accuracy. The present study contributes to the literature on machine learning and data mining by identifying the best classifier that accurately predicts and classifies patients as predicted diabetic and non-diabetic. For future studies, numerous opportunities for additional research would significantly extend the functionality of the current research.

References

1. Jothi, N., Husain, W.: Data mining in healthcare-a review. Proc. Comput. Sci. **72**, 306–313 (2015)
2. Aljumah, A.A., Ahamad, M.G., Siddiqui, M.K.: Application of data mining: Diabetes health care in young and old patients. J. King Saud Univ.-Computer Inf. Sci. **25**(2), 127–136 (2013)
3. Kandhasamy, J.P., Balamurali, S.J.P.C.S.: Performance analysis of classifier models to predict diabetes mellitus. Proc. Comput. Sci. **47**, 45-51 (2015)
4. Kavakiotis, I., Tsave, O., Salifoglou, A., Maglaveras, N., Vlahavas, I., Chouvarda, I.: Machine learning and data mining methods in diabetes research. Comput. Struct. Biotechnol. J. **15**, 104–116 (2017)
5. Zhou, F., Yu, T., Du, R., Fan, G., Liu, Y., Liu, Z., Xiang, J., Wang, Y., Song, B., Gu, X., Guan, L., Wei, Y., Li, H., Wu, X., Xu, J., Tu, S., Zhang, Y., Chen, H., Cao, B.: Clinical course and risk factors for mortality of adult inpatients with COVID-19 in Wuhan, China: a retrospective cohort study. The Lancet **395**(10229), 1054–1062 (2020)
6. Koh, H.C., Tan, G.: Data mining applications in healthcare. J. Healthc. Inf. Manag. **19**(2), 64–72 (2011)
7. Alasadi, S.A., Bhaya, W.S.: Review of data preprocessing techniques in data mining. J. Eng. Appl. Sci. **12**(16), 4102–4107 (2017)
8. Gopalan, R., Desai, M., Acharjya, D.P.: Customer classification in Indian retail sector-a comparative analysis of various machine learning approaches. Int. J. Oper. Quant. Manag. **26**(1), 1–28 (2020)
9. Acharjya, D.P., Abraham, A.: Rough computing-A review of abstraction, hybridization and extent of applications. Eng. Appl. Artif. Intell. **96**, 103924 (2020)
10. Kumari, N., Acharjya, D.P.: Data classification using rough set and bioinspired computing in healthcare applications - an extensive review. Multimedia Tools and Applications. Springer (2022). https://doi.org/10.1007/s11042-022-13776-1
11. Ahmed, P.K., Acharjya, D.P.: Knowledge inferencing using artificial bee colony and rough set for diagnosis of hepatitis disease. Int. J Healthcare Inf. Syst. Inf. **16**(2), 49–72 (2021)
12. Acharjya, D.P., Ahmed, P.K.: A hybridized rough set and bat-inspired algorithm for knowledge inferencing in the diagnosis of chronic liver disease. Multimed. Tools Appl. **81**(10), 13489–13512 (2022)
13. Kumari, N., Acharjya, D.P.: A decision support system for diagnosis of hepatitis disease using an integrated rough set and fish swarm algorithm. Concurr. Comput.: Pract. Exp. **34**(21), e7107 (2022)
14. Marinov, M., Mosa, A.S.M., Yoo, I., Boren, S.A.: Data-mining technologies for diabetes: a systematic review. J. Diabetes Sci. Technol. **5**(6), 1549–1556 (2011)
15. Zheng, T., Xie, W., Xu, L., He, X., Zhang, Y., You, M., Yang, G., Chen, Y.: A machine learning-based framework to identify type 2 diabetes through electronic health records. Int. J. Med. Inf. **97**, 120–127 (2017)
16. Soni, J., Ansari, U., Sharma, D., Soni, S.: Predictive data mining for medical diagnosis: an overview of heart disease prediction. Int. J. Comput. Appl. **17**(8), 43–48 (2011)
17. Ahmed, P.K., Acharjya, D.P.: A hybrid scheme for heart disease diagnosis using rough set and cuckoo search technique. J. Med. Syst. **44**(1), 1–16 (2020)

18. Vijayarani, S., Dhayanand, S.: Data mining classification algorithms for kidney disease prediction. Int. J. Cybernet. Inf. **4**(4), 13–25 (2015)
19. Kumar, D.A., Govindasamy, R.: Performance and evaluation of classification data mining techniques in diabetes. Int. J. Comput. Sci. Inf. Technol. **6**(2), 1312–1319 (2015)
20. Nass, L., Swift, S., Al Dallal, A.: Indepth analysis of medical dataset mining: a comparitive analysis on a diabetes dataset before and after preprocessing. KnE So. Sci. 45–63 (2019)
21. Orabi, K.M., Kamal, Y.M., Rabah, T.M.: Early predictive system for diabetes mellitus disease. In: Proceedings of Industrial Conference on Data Mining, Springer, Cham, pp. 420–427 (2016)
22. Priyam, A., Abhijeeta, G.R., Rathee, A., Srivastava, S.: Comparative analysis of decision tree classification algorithms. Int. J. Curr. Eng. Technol. **3**(2), 334–337 (2013)
23. Pradhan, M., Bamnote, G.R.: Design of classifier for detection of diabetes mellitus using genetic programming. In: Proceedings of the 3rd International Conference on Frontiers of Intelligent Computing: Theory and Applications, Springer, Cham, pp. 763–770 (2015)
24. Sharief, A.A., Sheta, A.: Developing a mathematical model to detect diabetes using multigene genetic programming. Int. J. Adv. Res. Artif. Intell. **3**(10), 54–59 (2014)
25. Islam, M.M., Ferdousi, R., Rahman, S., Bushra, H.Y.: Likelihood prediction of diabetes at early stage using data mining techniques. In: Computer Vision and Machine Intelligence in Medical Image Analysis. Springer, Singapore, pp. 113–125 (2020)
26. Durgesh, K.S., Lekha, B.: Data classification using support vector machine. J. Theor. Appl. Inf. Technol. **12**(1), 1–7 (2010)
27. Kumari, V.A., Chitra, R.: Classification of diabetes disease using support vector machine. Int. J. Eng. Res. Appl. **3**(2), 1797–1801 (2013)
28. Karthikeyan, R., Geetha, P., Ramaraj, E.: Rule based system for better prediction of diabetes. In: Proceedings of 3rd International Conference on Computing and Communications Technologies. IEEE, pp. 195–203 (2019)
29. Shuja, M., Mittal, S., Zaman, M.: Effective prediction of type ii diabetes mellitus using data mining classifiers and SMOTE. In: Advances in Computing and Intelligent Systems. Springer, Singapore, pp. 195–211 (2020)
30. Bhatti, S., Kehar, V., Memon, M.A.: Prognosis of diabetes by performing data mining of HbA1c. Int. J. Comput. Sci. Inf. Sec. **18**(1), 1–7 (2020)
31. Thaiyalnayaki, K.: Classification of diabetes using deep learning and svm techniques. Int. J. Curr. Res. Rev. **13**(01), 146–149 (2021)
32. Zhu, C., Idemudia, C.U., Feng, W.: Improved logistic regression model for diabetes prediction by integrating PCA and K-means techniques. Inf. Med. Unlocked **17**, 100179 (2019)
33. Joshi, R.D., Dhakal, C.K.: Predicting type 2 diabetes using logistic regression and machine learning approaches. Int. J. Environ. Res. Public Health **18**(14), 7346 (2021)
34. Pradhan, A.: Support vector machine-a survey. Int. J. Emerg. Technol. Adv. Eng. **2**(8), 82–85 (2012)
35. Park, H.A.: An introduction to logistic regression: from basic concepts to interpretation with particular attention to nursing domain. J. Korean Acad. Nurs. **43**(2), 154–164 (2013)

Prediabetes Prediction Using Response Surface Methodology and Probabilistic Neural Networks Model in an Ethnic South Indian Population

Raja Das, Shree G B Bakhya, Vijay Viswanathan, Radha Saraswathy, and K. Madhusudhan Reddy

Abstract Prevalence of prediabetes is increasing in India. Epidemiological studies have reported a strong association of anthropometric and biochemical parameters with the glycemic levels of prediabetes subjects. Identification of predictors of prediabetes using statistical and neural network models has been reported in this chapter. This chapter included 300 subjects from Timiri Block, Vellore, Tamilnadu, India, for neural network modeling. The subjects' biochemical and anthropometric variables are analyzed, and the glycemic levels of samples are predicted. Pearson correlation analysis and response surface method are performed using Minitab software. Artificial neural networks and probabilistic neural networks are served using Matlab software. Results obtained in all the models are validated using a testing data set. It shows an accuracy of 95% for the probabilistic neural network model. Salivary glucose, HbA1C, waist circumference, BMI, and LDL are found to be predictors of prediabetes.

R. Das (✉) · S. G. B. Bakhya · R. Saraswathy
Vellore Institute of Technology University, Vellore, Tamilnadu, India
e-mail: rajadasrkl@gmail.com

S. G. B. Bakhya
e-mail: bakhyashree@gmail.com

R. Saraswathy
e-mail: radhasaraswathy1960@gmail.com

V. Viswanathan
M V Hospital for Diabetes Prof M Viswanathan Diabetes Research Centre, Royapuram, Chennai, India
e-mail: vijayviswanathan92@gmail.com

K. M. Reddy
College of Computing and Information Sciences, University of Technology and Applied Sciences Shinas, Shinas, Sultanate of Oman
e-mail: madhu.katta@shct.edu.om

D. P. Acharjya and K. Ma (eds.), *Computational Intelligence in Healthcare Informatics*, Studies in Computational Intelligence 1132,
https://doi.org/10.1007/978-981-99-8853-2_5

1 Introduction

Prediabetes is defined as impaired fasting glucose of 100–125 mg/dl or an impaired glucose tolerance of 2-h plasma glucose levels between 140 and 199 mg/dl or HbA1C between 5.7% and 6.4% [1]. Generally, nondiabetic hyperglycemia, which does not satisfy the diagnostic criteria for diabetes, is termed prediabetes [2]. Rather than being a separate clinical entity, prediabetes is a crucial risk factor for type 2 diabetes and cardiovascular diseases [1]. Besides, the prediabetes state is generally asymptomatic [3]. The international diabetes federation report states that individuals at the prediabetes stage are more prone to macrovascular complications. The risk of cardiovascular disease (CVD) is higher in the prediabetes group than in normal individuals. It is because of impaired glucose tolerance and other known CVD risk factors like high blood pressure (BP), lipotoxicity, and abdominal obesity. Identifying predictors for prediabetes is essential to screen the high-risk group early and provide interventions to prevent the progression of diabetes complications.

Prediabetes prediction and screening for undiagnosed diabetes have been performed with different models in different populations. Most screening studies included questionnaires and diabetes risk scores for all the risk factors. Testing of Framingham risk score gave consistent results in specific people only [4–6]. A more recent model for the prediction of diabetes is the Indian diabetes risk score (IDRS) devised by the Chennai urban, rural epidemiology study (CURES). IDRS has reported age, abdominal obesity, family history of diabetes, and physical inactivity as predictors of prediabetes [7]. However, a robust epidemiology model is required as the screening based on a questionnaire might result in a few false positive and false negative results. Thus, this study has attempted to evaluate all the risk factors of the prediabetes group using statistical and neural network modeling. It could help in the early diagnosis of the high-risk group using predictors for three prediabetes. In addition, these models would help predict the risk factors directly related to the prediabetes condition in the ethnic group.

The rest of the chapter is organized as follows. Section 2 refers to brief introduction to prediabetes. Sample collection and materials are produced in Sect. 3. Statistical and neural network models are discussed in Sect. 4. Results analysis has been carried out in Sect. 5 followed by discussion in Sect. 6. The chapter is concluded in Sect. 7.

2 Brief Introduction to Prediabetes

Prediabetes represents a hyperglycemic condition with the plasma glucose above the normal range, but below that of the clinical diabetic condition [2]. It is generally characterized by impaired fasting glucose (IFG) and impaired glucose tolerance (IGT). IGT is usually measured by oral glucose tolerance testing (OGTT). HbA1C serves as a standard gold method. However, due to the invasive nature of the test, in this chapter, glucose levels are measured using saliva samples. Early diagnosis of diabetes is essential to provide timely interventions and prevent complications. Prediabetes, being intermediate hyperglycemia [8], requires more attention as the individuals in the prediabetic state are at a high risk of progression to diabetes. WHO fact sheet 2015 states that various lifestyle habits such as smoking, alcoholism, intake of high-calorie food, physical inactivity, and other demographic and anthropometric characteristics such as ethnicity, gender, family history of diabetes (FHD), obesity, and BP contribute to the risk of development of diabetes. Thus, this chapter presents an in-depth analysis of the study population's various clinical and biochemical characteristics and designs a mathematical model to predict prediabetes at the early stages. It might help to plan proper interventions for the high-risk group.

3 Materials and Methods

A total of 300 subjects are included in the study from Timiri block, Vellore district, comprising four villages: Pattanam, Vellambi, Palayam, and Mosur. It consists of 84 samples of Pattanam, 78 samples of Vellambi, 63 samples of Palayam, and 75 samples of Mosur. Institutional ethical clearance and permission from the department of health and family welfare, Government of Tamilnadu, and the directorate of public health (DPH) and preventive medicine, Chennai, is obtained to conduct the study. Informed consent from all the subjects who participated in the study was obtained. The study included samples above 20 years of age and without diabetes. Subjects with known diabetes, subjects diagnosed with thyroid or other diseases, pregnant women, and subjects aged below 20 were excluded from the study.

Blood samples, saliva samples, and urine samples were collected from all 300 subjects. Blood and urine samples are collected thrice for OGTT (during fasting, one hour after glucose consumption, and two hours after glucose consumption). In addition, 1 ml of saliva samples were contained in sterile 2 ml Eppendorf tubes after mouth rinse with drinking water in a fasting condition. The following methodology is carried out in this study.

3.1 Clinical Study

The step-wise approach of WHO is followed to collect data. The risk factor surveillance version 2.1 of chronic disease is considered for the following analysis.

1. *Demographic Analysis:* Age and gender of the subjects are analyzed based on the information of the subjects recorded in the WHO guidelines.
2. *Behavioral Measurements:* Lifestyle habits such as smoking, alcoholism, and smokeless tobacco usage are recorded for each subject, and the percentage of smokers, alcoholics, and smokeless tobacco users are calculated. The subjects' physical activity is measured based on the global physical activity questionnaire

(GPAQ) values and metabolic equivalent values according to the GPAQ guidelines.

3. *Physical Measurements:* The following physical measurements are collected.

   a. Blood glucose levels are estimated by using standardized protocol [9].
   b. Body mass index (BMI): The height and weight of the subjects are measured, and the following Eq. 1 is used to calculate the BMI.

$$BMI = \frac{Weight\ in\ kg}{(Height\ in\ m)^2} \tag{1}$$

   Classification of subjects based on the BMI is carried out with the Asian pacific standards for BMI from WHO global database on BMI.

   c. Blood Pressure: Systolic and diastolic BP is recorded twice for each subject in a sitting posture, and the average of the two readings is taken for analysis. Classifying subjects based on BP is carried out by following the JNC8 guidelines for BP.
   d. Waist Circumference: Waist circumference is measured for all the subjects. Classification of subjects as having a normal waist circumference or a higher waist circumference is done based on the WHO guidelines for waist circumference reports.

3.2 Biochemical Study

The biochemical study included the following tests.

1. *Oral glucose tolerance test:* Blood samples (2 ml) are collected thrice from all the subjects in ethylenediamine tetraacetic acid (EDTA) coated vacutainers (during fasting, one hour after the consumption of 75 g of glucose mixed in 250 ml of water and two hours after glucose consumption) for OGTT. OGTT is carried out with the standardized protocol [10]. According to Americans with disabilities act (ADA) (2014) guidelines, the cut-off for IFG is the normal fasting glucose $< 100\,\text{mg/dl}$; impaired fasting glucose between $100 - 125\,\text{mg/dl}$ and hyperglycemia $> 125\,\text{mg/dl}$. Similarly the cut-off for IGT is the normal glucose tolerance $< 140\,\text{mg/dl}$; impaired glucose tolerance between $140 - 199\,\text{mg/dl}$; hyperglycemia $> 199\,\text{mg/dl}$.
2. *Glycated hemoglobin (HbA1C) levels:* HbA1C test is performed for all the subjects to estimate the glycated hemoglobin levels, which would give the glycemic status of an individual for the past three months. Blood samples (3 ml) are collected in EDTA-coated vacutainers, and estimation of HbA1C is performed using the standardized protocol [9]. Classifying subjects based on HbA1C is based on ADA (2014) guidelines.

3. *Lipid profile:* Lipid profile of all the subjects are analyzed by measuring the serum high-density lipoprotein (HDL), low-density lipoprotein (LDL), triglyceride, and total cholesterol levels by using the standardized protocol [11]. Subjects are classified based on adult treatment panel (ATP) III guidelines (2008).
4. *Urine glucose test:* Qualitative analysis of urine glucose estimation is performed for all the subjects by the standardized protocol [12].
5. *Salivary glucose test:* 1 ml of saliva samples are collected in sterile 2 ml Eppendorf tubes after mouth rinse with drinking water in fasting condition from all the subjects. Salivary glucose estimation is performed using the modified protocol [13].

4 Computational Intelligence Techniques in Prediabetes Prediction

Computational intelligence techniques are employed as predictors for prediabetes. Pearson correlation, response surface methodology, artificial neural network (ANN), and probabilistic neural network (PNN) are considered for the analysis. A brief description of the techniques is furnished below.

4.1 Pearson Correlation

The most popular technique for analyzing numerical variables is the Pearson correlation, which assigns a value between 0 and 1. Here one refers to the total positive correlation, and 0 denotes the total negative correlation. According to the following interpretation, a correlation value of 0.7 between two variables denotes a significant and favorable association between them. A positive correlation means that if variable A rises, variable B will likewise increase, whereas a negative correlation means B declines if A increases. Given paired of data $\{(x_1, y_1), (x_2, y_2), \ldots, (x_n, y_n)\}$ consisting of n pairs, the Pearson correlation is defined in Eq. 2

$$\rho_{xy} = \frac{\sum_{i=1}^{n}(x_i - \overline{x})(y_i - \overline{y})}{\sqrt{\sum_{i=1}^{n}(x_i - \overline{x})^2}\sqrt{\sum_{i=1}^{n}(y_i - \overline{y})^2}} \tag{2}$$

4.2 Response Surface Methodology

A second-order polynomial empirical mathematical model may explain the correlation between the independent variables and the output responses. The second-order polynomial is defined in Eq. 3, where y is the predicted response, β_0 is the constant term, β_i is the linear term, β_{ii} is the quadratic term, β_{ij} is the interaction term, N is the number of variables, and X_i is the independent variable with i.

$$y = \beta_0 + \sum_{i=1}^{N} \beta_i X_i + \sum_{i=1}^{N} \beta_{ii} X_i^2 + \sum_{i=1}^{N-1} \sum_{j=1-i}^{N} \beta_{ij} X_i X_j \quad (3)$$

The proposed polynomial model is examined for consistency using the analysis of variance (ANOVA) method, taking into account the coefficient of determination R^2 and R^2_{adj}, as stated in the Eqs. 4 and 5 respectively. It is noted that df denotes the degree of freedom, and SS represents the sum of squares.

$$R^2 = \frac{SS_{residual}}{SS_{model} + SS_{residual}} \quad (4)$$

$$R^2_{adj} = \frac{SS_{residual}/df_{residual}}{(SS_{model} + SS_{residual})/(df_{model} + df_{residual})} \quad (5)$$

4.3 Artificial Neural Network

ANN is a highly flexible neurocomputing tool. It differs from conventional computing tools as it can draw mathematical mapping between input and output nodes. A typical ANN model has a set of input variables given in the input nodes and a set of output nodes. The input and output nodes are connected with the help of hidden nodes. This study initially constructed the ANN model for prediabetes prediction using the available clinical and biochemical data. Then, based on the results obtained in the Pearson correlation analysis and linear regression analysis, five input nodes are chosen (variables: salivary glucose, HbA1C, waist circumference, BMI, and LDL). In this study, the ANN model is constructed using five input nodes, four hidden nodes, and one output node.

The constructed model is tested for accuracy with a set of testing data sets. The modified methodology is followed for ANN model [14]. The input and output layers are fed with a multilayer perceptron (MLP) forward neural network. The construction of the ANN model using the input layer, hidden layer, and output layer is represented in Fig. 1.

Each interconnection between the nodes has a weight associated with it. Sigmoidal $S(.)$ is taken as the activation function for the hidden nodes, whereas an activation function of linear is taken for output nodes. The following Eq. 6 gives the net input of the jth hidden neuron, where $w1_{ji}$ is the weight between the ith node of input layer and jth node of hidden layer and $b1_j$ is the bias at jth node of hidden layer. The output of the jth hidden node is defined in Eq. 7.

$$y_j(x) = \sum w1_{ji} x_i + b1_j \quad (6)$$

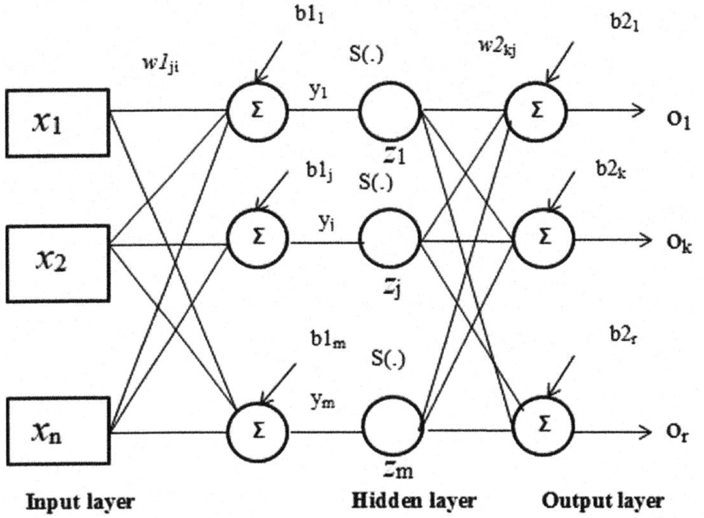

Fig. 1 Architecture of ANN model

$$z_j(x) = \frac{1}{1 + e^{-y_j(x)}} \tag{7}$$

Taking $w2_{kj}$ as the weight between the jth node of the hidden layer and the kth node of the output layer, $b2_k$ as the biasing term at the kth node of the output layer, the output value $o_k(x)$ of the kth node of the output layer is given by the sum of weighted outputs of the hidden nodes with an input vector x and the bias of the kth node which is represented in the Eq. 8

$$o_k(x) = \sum w2_{kj} z_j + b2_k \tag{8}$$

The predicted OGTT levels are compared with the experimental values, and the error percentage is calculated. In ANN, each node in a layer is not connected with the other nodes, while the adjacent layers are connected with the help of weighted coefficients. To reduce the error, the adjusted parameter is calculated by the difference between the actual and predicted output for the weighted coefficients. It is defined in Eq. 9. The error is back propagated through the network, and the process is repeated until the error becomes minimum. The ideal parameter, i.e., the minimum value of the square deviation of the actual output (o_k) and predicted output (t_k), is calculated by changing the weight values in the network, where r is the number of outputs.

$$E = \frac{1}{2} \sum_{k=1}^{r} (o_k - t_k)^2 \tag{9}$$

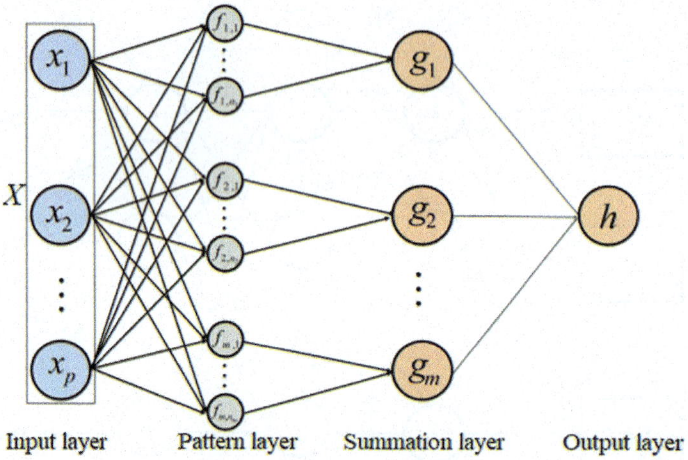

Fig. 2 Architecture of PNN model

The entire process is repeated for all the samples until the overall error value reduces below the minimum value. We could then claim that the network has learned the problem well enough to predict the desired output.

4.4 Probabilistic Neural Networks

PNN works on a statistical algorithm known as kernel discriminate analysis. In PNN, the process is organized into four different layers: the input layer, pattern layer, summation layer, and output layer, as represented in Fig. 2. The modified methodology is carried out to construct PNN model [15]. The node of the pattern layer computes the output by receiving the data from the input layer using Eq. 10.

$$f_k(x) = \frac{1}{(2\pi)^{\pi/2}\sigma^p} e^{-\frac{||X-X_k||^2}{2\sigma^2}} \tag{10}$$

The summation layer calculates the maximum likelihood pattern of the X and computes the average of the output of all neurons which belongs to the same category. It is defined in Eq. 11, where n_i is the number of samples in the ith population.

$$g_i(x) = \frac{1}{(2\pi)^{\pi/2}\sigma^p n_i} \sum_{k=1}^{n_i} e^{-\frac{||X-X_{ik}||^2}{2\sigma^2}} \tag{11}$$

The three groups considered for analysis are the normoglycemic group, prediabetes group, and undiagnosed diabetes. Suppose the probabilities for each group are identical, and the losses associated with making a wrong decision for each group are the same. In that case, the decision classifies the pattern X in agreement with Bayes' rule based on the output of all summation layer nodes.

$$h(X) = \mathbf{agmax}\{g_i(X)\}, \quad i = 1, 2, 3, \ldots, m \tag{12}$$

Where $h(x)$ denotes the estimated class of the pattern x and m is the total number of groups in the training samples. It includes determining the network size, the pattern layer neurons, and an appropriate smoothing parameter.

5 Results and Analysis

This study included 300 subjects from Timiri block, Vellore district, for screening of prediabetes and undiagnosed diabetes with an average age of 48.3 ± 14.90 years. The demographic information revealed the incidence of 66.3% of females and 33.7% of males in the study population. There is a significant percentage of combined IFG and IGT (prediabetes) and undiagnosed diabetes, i.e., 23.3% and 6.6%, respectively. Matlab and Minitab software were used to construct the models mentioned above and to assess the accuracy of the constructed models by using a testing data set.

5.1 Pearson Correlation Analysis

Pearson correlation analysis is performed between biochemical parameters such as HbA1C, salivary glucose, LDL, HDL, total cholesterol, triglycerides, one-hour OGTT, two-hour OGTT, BP, and BMI. The correlation analysis showed a significant positive correlation between the various biochemical parameters in the study population, as depicted in Table 1.

Further, to obtain the correlation between the variables, Pearson correlation analysis is used. As a result, a significant positive correlation was observed between various biochemical variables with R^2 values and respective p values as represented in Table 1. Therefore, the variables which exhibited a positive correlation in Pearson correlation analysis are included in linear regression analysis.

5.2 Response Surface Methodology Analysis

Response Surface Methodology (RSM) is performed to generate a non-linear regression model for predicting glycemic levels with the various factors influencing the glycemic levels of prediabetes and undiagnosed diabetes subjects. Initially, a factorial design experiment is performed to identify the factors influencing glycemic

Table 1 Pearson correlation analysis for various biochemical parameters

Variables	Pearson correlation coefficient	*p* value
Salivary glucose versus LDL	0.136	0.032
Cholesterol versus LDL	0.853	0.000
Triglyceride versus LDL	0.252	0.000
Triglyceride versus total cholesterol	0.421	0.000
Systolic B.P versus LDL	0.172	0.006
Systolic B.P versus total cholesterol	0.188	0.003
Diastolic B.P versus LDL	0.170	0.007
Diastolic B.P versus total cholesterol	0.231	0.000
Diastolic B.P versus HDL	0.607	0.000
HbA1C versus salivary glucose	0.133	0.035
HbA1C versus LDL	0.223	0.000
HbA1C versus total cholesterol	0.210	0.001
HbA1C versus triglyceride	0.295	0.000
HbA1C versus systolic B.P	0.200	0.001
HbA1C versus diastolic B.P	0.162	0.010

levels. Then, the elements with p values less than 0.05 are selected for constructing a non-linear regression model using RSM. The factorial design represented five factors, namely, HbA1C, salivary glucose, LDL, BMI, and waist circumference, as significant factors influencing the glycemic levels of prediabetes and undiagnosed diabetes subjects are presented in Table 2. Other variables were not important. Therefore, these five variables are used to construct a non-linear regression model using RSM.

RSM is a collection of statistical and mathematical techniques which has broad applications in the design and optimization of models and for the improvement of an existing model. The factorial method is used to identify the significant factors. Further, RSM is used to determine the factors that influence glycemic levels. A second-order polynomial regression model is created for the five essential variables in the factorial test. Based on the adjusted R^2 value and p values, the variables which showed significant association is chosen, and a non-linear regression equation is generated to predict the glycemic levels of the study subjects.

According to the non-linear regression analysis, salivary glucose, HbA1C, waist circumference, BMI, and LDL are predictors for prediabetes. The computation is shown in Table 3. A testing data set ($n = 50$) gave an accuracy of 80% in predicting glycemic levels in a non-linear regression model using RSM. The fasting blood glucose (FBG) is defined in Eq. 13

Table 2 Factorial fit

Factors	Coefficient	SE coefficient	T	p
Constant	−5.7	163.63	−0.03	0.972
HbAIC	125.0	45.62	2.74	0.007*
Salivary glucose	−204.4	141.55	−1.44	0.015*
HDL	−212.8	163.13	−1.30	0.193
LDL	−193.1	189.89	−1.02	0.031*
Cholesterol	206.3	281.95	0.73	0.465
Triglyceride	30.9	63.97	0.48	0.630
BMI	−167.9	279.02	−0.60	0.054*
Systolic BP	151.1	112.49	1.34	0.181
Diastolic BP	−64.0	85.35	−0.75	0.454
Hip circumference	−375.8	184.56	−2.04	0.043
Waist circumference	422.9	210.09	2.01	0.045*

* significant p values

Table 3 Non-linear regression model for prediabetes prediction using RSM

Predictor	Coefficient	SE coefficient	T	p
Constant	159.325	33.503	4.756	0.000
HbA1C	−24.755	6.580	−3.762	0.000
Salivary glucose	−2.370	2.065	−1.147	0.052
LDL	0.771	0.1639	4.705	0.000
BMI	−0.675	0.4604	−1.466	0.144
Waist circumference	−1.567	0.327	−4.790	0.000
HbA 1C × HbA1c	1.549	0.361	4.286	0.000
HbA1C × Salivary glucose	0.725	0.174	4.150	0.000
HbA1C × LDL	−0.100	0.024	−4.141	0.000
HbA1C × Waist circumference	0.315	0.053	5.898	0.000
Salivary glucose × LDL	−0.039	0.013	−2.877	0.004
Salivary glucose × BMI	0.089	0.065	1.366	0.173

$S = 21.16$; $R^2 = 69.3\%$; $R^2(adj) = 68.1\%$

Table 4 Models used for prediabetes prediction

Method	Testing data set	Error (%)	Accuracy (%)
RSM	50	20	80
ANN	50	10	90
PNN	50	5	95

$$\begin{aligned}FBG = {} & 159.325 - (24.755 \times HbA1C) - (2.370 \times salivary\ glucose) \\ & -(0.771 \times LDL) - (0.675 \times BMI) + (1.549 \times HbA1C^2) - \\ & (1.567 \times waist\ circumference) - (0.1 \times HbA1C \times LDL) + \\ & (0.315 \times HbA1C \times waist\ circumference) + \\ & (0.725 \times HbA1C \times salivary\ glucose) - \\ & (0.039 \times salivary\ glucose \times LDL) + \\ & (0.089 \times salivary\ glucose \times BMI)\end{aligned} \tag{13}$$

5.3 *Artificial Neural Networks*

ANN is used to predict the subjects' IGT and IFG levels to identify the high-risk group (prediabetes and undiagnosed diabetes). Five input variables, salivary glucose, HbA1C, waist circumference, BMI, and LDL, predict the study participants' IGT and IFG. The testing data ($n = 50$) showed a 90% probability of getting true positive results in identifying the prediabetes and undiagnosed diabetes group.

5.4 *Probabilistic Neural Networks*

PNN is used to validate the results obtained in the other two models. PNN model showed a higher probability of getting 95% of true positive results in predicting the IFG and IGT of the testing group to identify as either prediabetes or undiagnosed diabetes. The comparative analysis is presented in Table 4. Besides, PNN is recognized as a significantly best-fit model for prediabetes prediction. Figure 5 depicts the regression analysis between the actual glycemic levels and the predicted glycemic levels using the PNN model. An R^2 value of 0.9056, ($p < 0.05$) for the regression analysis represents a significant positive correlation between actual and predicted glycemic levels. Thus PNN could be used as a model for prediabetes prediction.

6 Discussion

OGTT depicted 210 subjects with normoglycemia (70%), 70 subjects with combined impaired fasting glucose and impaired glucose tolerance (23.3%), and 20 subjects with hyperglycemia (6.6%). All the other biochemical parameters, namely HbA1C levels, lipid profile, Urine glucose, and salivary glucose levels, are considered input variables for the identification of predictors of prediabetes using the statistical models and neural networks model. A significant percentage of prediabetes and undiagnosed diabetes is observed in this study population. A similar study has reported a prevalence of 8.3% of prediabetes subjects from Tamilnadu [16]. Other clinical parameters, namely behavioral and physical measurements, are input variables for prediabetes prediction. Variables, namely salivary glucose, HbA1C, waist circumference, BMI, and LDL, significantly affect glycemic levels in the study of factorial design. The RSM model uses the above factors to obtain a non-linear regression expression for glycemic levels.

6.1 Residual Plots for Predicted Glycemic Levels Using RSM

Figure 3 represents the residual plots of the predicted OGTT levels using the non-linear regression model. Plot A represents the graph of the predicted OGTT levels against the actual OGTT levels. Besides, a strong correlation between the predicted and the actual OGTT values has been observed. Plot B represents the residual versus fitted values of the RSM model. This plot represents the correlation between the observed and predicted OGTT values with minimum error. It also describes the dependency between the residual and fitted values.

Plot C represents the histogram of the residuals. It represents the deviation of the predicted OGTT levels using non-linear regression and the observed variation frequency. Here, it is observed that the predicted values represent minimum error as the error plot is usually distributed with the maximum frequency at zero. The normality assumption of the non-linear regression model is likely to be true in this study. Plot D represents the residual values for each observed value. Again, it is observed that the residual values almost approach zero, indicating a minimum error in the predicted OGTT levels. Thus a non-linear regression model has been designed using RSM and validated for prediabetes prediction in this study. A similar study has reported the accuracy of predicting fasting glucose levels using RSM and data mining [17].

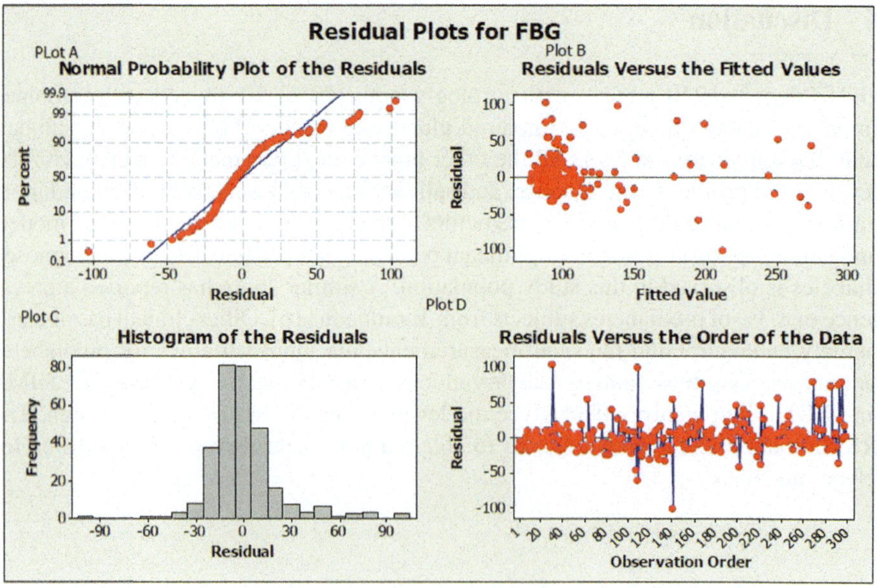

Fig. 3 Residual plots for predicted glycemic levels using RSM

6.2 *Regression Plot for Prediabetes Prediction Using ANN*

In this study, the ANN model is applied over the entire data set $n = 300$ subjects. The 300 subjects are divided into a training data set of 250 and a validation and testing data set of 50. Initially, an ANN model is created and trained with the training data set. Finally, the developed ANN model is tested for its accuracy using a testing data set of 50 subjects. The first three plots in Fig. 4 represent the regression analysis for the training, validation, and testing data. In each plot, exact linearity is observed, represented by the solid line in Fig. 4. Similarly, since the R value is approaching 0.9 in all the cases, it reflects an exact linear relationship between the output values and the targets. Thus ANN model could be used for prediabetes prediction.

Figure 5 represents the regression plot for assessing the accuracy of the ANN model for prediabetes prediction. Using the predictors, the ANN model is designed to predict the glycemic levels of the subjects. It showed 90% accuracy in predicting the glycemic levels of the samples. The PNN model exhibited 95% accuracy in predicting the glycemic levels of the study subjects. The accuracy of the constructed models is validated using a testing data set. The residual curves for the RSM model, regression analysis for the ANN model, and percentage of accuracy observed for the PNN model are significant. This study identifies the predictors of prediabetes, which could be used for the early prediction of the glycemic levels of the study subjects in the ethnic population. Simultaneously, early screening helps for better management of the disease. A hospital-based study has reported the prediction of prediabetes by

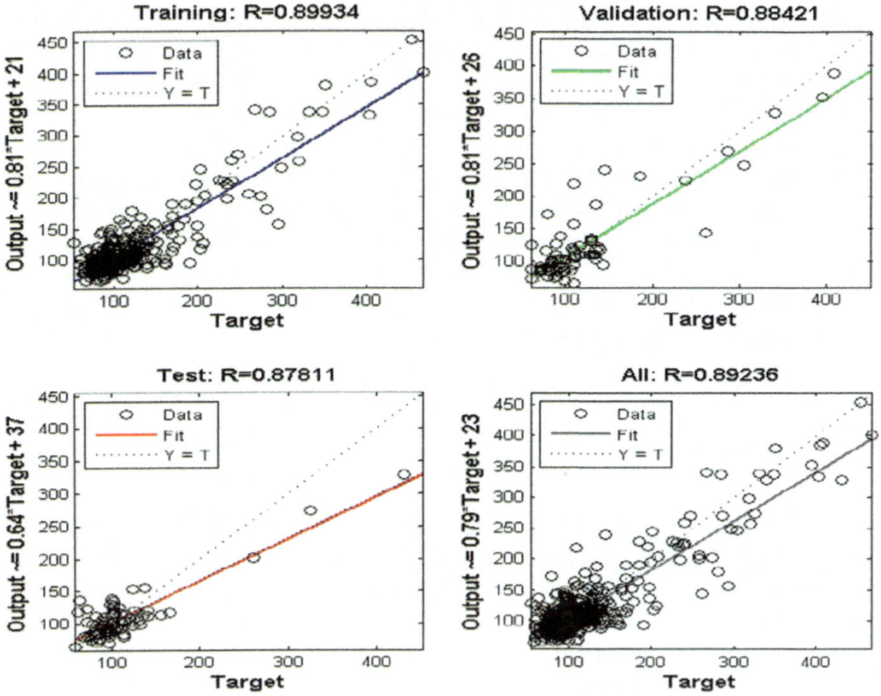

Fig. 4 Regression plot for training, testing, and validation data of ANN

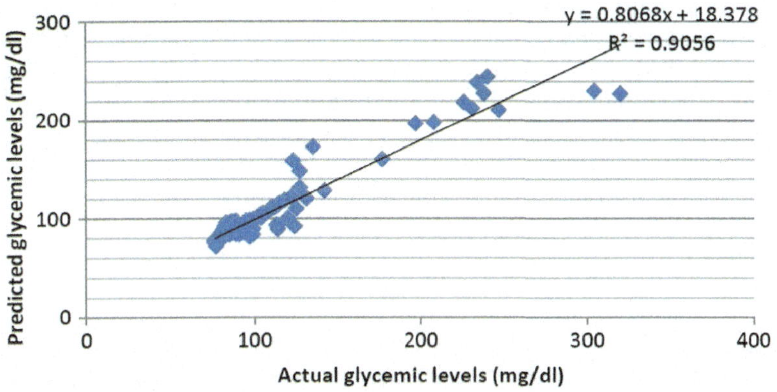

Fig. 5 Regression plot for assessment accuracy of ANN model for prediabetes prediction

using mathematical models among 1252 patients [18]. A linear regression trend line prediction model is devised using clinical data and HbA1C levels. It is used to predict prediabetes subjects.

A multinomial logistic model showed that variables such as age, gender, BMI, hypertension, level of physical activity, and hypercholesterolemia are potential risk factors for prediabetes and undiagnosed diabetes among the subjects studied from Florida [19]. Therefore, the study has attempted to predict prediabetes in a testing data set using these factors as predictors. Our results follow these studies, where an attempt has been made to indicate prediabetes in the chosen ethnic group with the help of mathematical models. However, this study has attempted to devise three different models, namely, linear regression and neural networks models, using ANN and PNN to analyze the robustness of all the proposed models in predicting prediabetes. Among the three models, the PNN model is more accurate, with a minor error in predicting prediabetes in the study population.

7 Conclusion

As the Prevalence and intensity of a metabolic condition like prediabetes increases, simultaneously in the clinical settings, new machine learning models have been created which would be used for the early screening of the disease. These modeled systems are proven accurate in screening for prediabetes and undiagnosed diabetes. Modeling also would require the identification of predictors of prediabetes and undiagnosed diabetes. These predictors would be used for the estimation of the glycemic levels of an individual through the mathematical models designed. The predictors vary in different ethnicities based on their lifestyle habits. This study identified salivary glucose, HbA1C, waist circumference, BMI, and LDL as potential prediabetes predictors. Statistical and neural network models were tested using testing data for accuracy. The PNN model gave maximum accuracy in predicting prediabetes and undiagnosed diabetes groups. These models could be used for early diagnosis of prediabetes and undiagnosed diabetes in the ethnic group in the future.

References

1. William, T.: C: Standards of medical care in diabetes: diabetes care. Am. Diabet. Assoc.**39**(1), S1–S112 (2016)
2. Rhee, S.Y., Woo, J.T.: The prediabetic period: review of clinical aspects. Diabet. Metab. J. **35**(2), 107–116 (2011)
3. Jali, M.V.: Prediabetes-Early detection and interventions. Medicine **18**, 633–644 (2008)
4. Empana, J.P., Ducimetière, P., Arveiler, D., Ferrieres, J., Evans, A., Ruidavets, J.B., Yarnell, J., Bingham, A., Amouyel, P., Dallongeville, J.: Are the Framingham and PROCAM coronary heart disease risk functions applicable to different European populations? The PRIME Study. Eur. Heart J. **24**(21), 1903–1911 (2003)
5. Game, F.L., Jones, A.F.: Coronary heart disease risk assessment in diabetes mellitusa comparison of PROCAM and Framingham risk assessment functions. Diabet. Med. **18**(5), 355–359 (2001)

6. Knuiman, M.W., Vu, H.T.: Prediction of coronary heart disease mortality in Busselton, Western Australia: an evaluation of the Framingham, national health epidemiologic follow up study, and WHO ERICA risk scores. J. Epidemiol. Commun. Health **51**(5), 515–519 (1997)
7. Mohan, V., Deepa, R., Deepa, M., Somannavar, S., Datta, M.: A simplified Indian Diabetes Risk Score for screening for undiagnosed diabetic subjects. J. Assoc. Phys. India **53**(9), 759–763 (2005)
8. Free, H.M., Collins, G.F., Free, A.H.: Triple-test strip for urinary glucose, protein, and pH. Clin. Chem. **6**(4), 352–361 (1960)
9. Braga, F., Dolci, A., Montagnana, M., Pagani, F., Paleari, R., Guidi, G.C., Mosca, A., Panteghini, M.: Revaluation of biological variation of glycated hemoglobin (HbA1c) using an accurately designed protocol and an assay traceable to the IFCC reference system. Clin. Chim. Acta **412**(15/16), 1412–1416 (2011)
10. Heine, R.J., Hanning, I., Morgan, L., Alberti, K.G.M.: The oral glucose tolerance test (OGTT): effect of rate of ingestion of carbohydrate and different carbohydrate preparations. Diabet. Care **6**(5), 441–445 (1983)
11. Argmann, C.A., Houten, S.M., Champy, M.F., Auwerx, J.: Lipid and bile acid analysis. Curr. Protoc. Mol. Biol. **75(1)**, 29B.2.1–29B.2.24 (2006)
12. Free, H.M., Collins, G.F., Free, A.H.: Triple-test strip for urinary glucose, protein, and pH. Clin. Chem. **6**(4), 352–361 (1960)
13. Balan, P., Babu, S.G., Sucheta, K.N., Shetty, S.R., Rangare, A.L., Castelino, R.L., Fazil, A.K.: Can saliva offer an advantage in monitoring of diabetes mellitus?-A case control study. J. Clin. Exp. Dent. **6**(4), e335 (2014)
14. Vickram, A.S., Das, R., Srinivas, M.S., Rao, K.A., Jayaraman, G., Sridharan, T.B.: Prediction of Zn concentration in human seminal plasma of Normospermia samples by artificial neural networks. J. Assist. Reproduct. Genetics **30**(4), 453–459 (2013)
15. Badarinath, A.R.S., Das, A.R., Mazumder, S., Banerjee, R., Chakraborty, P., Saraswathy, R.: Classification of PCR-SSCP bands in T2DM by probabilistic neural network: a reliable tool. Int. J. Bioinform. Res. Appl. **11**(4), 308–314 (2015)
16. Anjana, R.M., Deepa, M., Pradeepa, R., Mahanta, J., Narain, K., Das, H.K., Adhikari, P., Rao, P.V., Saboo, B., Kumar, A., Bhansali, A., John, M., Luaia, R., Reang, T., Ningombam, S., Jampa, L., Budnah, R.O., Elangovan, N., Yajnik, C.S.: Prevalence of diabetes and prediabetes in 15 states of India: results from the ICMR-INDIAB population-based cross-sectional study. Lancet Diabet. Endocrinol. **5**(8), 585–596 (2017)
17. Yamaguchi, M., Kaseda, C., Yamazaki, K., Kobayashi, M.: Prediction of blood glucose level of type 1 diabetics using response surface methodology and data mining. Med. Biol. Eng. Comput. **44**(6), 451–457 (2006)
18. Huang, C.L., Iqbal, U., Nguyen, P.A., Chen, Z.F., Clinciu, D.L., Hsu, Y.H.E., Hsu, C.H., Jian, W.S.: Using hemoglobin A1C as a predicting model for time interval from pre-diabetes progressing to diabetes. PloS One **9**(8), 1–7 (2014)
19. Okwechime, I.O., Roberson, S.: Prevalence and predictors of pre-diabetes and diabetes among adults 18 years or older in Florida: A multinomial logistic modeling approac. PloS One, **10(12)**, 1–17, e0145781 (2015)

Localization and Classification of Brain Tumor Using Multi-layer Perceptron

Ajay Kumar and Yan Ma

Abstract A brain tumor is a group of structures created through the gradual accumulation of unusual cells. It occurs when cells in the brain develop abnormally. It has recently become a leading cause of death for many people. Because brain tumors are among the most dangerous cancers, prompt detection and treatment are required to preserve life. Because of the growth of tumor cells, detecting these cells is a difficult task. Therefore, it is essential to contrast MRI findings with brain tumor therapy. Abnormal brain areas are very challenging to visualize using straightforward imaging methods. Ensemble approaches have been the most significant development in information mining and machine intelligence over the past ten years. They combine several models into one, typically more realistic than the sum of its components. This chapter offers a framework for identifying and categorizing different tumor kinds. Based on the examination of sizable datasets, this study presents a strategy for identifying various tumor forms of the brain. The precise structure of the brain may be seen in MRI pictures and thus taken into consideration for analysis without the need for surgery. The MRI scan reveals the brain's structure, which aids in further processing and tumor diagnosis. The field of artificial neural networks, frequently referred to as multi-layer perceptrons, is the most practical variant of neural networks used for brain tumor analysis. The study uses 212 samples of brain MR images for classifying brain tumors using the multi-layer perceptron model. The classification accuracy achieved by the model for classifying brain tumors is 98%.

A. Kumar (✉)
Manipal University Jaipur, Jaipur, India
e-mail: dr.ajaykr0088@gmail.com

Y. Ma
College of Information, Mechanical and Electrical Engineering, Shanghai Normal University, Shanghai 200234, China
e-mail: ma-yan@shnu.edu.cn

D. P. Acharjya and K. Ma (eds.), *Computational Intelligence in Healthcare Informatics*, Studies in Computational Intelligence 1132,
https://doi.org/10.1007/978-981-99-8853-2_6

1 Introduction

The identification of brain tumors is one of the most significant issues with the processing of medical pictures. A "brain tumor" is a condition where the number of cells in the human brain increases unnaturally. It often develops from brain cells, blood vessels, and protruding brain nerves. The tumors are classified as benign or malignant. Slow-growing tumors are benign tumors. Benign tumors do not invade the surrounding brain tissue. These tumors will only exert pressure that might be destructive. Rapidly growing tumors are referred to as malignant tumors. These tumors can spread throughout the brain [1, 2]. Tumors can destroy normal brain cells by causing inflammation, putting pressure on brain areas, and increasing pressure in the skull. Globally, brain tumors have become a significant cause of mortality and impairment. With the advancement of image recognition, early detection of brain tumors [3] is now possible. Prior diagnosis of patients who have survived a brain tumor is facilitated by medical image processing [2]. The primary objective highlights how deep learning and machine learning methodologies have influenced medical image processing.

Machine learning is essential in tumor prognosis. This chapter has built and suggested a model for detecting and classifying brain tumors in this proposed study. The image is processed and smoothed as part of the procedure. The segmentation is performed using morphological operations followed by masking, which improves classification accuracy. Different feature extraction methods are applied to extract the feature from the masked image, and the multi-layer perceptron is used for classification.

The rest of this book chapter is organized as follows: Sect. 2 discusses the related work relating to brain tumor classification. Section 3 briefly highlight the various neural network architectures. Various phases of brain tumor detection are discussed in Sect. 4. Experimental results are produced in Sect. 5. Conclusion is presented in Sect. 6.

2 Background Literature

This section briefly highlights the background literature on brain tumor classification using computational intelligence techniques. Image segmentation plays a vital role in disease detection. The notion of fuzzy c-means (FCM) segmentation is introduced to discriminate between brain areas with tumors, and those without them [4]. Wavelet attributes can also be extracted using a multi-layer discrete wavelet transform (DWT). Finally, a deep convolutional neural network (DNN) is employed to classify brain tumors accurately. The accuracy percentage for the DNN-based brain tumor classification study was 96.97%.

Another work presents a novel bio-physiomechanical tumor development simulation to study patient tumor progression systematically. It Combines both discrete and continuous methods in a model of tumor development [5]. Despite its lengthy computing time, this approach is mainly used for segmenting brain tissue. Similarly, the brain tumor is detected and segmented using unique multi-fractal feature extraction and enhanced AdaBoost classification approaches. The texture of brain tumor tissue is extracted using the multi-fractal feature extraction method. The intricacy is very high in this procedure [6]. Likewise, the local independent projection-based classification (LIPC) method is used to classify the brain component [7]. Literature also found a graph-cut-based segmentation method with a seeded tumor segmentation strategy using cutting-edge cellular automata (CA) technology. This approach uses the volume of interest and seed selection for effective segmentation of brain tumors [8]. But, the accuracy is moderate, and the complexity is high.

A novel brain tumor segmentation method, called multimodal brain tumor segmentation, is introduced in the literature [9]. The challenge relating to brain tumor segmentation and classification is high. First, an overview of brain tumor segmentation is presented in the literature. The validity, robustness, and accuracy of each approach are assessed [10]. Similarly, hybrid feature selection with ensemble classification is also used to diagnose brain tumors [11]. Next, the fuzzy-based control theory is used to segment and classify brain tumors. Although the performance is outstanding, the accuracy is subpar [12]. Likewise, an adaptive histogram equalization to boost picture contrast is used in the tumor classification [13]. The tumor is subsequently separated from the rest of the brain image using FCM. Finally, the fuzzy with k-nearest neighbor (KNN) classification is employed to identify the anomaly in the brain MRI image. The challenge is very high, and the accuracy is subpar. This chapter suggested a method that performs a unique automated brain tumor classification using an artificial neural network.

3 Foundations of Neural Network

Neural network architecture and technology resemble the human brain. Typical applications of neural networks include vector quantization, approximation, data clustering, pattern matching, optimization operations, and classification methods. A neural network is divided into three different sorts by its connections. Neural networks can be classified as either feedback, feed-forward, or recurrent. Further, the feed-forward neural network has two variations: single-layer and multi-layer. In a single-layer network, the hidden layer is invisible. It just has an input layer and an output layer.

In contrast, the multi-layer consists of three layers: input, hidden, and output. The recurrent network is a feedback system based on closed loops. In addition, the proposed method employs a feed-forward neural network. Recurrent, radial basis and feed-forward neural networks are three distinct kinds of neural networks that have been used and compared for operational excellence [14]. The subsections below describe these networks briefly. An overview of neural network architecture is presented in Fig. 1.

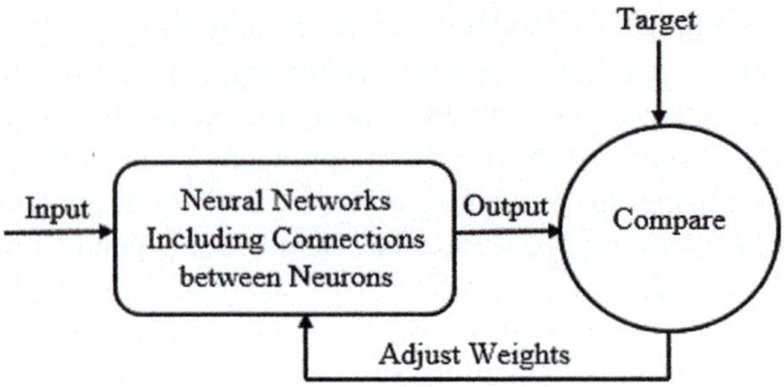

Fig. 1 Neural network architecture

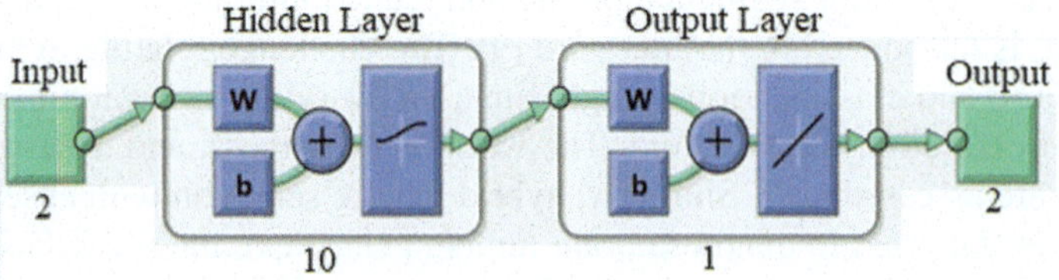

Fig. 2 Feed forward neural network

3.1 Feed-Forward Networks

In feed-forward neural networks, data only travels in one direction. It moves from the input node to the output node via the hidden nodes. Every stratum's output is correlated with the input of the layer below it. In a feed-forward network, various transfer functions are utilized for computation. As illustrated in Fig. 2, the feed-forward network contains a hidden layer with a sigmoid function, followed by an output layer with a linear function. A sigmoid function can be employed in the output layer of the network to confine its output. The network doesn't contain any loops or cycles.

3.2 Recurrent Networks

Recurrent networks have feedback connections from the output to the intermediate layer. Examples of recurrent networks are the Hopfield network and the Elman network. Elman networks are two-layer architecture. Furthermore, it organizes a criticism association from the yield of the primary layer to its information. This repetitive association permits the Elman system to identify and create time-changing examples. This system has a tansig move work in its concealed layer and a purlin move work in its yield layer, as appeared in Fig. 3.

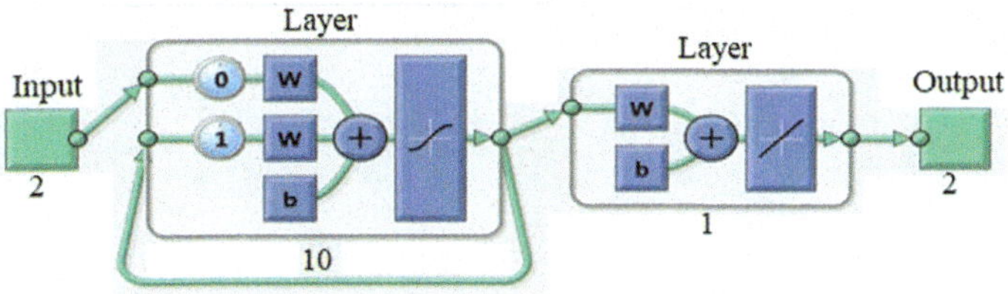

Fig. 3 Elman neural network

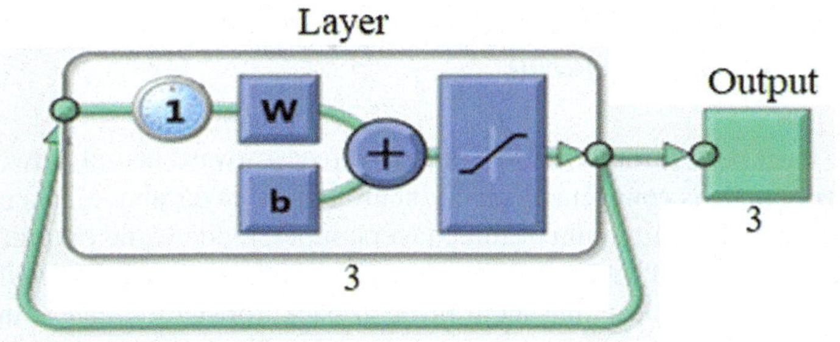

Fig. 4 Hopfield neural network

Hopfield networks utilize the immersed straight transfer work. This system can have at least one info vectors which is mentioned as starting condition of the system. In the wake of giving the underlying conditions, the network produces a yield which is then bolstered back to turn into the contribution as appeared in Fig. 4. This process is rehashed again and again until the yield settles.

3.3 *Radial Basis Network*

Radial basis network systems may require more neurons than feed-forward networks. However, regularly these can be structured in a shorter time than the standard feed-forward systems. These systems work best when many preparing vectors are accessible. However, these may have higher memory and calculation costs. The vector separation between the info vector and the weight vector, repeated by the inclination, is shown in Fig. 5. The net contribution to the radbas transfer function in these systems.

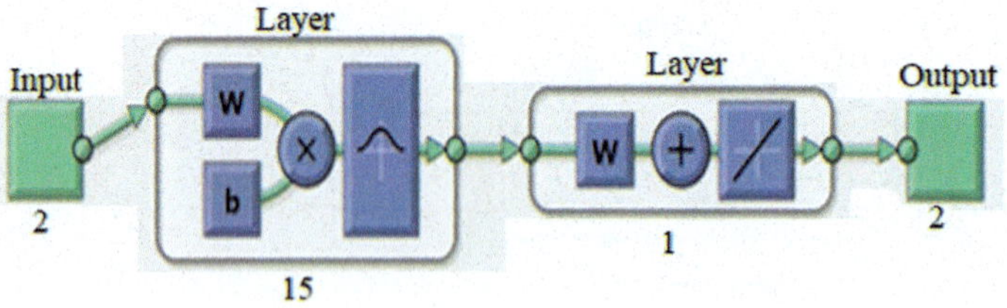

Fig. 5 Radial basis network

3.4 Multi-layer Perceptron

The multi-layer perceptron is a fully connected feed-forward neural network. In this system, every node is connected to every neuron in the layer above. In a multi-layer perceptron, there is a minimum of three layers: input, hidden, and output. It spreads in both the forward and backward directions. The activation function multiplies the weighted inputs before sending them through backpropagation and is modified to reduce loss. Simply put, weights are learned values from neural networks by computers. Based on the difference between training inputs and anticipated outcomes, they self-adjust. Nonlinear activation functions are initially used as the output layer activation function, followed by SoftMax. The general architecture of multi-layer perceptron is presented in Fig. 6.

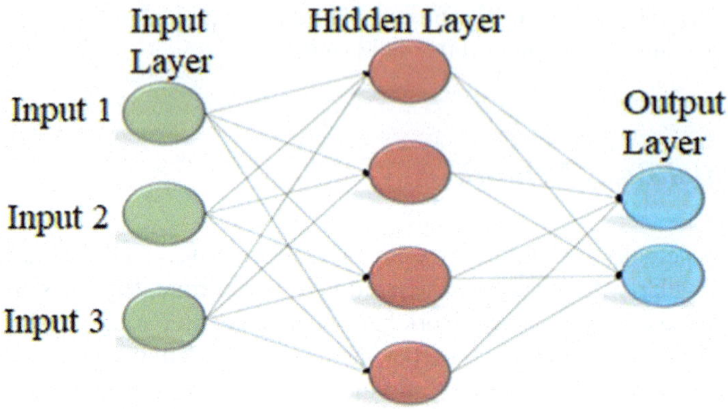

Fig. 6 Multi-layer perceptron

4 Phases of Brain Tumor Detection

Various phases of brain tumor detection are discussed in the section. Preprocessing, segmentation, feature extraction, and classification are the four stages of the brain tumor detection procedure, as shown in Fig. 7.

The various imaging modalities accessible in recent days include X-ray, mammography, CT scans, and magnetic resonance imaging (MRI) [15]. MRI imaging can capture the whole architecture and composition of the brain or skull. In addition, an MRI scan can be used to learn more about the blood flow inside the brain [16]. As a result, MRI methods have emerged as a crucial tool for establishing diagnoses, spotting abnormalities, and tracking the development of disease [17]. In addition, these MRI images are further processed to detect the brain tumor.

Image Preprocessing: The prime objective of image preprocessing is to reduce unwanted noise and enhance the image. Techniques for image preparation can enhance a picture's qualities. The quality of the uncorrupted image, processing time, computing cost, and noise-reduction strategies are only a few factors affecting picture improvement. These are taken care of in the image preprocessing phase [18]. The next phase leads to image segmentation. The primary objective of segmentation is to separate related objects in an image. In this phase, an image is divided into several areas in the segmentation stage [19]. These zones are comparable in terms of texture, color, intensity, contrast, and gray level. Various segmentation techniques include threshold-based segmentation, edge-based segmentation, segmentation using clustering, and segmentation using region.

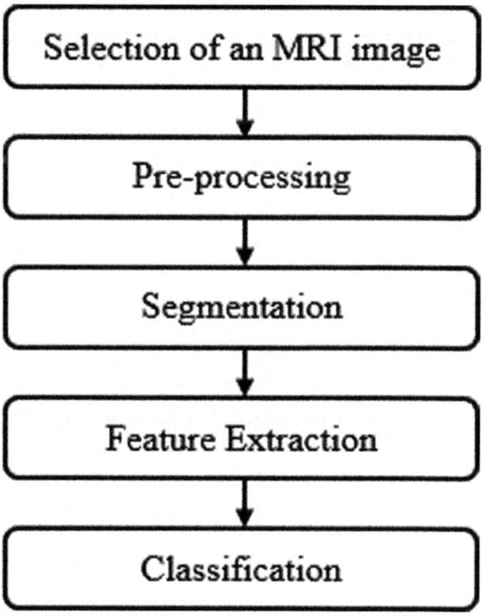

Fig. 7 The stages of brain tumor detection

Segmentation: Threshold-based segmentation is beneficial for images with varying brightness levels. Based on the brightness of their pixels, images may be divided into various categories. Global, local, and adaptive thresholding are the three different types of thresholding techniques. Similarly, depending on a dramatic shift in the brightness of pixels close to the borders, an edge-based segmentation technique is used to split the image. By utilizing this technique, a binary picture is produced. This graphic demonstrates the component limitations. Watershed segmentation, a gradient-based strategy, and gray histogram approaches were the methodologies used in this study. Likewise, clustering is the most used technique for segmenting MRI pictures. This method divides pixels into classes without the aid of historical data. The same class is given to the pixels with the highest potential [20, 21]. K-means, fuzzy C-means, and hierarchical clustering are examples of approaches used for clustering. Furthermore, images are segmented using regions. It splits the images into various areas based on a certain system. Nature is comparable in these areas. This method employs regional growth, region splitting, and region merging strategies.

Feature Extraction: When a computation input is too significant and redundant to process, it is condensed into a feature vector, which is a smaller collection of features. The process of turning raw data into a group of features is known as feature extraction. In this stage, the primary characteristics needed for image categorization are extracted. First, the segmented brain MRI extracts texture characteristics from the segmented image. Further, the texture property of the image is demonstrated. These characteristics are extracted using the gray level co-occurrence matrix (GLCM), a dependable and effective method [20]. The GLCM texture analysis method is highly competitive because it uses fewer gray levels, which minimizes the size of the GLCM and lowers the processing cost of the algorithm while retaining high classification rates. In addition, it finds essential details regarding the surface's structure in the texture. For example, gray tone spatial dependencies-based textural qualities are often helpful for image classification. The following are the GLCM texture characteristics that are extracted.

1. Power: The power E measures the number of repeated pixel pairs, a measure of textural consistency. The range lies between [0, 1].

$$E = \sum_{i=0}^{N_g-1} \sum_{j=0}^{N_g-1} p(i, j)^2 \tag{1}$$

2. Variance: It evaluates the overall picture intensity contrast, Con, between a pixel and its neighbors [6]. The range lies between [0, 1].

$$Con = \sum_{n=0}^{N_g-1} n^2 \sum_{i=0}^{N_g-1} \sum_{j=0}^{N_g-1} p(i, j)^2 \tag{2}$$

3. Similarity: This gauges how connected a pixel is to its neighbors throughout the image. The range lies between [−1, 1].

$$C = \frac{1}{\sigma_x \sigma_y} \sum_{i=0}^{N_g-1} \sum_{j=0}^{N_g-1} (i, j) p(i, j)^2 - \mu_x \mu_y \tag{3}$$

4. Cohesion: This gauges how well the distribution of elements in the GLCM aligns with its diagonal. The range lies between [0, 1].

$$H = \sum_{i=0}^{N_g-1} \sum_{j=0}^{N_g-1} \frac{p(i, j)}{(1 + mod(i, j))} \tag{4}$$

An MRI scan of the brain is taken, preprocessed, and segmented to determine if it is normal or abnormal. Next, the segmented picture is subjected to feature extraction, and texture properties are recovered using GLCM [20]. Finally, machine learning techniques identify the MR brain picture as normal or abnormal.

The machine learning algorithm's main objective is to learn and make sensible conclusions automatically. As mentioned above, the feature set produced by the method is put into a multi-layer perceptron for classification (MLP). A set of input data is transformed into pertinent output data using the MLP artificial neural network model. It is known as a feed-forward neural network. In the feed-forward network, there are no cycles, and the network output is simply dependent on the current input instance. A neuron with a nonlinear activation function makes up each node in the MLP. It is based on the supervised learning method. The connection weights are modified based on the degree of inaccuracy between the intended output and the expected result after processing each data item. The goal of the learning process is to decrease error by increasing the current weight values assigned to each edge. Because the weights are reversed, the model is known as back diffusion.

5 Experimental Results

In this study, effective automated brain tumor detection is achieved using an artificial neural network. The simulation is run using Python. Calculated accuracy is compared to other cutting-edge methods. The research employed 212 brain MRI images. Each image contains the texture-based characteristics that are extracted. Energy, contrast, correlation, and homogeneity are a few texture-based attributes extracted using GLCM. The MLP with a 66% split is used for training categorization. The remaining instances of 34% percent are utilized for testing. The comparative results is presented in Table 1.

Table 1 Assessment of investigational observations

ML algorithm	Sample count	Accuracy
Multi-layer perceptron	212	98.2
Radial basis function	212	76.4
Recurrent neural network	212	95.3
Naive Bayes	212	97.6

6 Conclusion

This research aims to create a rapid, accurate, and easy-to-use system for automatically classifying brain tumors. It is simple to understand this study on machine learning-based brain tumor detection. It uses co-occurrence GLCM to extract information from the images. Besides, it uses the Python programming language. A database called picture net is utilized for categorization. In this proposed piece, the textural elements of the image are evaluated for energy, contrast, correlation, and homogeneity. Maximum accuracy of 98.2% is attained by employing MLP for classification on 212 samples of brain MRI. The MLP is also compared with radial basis function, recurrent neural network, and Naive Bayes method. A more extensive data set, additional models that perform better, and the extraction of intensity-based characteristics and texture-based features might increase this accuracy.

References

1. Bhagat, M.J.V., Dhaigude, N.B.: A survey on brain tumor detection techniques. Int. Res. J. Eng. Technol. **4**(3), 1795–1796 (2017)
2. Kapoor, L., Thakur, S.: A survey on brain tumor detection using image processing techniques. In: Proceedings of 7th IEEE International Conference on Cloud Computing, Data Science & Engineering-Confluence, pp. 582–585 (2017)
3. Sapra, P., Singh, R., Khurana, S.: Brain tumor detection using neural network. Int. J. Sci. Modern Eng. **1**(9), 83–88 (2013)
4. Mohsen, H., El-Dahshan, E.S.A., El-Horbaty, E.S.M., Salem, A.B.M.: Classification using deep learning neural networks for brain tumors. Future Comput. Inform. J. **3**(1), 68–71 (2018)
5. Bauer, S., May, C., Dionysiou, D., Stamatakos, G., Buchler, P., Reyes, M.: Multiscale modeling for image analysis of brain tumor studies. IEEE Trans. Biomed. Eng. **59**(1), 25–29 (2011)
6. Islam, A., Reza, S.M., Iftekharuddin, K.M.: Multifractal texture estimation for detection and segmentation of brain tumors. IEEE Trans. Biomed. Eng. **60**(11), 3204–3215 (2013)
7. Huang, M., Yang, W., Wu, Y., Jiang, J., Chen, W., Feng, Q.: Brain tumor segmentation based on local independent projection-based classification. IEEE Trans. Biomed. Eng. **61**(10), 2633–2645 (2014)
8. Hamamci, A., Kucuk, N., Karaman, K., Engin, K., Unal, G.: Tumor-cut: segmentation of brain tumors on contrast enhanced MR images for radiosurgery applications. IEEE Trans. Med. Imaging **31**(3), 790–804 (2011)

9. Ghaffari, M., Sowmya, A., Oliver, R.: Automated brain tumor segmentation using multimodal brain scans: a survey based on models submitted to the BraTS 2012–2018 challenges. IEEE Rev. Biomed. Eng. **13**, 156–168 (2019)
10. Liu, J., Li, M., Wang, J., Wu, F., Liu, T., Pan, Y.: A survey of MRI-based brain tumor segmentation methods. Tsinghua Sci. Technol. **19**(6), 578–595 (2014)
11. Huda, S., Yearwood, J., Jelinek, H.F., Hassan, M.M., Fortino, G., Buckland, M.: A hybrid feature selection with ensemble classification for imbalanced healthcare data: a case study for brain tumor diagnosis. IEEE Access **4**, 9145–9154 (2016)
12. Karuppathal, R., Palanisamy, V.: Fuzzy based automatic detection and classification approach for MRI-brain tumor. ARPN J. Eng. Appl. Sci. **9**(12), 42–52 (2014)
13. Janani, V., Meena, P.: Image segmentation for tumor detection using fuzzy inference system. Int. J. Comput. Sci. Mobile Comput. **2**(5), 244–248 (2013)
14. Joshi, D.M., Rana, N.K., Misra, V.: Classification of brain cancer using artificial neural network. In: Proceedings of 2nd IEEE International Conference on Electronic Computer Technology, pp. 112–116 (2010)
15. Rajeshwari, S., Sharmila, T.S.: Efficient quality analysis of MRI image using preprocessing techniques. In: Proceedings of IEEE Conference on Information & Communication Technologies, pp. 391–396 (2013)
16. Sun, L., Zhang, S., Chen, H., Luo, L.: Brain tumor segmentation and survival prediction using multimodal MRI scans with deep learning. Front. Neurosci. **13**, 810 (2019)
17. Bhanothu, Y., Kamalakannan, A., Rajamanickam, G.: Detection and classification of brain tumor in MRI images using deep convolutional network. In: Proceedings of IEEE International Conference on Advanced Computing and Communication Systems, pp. 248–252 (2020)
18. Borole, V.Y., Nimbhore, S.S., Kawthekar, D.S.S.: Image processing techniques for brain tumor detection: a review. Int. J. Emerg. Trends Technol. Comput. Sci. **4**(5), 28–32 (2015)
19. Kadkhodaei, M., Samavi, S., Karimi, N., Mohaghegh, H., Soroushmehr, S.M.R., Ward, K., All, A., Najarian, K.: Automatic segmentation of multimodal brain tumor images based on classification of super-voxels. In: Proceedings of 38th Annual International Conference of the IEEE Engineering in Medicine and Biology Society, pp. 5945–5948 (2016)
20. Chowdhary, C.L., Acharjya, D.P.: Segmentation and feature extraction in medical imaging: a systematic review. Procedia Comput. Sci. **167**, 26–36 (2020)
21. Chowdhary, C.L., Acharjya, D.P.: Segmentation of mammograms using a novel intuitionistic possibilistic fuzzy c-mean clustering algorithm. In: Nature Inspired Computing, pp. 75–82. Springer, Singapore (2018)

Computational Intelligence in Analyzing Health Data

Information Retrieval from Healthcare Information System

Nimra Khan, Bushra Hamid, Mamoona Humayun, N. Z. Jhanjhi, and Sidra Tahir

Abstract The term medical information retrieval refers to the collection of the dataset from different sources like hospitals, organizations, and healthcare research centers and its use for subsequent experiments to improve the treatments for complex medical conditions. Such information retrieval systems are designed to enhance the healthcare system, speed up disease diagnosis, and offer patients better alternatives. Today, the Internet has interconnected the entire world, making it incredibly simple for institutions conducting medical research to exchange test results and medical data. With the aid of gathered medical data, research might be carried out at subsequent levels of studies. Even countries can exchange medical data and combine performing medical research using the acquired data. This chapter attempts to examine medical information retrieval's significance, its techniques, and its utility in health care. The rapid development of Internet of Things (IoT) makes it possible to connect various smart devices over the Internet and provides more approaches for data interoperability for application needs. These IoT applications, which could be employed in information-intensive industries, are supported by more recent studies in the healthcare services sector. This work initially recommends a semantic data model to store and analyze IoT data. Further, a resource-based data access technique is designed to increase access to IoT data resources by collecting and using IoT data anywhere. Finally, an IoT-based emergency medical service solution is presented. In order to

N. Khan · B. Hamid · S. Tahir
University Institute of Information Technology, University of Arid Agriculture, Rawalpindi, Pakistan
e-mail: bushrakiani@uaar.edu.pk

S. Tahir
e-mail: stahir@uaar.edu.pk

M. Humayun
Department of Information Systems, College of Computer and Information Sciences, Jouf University, Sakakah 42421, Saudi Arabia
e-mail: mahumayun@ju.edu.sa

N. Z. Jhanjhi (✉)
School of Computer Science and Engineering, Taylor's University, 47500 Subang Jaya, Malaysia
e-mail: noorzaman.jhanjhi@taylors.edu.my

D. P. Acharjya and K. Ma (eds.), *Computational Intelligence in Healthcare Informatics*, Studies in Computational Intelligence 1132,
https://doi.org/10.1007/978-981-99-8853-2_7

facilitate emergency medical services, flexibility is needed. The result shows that the resource-based strategy to accessing IoT data is effective in a context with distributed heterogeneous data for supporting quick, all-encompassing data access on a cloud and mobile computing platform. These IoT applications might be employed in information-intensive businesses like healthcare services. However, the variety of IoT items resulted in a problem with the data in a situation where there is a dispersed heterogeneous dataset to enable quick and widespread data access on a cloud and mobile computing platform.

1 Introduction

For researchers and medical institutions, the main problems with the Medical Information Retrieval System (MIRS) is that they should possess the necessary information to comprehend the data and should be equipped with tools for data conversion as needed by the healthcare system. The method to enhance medical information retrieval and medical care is to obtain the necessary beneficial information from the massive data readily available in the healthcare industry. The MIRS suggests the collection of data from research offices, institutions, and medical facilities from research organizations, such as emergency rooms. Further, the use of information for additional research to find new treatments for complex health issues. A basic model for MIRS is depicted in Fig. 1.

This particular data recovery architecture seeks to improve the provision of human services, quickly identify infections, and provide patients with better substitute

Fig. 1 A basic model for a medical information retrieval system that collects medicines, clinical data, and stores in a centralized database

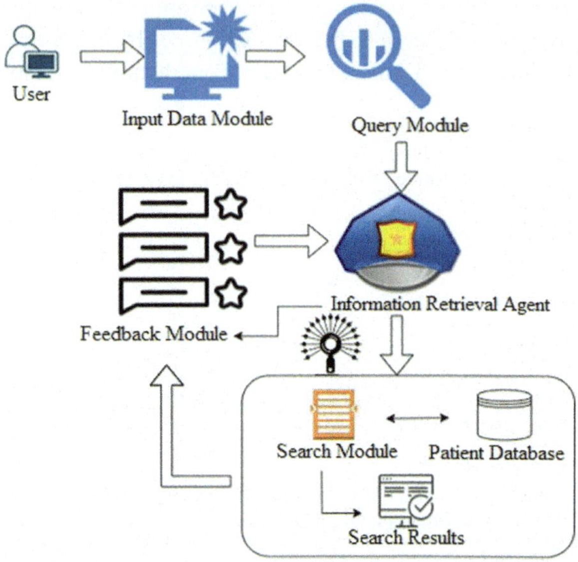

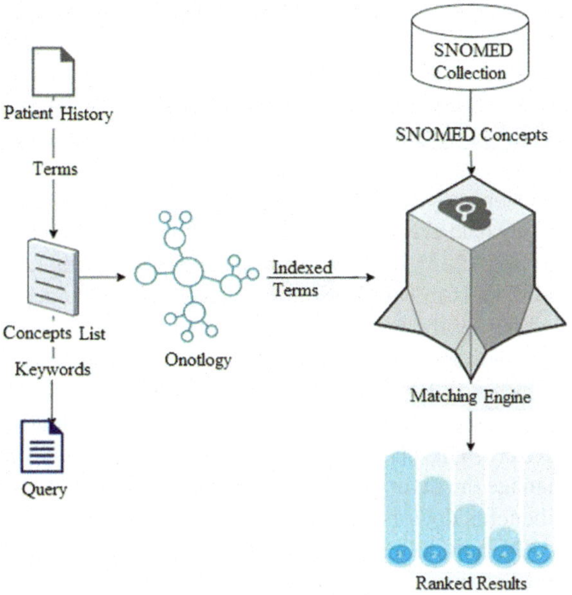

Fig. 2 An implementation of a healthcare information retrieval system

treatment options. The world is now connected via the Internet, making it incredibly simple for therapeutic research organizations to exchange medical information and test results. They are not compelled to conduct the same trials that have just occurred across the nation. The next stage of trials might be completed by the examination of research associations with the help of gathered therapeutic data from the scientists of one nation. Indeed, even countries can collaborate on medical research projects and share medical data. This data recovery framework type aims to use the recovered data to enhance the human services framework. However, this investigation aims to consider the value of information on medicine. An example of retrieving information about patient data is shown in the following Fig. 2.

Besides, retrieval and its methods are also considered in the healthcare industry. It is clear that the investigations face a few challenges because of the dialects of the various countries. IoT technology advancements offer tremendous possibilities for improved and more convenient access to healthcare services. Using these technological methods, physicians participate in healthcare servicing medical facilities that have access to several data sources. As a result, the rising population needs and disparate systems between the accessibility and availability of health services in rural and urban areas. One of the most fundamental difficulties in providing health care is the inadequate sharing of medical data and other types of information.

In China, a lot of work has been done to address the issue of hospitals sharing clinical data. There is now a comprehensive clinic information portal to communicate data among hospital information systems. Some districts have created resident health document systems enabling residents to save and access their electronic health records in a cloud-based system. These programmers improve the clinic's data environment, making it easier for medical researchers to obtain patient information. Still, they are insufficient to support diagnosing, especially in emergency medical services, where more data must be quickly accessed across organizations to coordinate group activities. Although an application-level standard called Health Level 7 (HL7) is available for network protocol for exchanging clinical data [1], it is still challenging for use in actual practice.

Over ten years, more research has concentrated on using IoT technology for collecting data everywhere, process it quickly, and transfer it wirelessly in the healthcare industry [2, 3]. Further, Ambient Assisted Living (AAL) intended to support elderly individuals in carrying out everyday tasks as independently as feasible is also carried out by the researchers [4]. In the literature, IoT is described as a Web resource management through web service interoperability [5]. Besides, Universal Resource Identifier (URI) could be used to retrieve web services [6, 7]. It can be adaptable to resource representation and wireless communication applications for ubiquitous services. This chapter suggests a method using Universal Data Access (UDA)-IoT to manage the diversity of IoT-based data utilized in medical services. Simultaneously, it benefits doctors and patients.

The remaining sections of the chapter are arranged as follows. Section 2 presents the challenges in information retrieval systems from health care. Section 3 describes the use of UDA-IoT. It throws light on identifying and addressing the access issues with distributed heterogeneous IoT data which is everywhere in enterprises. Section 4 discusses the use of the UDA-IoT methodology with a case study for medical emergency services. The conclusion is presented in Sect. 5.

2 Challenges of Information Retrieval in Health Care

Verifying the accuracy of the data gathered for the healthcare system is crucial. Inaccurate data could slow down the research process and potentially cause severe findings to appear in the future. Even the study process is highly challenging. The issues researchers may encounter when correcting the current datasets are depicted in Fig. 3.

Valuable reports, radiology reports, pathology reports, and release rundowns are examples of significant restorative information. Reports are organized in a literary style. It is advisable to frequently explore databases with free-content beneficial stories to find relevant information for clinical and research needs. The following Table 1 presents the most typical healthcare system information retrieval challenges.

2.1 The Medical Healthcare Benefits and Challenges

The healthcare data collected worldwide has several advantages and leads to challenges. The data generated through several examinations must address several topics.

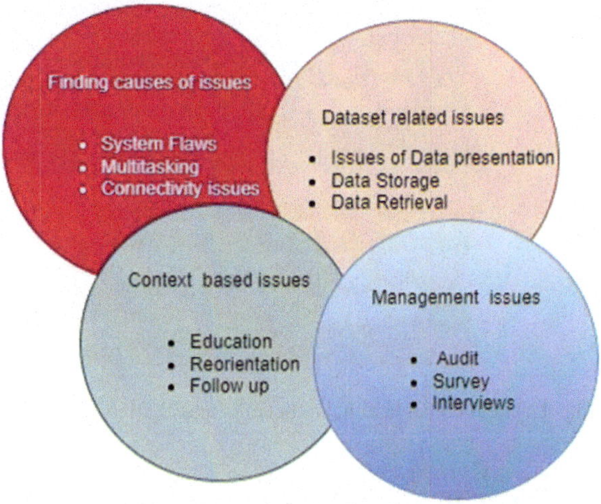

Fig. 3 The challenges of correction in patient information datasets

Table 1 Healthcare system information retrieval challenges

Challenges	Description
Inadequate information	Inexperienced researchers were unable to understand the format of information. As a result, processing enormous datasets for subsequent research is quite tricky
Modification of medical data	One of the most significant challenges is the potential loss of many months or years of research if a method or adjustment is reversed
Correction in medical information data	The researchers' language is the key challenge in combining research with the information gathered from previous studies for a given condition
Accuracy of information	It is exceedingly challenging to determine if the data gathered is entirely accurate or not

Because the languages of the various nations vary, the remedial terms are accepted worldwide. Thus, synchronizing the recovered data is challenging today and again. The fact that it goes beyond creative is one of the biggest tests. For such a data recovery system, there are numerous audit associations which gather online information frameworks, and Internet users interface clinical data and handle them carefully. The computation shown significantly enhances data recovery displayed on restorative accounts. Figure 4 depicts the several medical healthcare benefits and challenges.

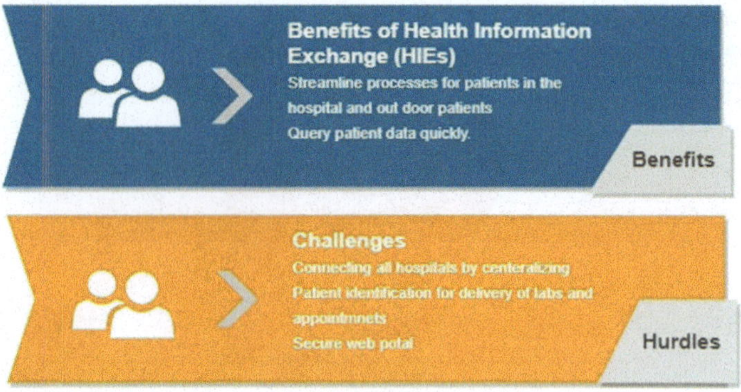

Fig. 4 The medical healthcare benefits and challenges

2.2 *IoT-Based Medical Services Data Model*

The provision of healthcare services is a dynamic process that involves processing largely before, during, and after treatment. Activities related to healthcare services both within and outside hospitals is depicted in Fig. 5. It includes the provision of drugs and equipment and the processing of insurance documents. Various departments, patient types, professional staff members, and doctors are all involved in the delivery of health care. Doctors must access patients' medical histories, and this type

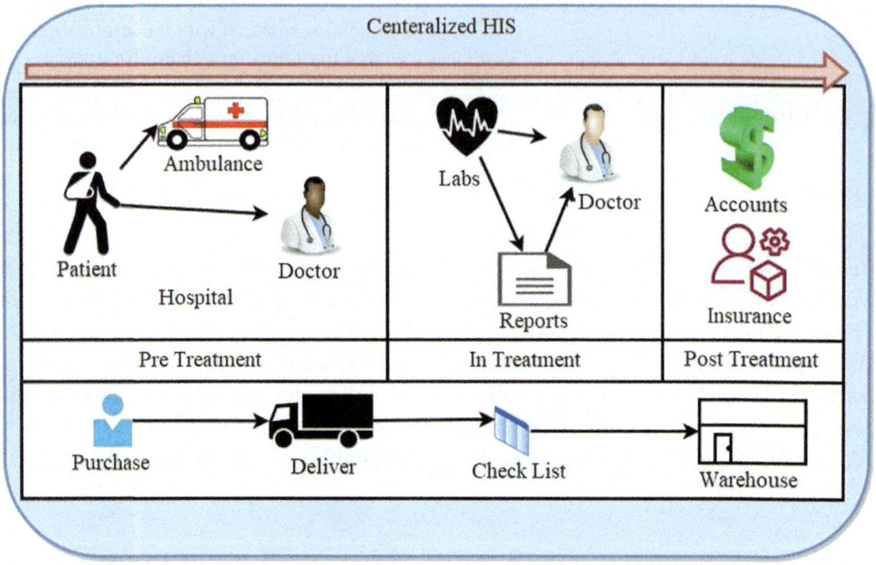

Fig. 5 Services related to health care

of data can be kept distributed. The location to collect the data and the equipment's busy/free operating status may require accessing data from the equipment. IoT technology is now frequently used in delivering healthcare services [8]. For example, ambulances and equipment are connected to the global positioning Radio Frequency Identification (RFID) and the Global Positioning System (GPS), making it easier to find. Drugs have barcode labels so that patients can be administered with greater accuracy. As a result, sharing medical information when processing medical services becomes crucial and challenging for doctors and management at medical centers as it calls for close collaboration. IoT notes link many items to one another to provide patients with healthcare services. Patients' and doctors' ID cards connect ambulances to the IoT, and RFID tags are used to find expensive medical equipment and instruments and scan medications for scanning into hospital information systems. A data model that is flexible and semantic is necessary to help clinicians and managers access data resources effectively. Additionally, data must be easily accessible at all times and locations for healthcare services.

2.3 *Frequently Attainable Metadata Model for IoT Data*

The UDA-IoT model in medical services must do the following duties after summarizing the features of medical services. The model must access enormous amounts of data and must support heterogeneous format of data. Besides, it must assist in the creation of real-time application systems. In order to achieve the requirements mentioned above, a consistent metadata schema to describe IoT notes is presented, in which data are heterogeneous in format, and it makes data sharing and interoperating simpler. It is depicted in Fig. 6. Three levels make up the data structure including the value, the annotation, and the semantic reason. Value is the term used to describe the collection of data that shows a patient's attribute or a fact regarding healthcare services. An annotation is a name for the data retrieval caption. Semantic explanation refers to the standard definition of the data used for data sharing.

This data model focuses more on self-explanation of data value for ubiquitous data access than on the specification of data structures for data organization, in contrast to standard data structures like relational data structures. Each IoT data point is defined and notated in XML format, enabling web access and self-description. Each piece of data is described by ontology, allowing semantic-level explanation, and is then discussed with additional examples of equivalent concepts. Another advantage of this data model is that it may be more easily customized for large data applications. The IoT application data is constantly increasing. To better meet the needs of large data in IoT and to increase the efficiency of data storage using NoSQL databases, the ontology incorporates the description of data connection, which is the specification of the data structure in the traditional data model. Furthermore, each IoT note is appended with time and location information for IoT applications in medical services.

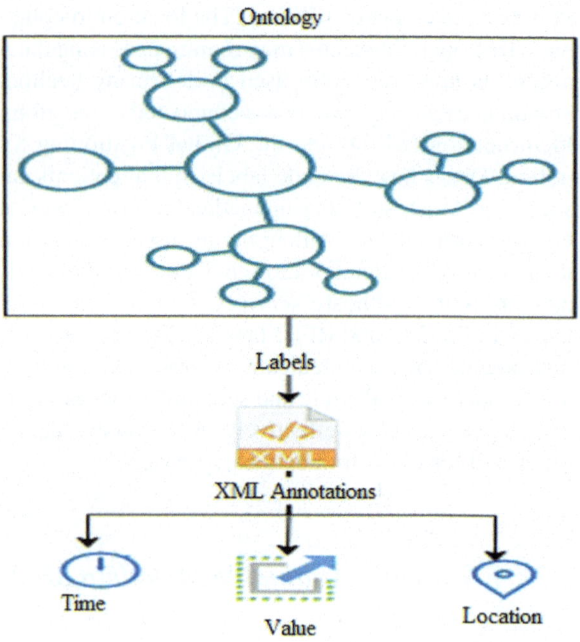

Fig. 6 IoT metadata model note

For systems that use real-time applications, the time tag is helpful to position the tag and would be crucial in smart medical services for the dynamic positioning of pricey medical devices and equipment.

3 UDA-IoT Ubiquitous Data Accessing for Information System

IoT physical entity and information entity mapping physical entities are used in IoT application systems as sensors to connect the information systems. They are used as the physical entities' representations in information systems. Physical entities cannot be transferred directly through information systems; instead, models of the physical entities can be transferred through the information system. In order to realize the physical interactions, entities are transferred. For instance, if someone is curious about whether the patient is in the medical ward of room number 3203 in the hospital, they might access the state of the physical entity's representation of "medical ward room 3203" in computerized databases. It is formally referred to as Entity-oriented Resource (EoR). EoR is the informational equivalent of physical item systems. It is defined as

$$EoR :=< URI, AttrSet :=< Attributes >, Persistence$$
$$:=< Driver, Address, Authentication >>$$

The EoR corresponding to the representation of the physical entity is kept in the URI, and the only address that application systems can use to access physical things. The EoR's attribute set is called AttrSet. Persistence notifies the data storage layer of the EoR's location, sources, and access points. It is to be highlighted that an EoR corresponds to distinct URIs, although the EoR's content can refer to many persistences. Once the EoR is defined, applications disclose their URI to the application layer. Persistence, on the other hand, is invisible to the application layer. Some physical entities in IoT applications are made up of other physical entities. For example, the ambulance "Sprinter-324EoR"'s is made up of other EoRs like a stretcher, a breathing machine, and a medical monitor. Information system representation of composed physical elements is referred to as composited EoR (cEoR). It changes with time. Depending on the company's needs, they might be integrated or divided. For instance, "Sprinter-324 Ambulance" can be broken down to merely contain a stretcher and an oxygen tank, leaving out the breathing apparatus and medical monitor. There are two ways to satisfy a composition such as via reference and through aggregation.

According to customer needs, things composed through reference cannot be disassembled, whereas entities composed through aggregation can be decomposed or recomposed. For instance, the physical entity "driver" of "Sprinter-324 Ambulance" is tied to it by reference and cannot be separated from the ambulance. It is defined as

$$
\begin{aligned}
cEoR &:=< URI, Composition \\
&:=< EoR/cEoR, Type\ reference/Aggregation > AttrSet \\
&:=< attributes >>
\end{aligned}
$$

The "Composition" refers to adding new entity resources to the composited resources. It is important to remember that the composed resources might either be composited or traditional entity resources. Figure 7 uses an application example to explain entity relationships (EoRs and cEoRs) and the concept of EoRs and cEoRs. According to the reference, the cEoR "Sprinter-324 Ambulance" in Fig. 7 comprises nurses in addition to an oxygen tank, breathing apparatus, stretcher, and medical monitor. The "Nurse's" primary resource is the ID card. The whole information system representation of the "Sprinter-324 Ambulance" which includes a breathing oxygen tank, machine, stretcher, and medical access is made available to IoT application systems when they need a physical item of type "/Ambulance/S324". The entity resource of an ambulance can be altered through recomposition or deconstruction when its components malfunction. When the state of the breathing machine "./ER/x-117" in the ambulance "./Ambulance/S324" is out of commission, the resource "Sprinter-324 Ambulance" can be reconfigured to an aggregate breathing machine "./ER/x-102" from "./Room/3309" instead of breathing machine "./ER/x-117". EoR and cEoR in IoT applications provide a versatile method for mapping a physical thing to an information entity. One copy of the entity resource data is used by just one physical entity in an entity resource model. The transition is started when the physical entity's condition changes in the resource model to maintain similarity.

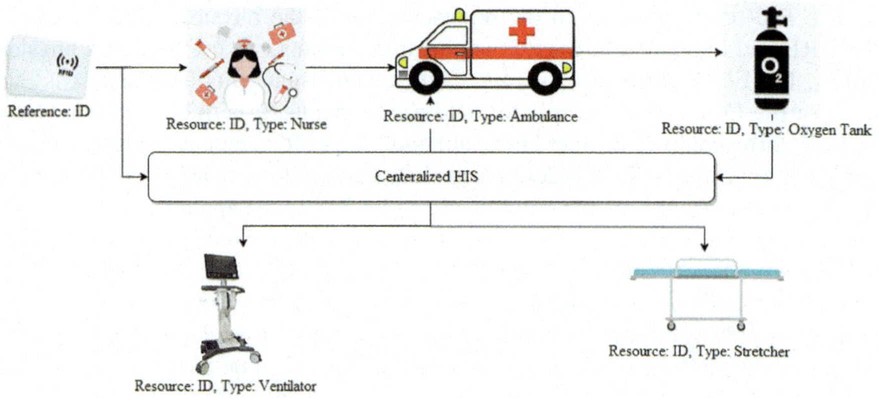

Fig. 7 EoR and cEoR

The transition does not need to spread across many information systems because it only affects a version of the mentioned resource model. Maintaining synchronization between the physical entity, its information representation, and ensuring timely information flow can be effective. Map-making is putting the business function between the transitory resources. EoR and cEoR models can perfectly describe the attributes of IoT physical entities. Still, the functional characteristics of IoT physical entities, which are crucial and necessary to IoT applications like "assigning ambulances approaching the accidental location", "dispatching doctors to the ambulances", and "starting the rescuing preplan" cannot be described in the entity resource models. These activities are information services, responding to the physical unions of operations on information entities in sets. For example, one of them is designating an ambulance approaching the saving spot. Information service that includes the used ambulance's URI and transferring the accident scene is required. It is because the ambulance is a specific type of cEoR. Following the definition of the cEoR of the ambulance, the driver, nurse, medical professional, stretcher, oxygen tank, and medical monitor will all turn on simultaneously. In addition, information services align with changes in the entity assets. For information service, the driver's status could shift from "driver" to "assigning ambulance". The status of the doctor shifts from "working" to "out-working", and from "waiting" to "working". It is depicted in Fig. 8. In conclusion, Transition-oriented Resources (ToR), are defined below as business functions.

$$
\begin{aligned}
ToR &:=< URI, Input :=< EoR/cEoR >, Output \\
&:=< EoRs/cEoRs >, Pre-condition, Effect \\
&:=< ToRs >>
\end{aligned}
$$

The URI is the ToR's accessible address. Through the HTTP protocol, the post method can be used to launch a ToR servicing. A ToR, the DELETE method, allows the execution to be stopped. The entity resources that the transition resource activates

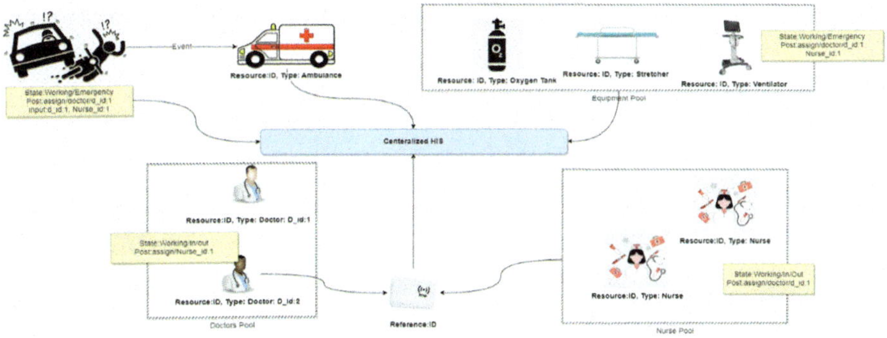

Fig. 8 A transitive resource representation

are referred to as input. The created entity resources are referred to as output after the transition. The servicing of resources is done. Preconditions are the requirements before the change resource servicing may begin. For example, the desired state of the physical entities is affected. Similarly, the resource servicing for follow-up is affected; for instance, if one doctor retires, another doctor must fill the void.

3.1 Accessing UDA-IoT Data and Cloud Platform

Healthcare data is gathered due to IoT and electronic medical record applications in medical services. This research uses a cloud platform with multitenant data management rather than the conventional data distribution architecture as depicted in Fig. 9 to handle massive healthcare data. The data management architecture consists of three levels. Tenant database layer, which houses multitenant databases, is the lowest tier. The data access control layer is the middle layer. Resource control mechanisms to arrange distributed healthcare data is employed in the middle layer. The business layer is the upper layer. It manages business operations and workflow to organize data exchange and interoperation. The three data management levels are described in more detail below. The layer that connects to physics databases is the multitenant layer for cloud applications.

Isolated databases and shared databases make up the two main parts. Applications using big data use data from multiple users. A patient's information might be dispersed to whole distinct hospital databases, similar to what happens in healthcare applications because a patient might visit many hospitals to see various doctors. In the cloud platform, databases from several hospitals should be segregated due to organization information policies and patient privacy. While this is happening, shared databases are created to hold standard data definitions for data access due to the growing need for data sharing in the healthcare industry. The crucial element is the resource layer in a heterogeneous data application environment for data

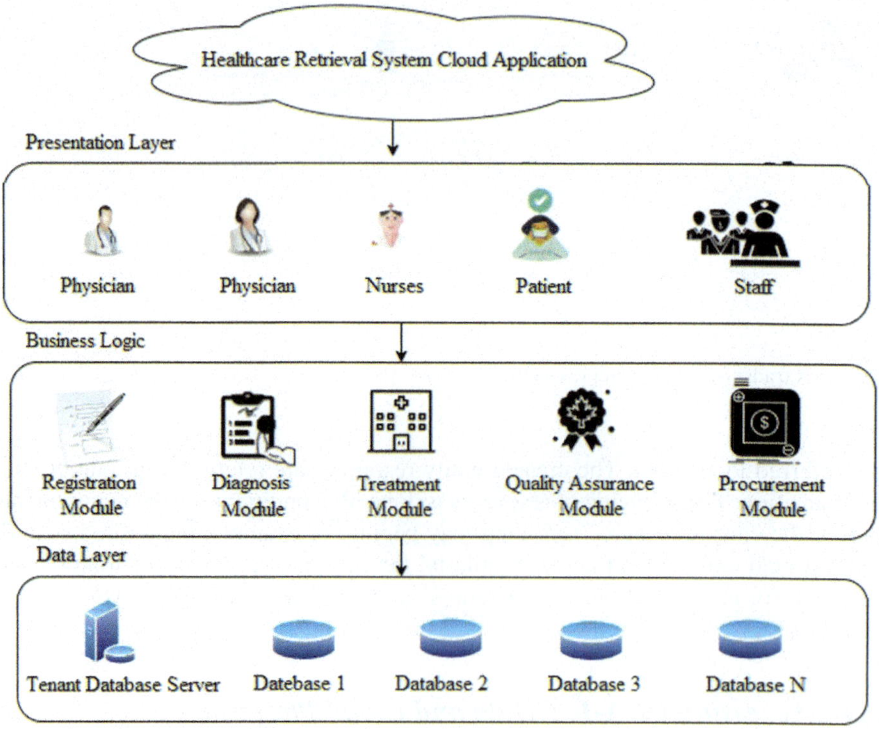

Fig. 9 Multitenant cloud platform for data storage

access control. The necessary information may be kept in various forms and tenant databases. Data are designated as resources in the resource layer for resource control mechanism retrieval, facilitation, and access capabilities. The resource access control mechanism translates the user's request for data access into tasks for resource retrieval from several tenants.

The resource access control mechanism translates the user's request for data access into tasks for resource retrieval from several tenants. The reasoning behind business operations is explained by the business layer, which also manages data access. The resource management layer receives the request for data access via the interface from restful web services. The following design can be used to model the ubiquitous IoT data access process:

1. Cloud applications send requests for accessing data.
2. Once the access rights have been verified, the business layer submits the request to the restful web service.
3. The resource control method makes it easier to send a request to access a resource to a database and for the database to process the request.

4. The bridging database layer retrieves information from separate and connected databases. IoT data is categorized as data resources for the pervasive IoT data accessing procedure. Any database that contains the particular URI of the resources for universal access can store them.

4 A Case Study on UDA-IoT Methodology

Emergency medical care is a decision-making process that necessitates close collaboration, such as an accident rescue. Many ambulances can be assigned and planned to transport the injured patients to various local hospitals. Emergency medical services must act as rapidly as possible, using accident rescue as an example. Most of the time, decision makers do not have much time for discussion. During the execution of a decision, actions frequently need to be adjusted for changing circumstances. Therefore, it is essential for decision makers in emergency medical systems to immediately organize, contact involved actors, and use the available resources.

Figure 10 illustrates how UDA-IoT may be applied to assist emergency medical services. A few examples of the types of entities and data resources that are stored on cloud servers include ambulances, nurses, doctors, and patient medical records. A medical service process is loaded during Emergency Events (EE), and several

Fig. 10 An example of a UDA-IoT scenario for an emergency medical service

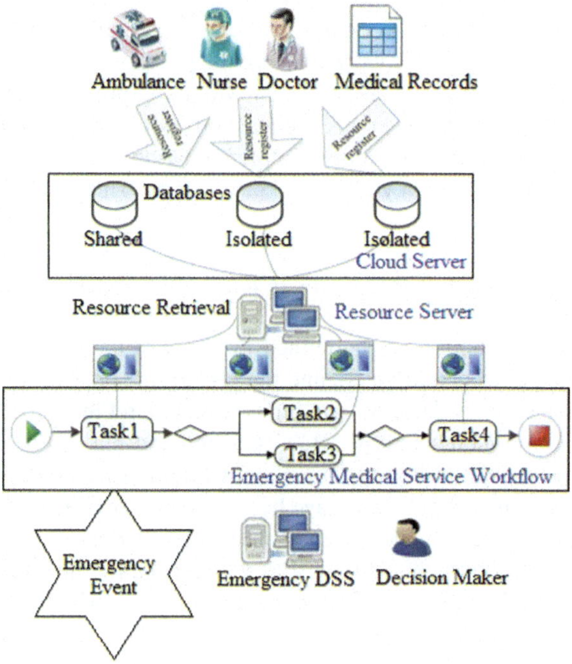

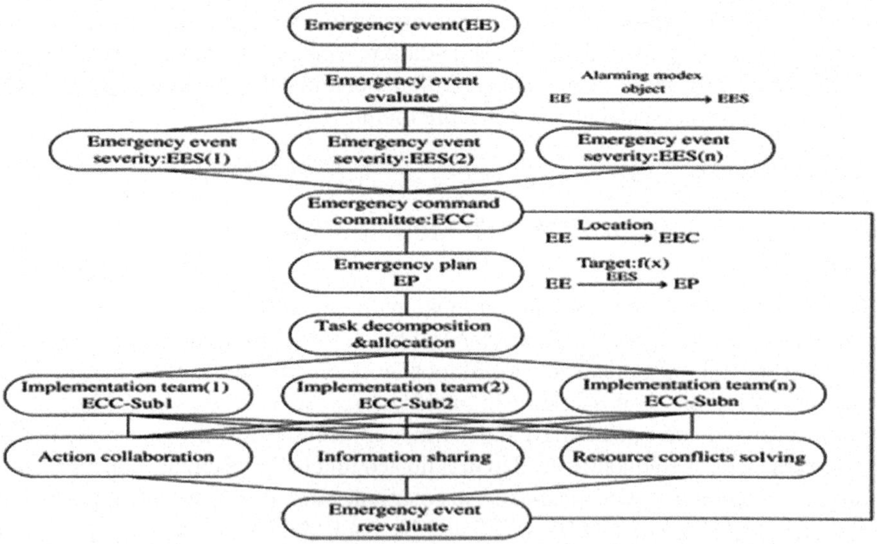

Fig. 11 Process of emergency decision-making

actions are necessary. Resources are required for these rescue missions. Because EE contain the features of information ambiguity, an emergency decision support system (DSS) is meant to be developed dynamically, dependent on the evolution of the events and the condition of the resources to be used. It adopts a descriptive decision technique in this work to solve the problems with emergency decision-making. The strategy focuses on the decision-making process instead of weighing and choosing the possibilities. Focus is placed on cooperation and information sharing within task groups during planning, implementation, and execution to encourage information sharing through widespread data access. These definitions apply to the denotations. The EE is defined as

$$Emergency\ Event\ EE = (Time,\ Location,\ Object,\ Target)$$

The notion EE is defined in terms of the circumstances surrounding the event, the objects it affected, and the goal of the decision-making process. The process of emergency decision-making is depicted in Fig. 11. The decision-making of the work group members will affect the time and place. For instance, if an EE occurred in Shanghai, task group members would benefit more from traveling there. Similarly, task members would have been on call if the EE had occurred at night.

The Emergency Event Severity (EES) class is defined below. The outcome of EE evaluation is known as EES, which refers to the alarm mode. For instance, calling "110" would be the alert mode in China, which would engage the authorities in interfering with the EEs:

$$EE(x, Object) = EE \overset{Alarmingmode:xobject}{\longrightarrow} EES$$

Making and directing decisions in an emergency group is known as the Emergency Command Committee (ECC). Separate division and cooperation are crucial in emergency management. Various EE types connect to different governing bodies and professions. In the interim, it would be effective if the ECC members were chosen from the local organizations where EEs took place. The ECC is defined below

$$(head, (ecc - sub1, ecc - sub2, \cdots)) = EE \overset{Location}{\longrightarrow} ECC$$

It is preferable to create different strategies based on the EE severity class to respond quickly and effectively to emergency situations. The target uses the utility function $f(x)$ while making decisions. Likewise, Emergency Plan is denoted as EP and is defined below

$$EE \overset{Target:f(x)EES}{\longrightarrow} EES$$

4.1 Emergency Medical DSS Ubiquitous Data Accessing Implementation

the UDA-IoT implementation in emergency medical DSS is depicted in Fig. 12. In this approach, the restful services management modules maintain both the physical resources-such as work groups and vehicles—and the data resources, such as the patients' medical records. These tools are designed for the region where the emergencies occur.

Resources are invested in an emergency decision-making platform to enable a dynamic decision-making process based on the location and severity level of the issue when it happens. Online information exchange is essential for decreasing information uncertainty in this emergency decision-making process. As a result, the DSS architecture's core component is the ubiquitous access information resource component. Two categories of resources—entity resources and data resources—are managed following our suggested universal IoT data model. Entity resources point to the standard physical resources used in medicine, such as staff, drugs, ambulances, and equipment. Data resources are the data that are important to decision-making and are produced by IoT and are kept in databases, such as the patient's digital medical records. More complex materials can be made by combining different components. Resources are encapsulated in restful web services by the resource control mechanism, which application systems can register for, access, and use.

An EE is divided into several subtasks for medical decision-making, each corresponding to a different type of rescue intervention. It requires the employment of many physical assets required to implement the emergency plans. As a result, making emergency decisions in IoT medical systems involves fusing physical resources with rescue planning. Besides, real-world objects, events, and business processes exist

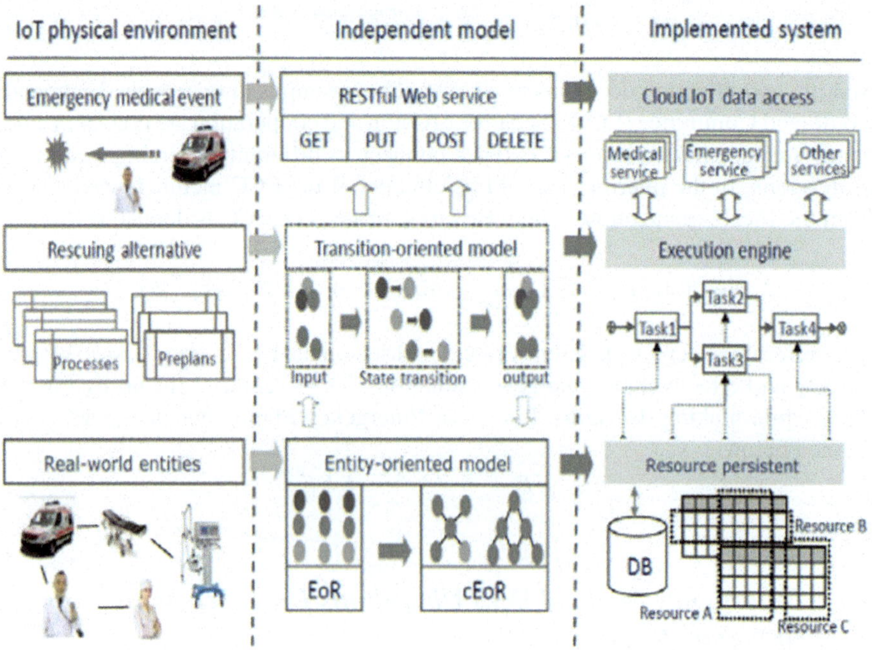

Fig. 12 UDA-IoT implementation in emergency medical DSS

and are mapped into the information environment when IoT application systems are installed.

The physical environment entity model, transition model, and PIM model, including entity model, transition model, and restful interface, are mapped to the actual IoT items. Then, according to the needs of the business, EoR resources are integrated with cEoR resources. For instance, a cEoR resource consists of EoRs for an oxygen tank, stretcher, breathing machine, and medical monitor. According to the phrase's definition, the rescue operation and the preplans can be transferred to transitive resources.

The business process or preplans commonly include many transition resources since they frequently integrate several merged operations of different information entities. Additionally, each transition resource provides the standard restful service interface for an application system, and the business process can be completed by gaining access to restful services. Interfaces are applied sequentially. PIM must be transformed into the Platform Specific Model (PSM) to create application systems. Information entities' resources are transferred to the reference of the data sources. As seen in the example below, an EoR resource could be a reference to an entire database table or many fields inside a single database table

$$implement(EoR) := reference(table(column))$$

An example of a cEoR resource is the fusion of several databases. The following list of possible data sources is presented in a cEoR resource as below

$$implement(cEoR) := reference(datasource(EoR))$$

The activities in the application systems receive the transition resources depending on EoR/cEoR. Each task includes the ability to control the EoR/cEoR state. The following is the list of sequentially completed subtasks which completes the business process defined as below

$$\begin{aligned} implement(ToR) &:= Task(Transition(EoRs/cEoRs)) \\ implement(Process) &:= Invoke(ToRs) \end{aligned}$$

Each business process uses the HTTP protocol to offer an interface for application systems to access its services. The interface of the emergency medical DSS using UDA-IoT is shown in Fig. 9. Different types of information are timely retrieved and displayed on the same screen, allowing decision-makers to gain a complete picture of how the emergency event is developing and the current state of the rescuing resources to act quickly.

4.2 Discussion

The case study illustrates how emergency decision-making is a continuous process that changes over time. Decision-makers must gather materials to ensure an efficient and quick rescue. UDA-IoT employs IoT to find ambulances and healthcare facilities. Mobile workers at EE are employed making use of a resource model. UDA-IoT on-demand data from hospitals serve as the foundation for sharing patient data. The emergency medical rescuing process may involve a large number of personnel. Complexity exists in the administration of numerous resources. It uses a cloud computing platform to coordinate data amongst several organizations [9, 10]. Figure 13 shows the emergency medical DSS's programmed interface.

5 Conclusion

Innovative IoT applications for health care create challenges when attempting to access heterogeneous IoT data, especially in mobile contexts with real-time IoT application systems. These applications help doctors and managers by providing them with access to various data sources. Accessing IoT data might be challenging due to extensive data collection from IoT devices. The analysis made in this chapter yielded three important conclusions. Firstly, it provides a platform for accessing vast volumes of data sources in a mobile application environment. It has been concluded

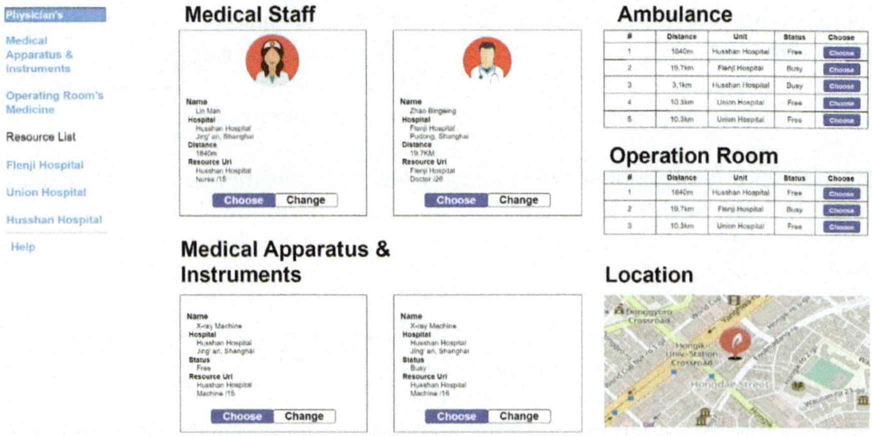

Fig. 13 Emergency medical DSS's programmed interface

that the IoT is useful in industrial applications that require a lot of data. More data collection is made possible by IoT, which is essential for commercial applications like health care. Using the information obtained by the manager, analysts may perform better business analytics IoT gadgets. Second, from a methodological perspective, the viability of ubiquitous access to heterogeneous IoT data is demonstrated. In many IoT applications, many smart objects are constantly moving and hence ubiquitous data access is crucial for IoT data analysis. Resource-based data models can offer URI-based cross-platform data access for IoT applications. Lastly, the use of IoT in UDA-applications in emergency medical care is highlighted. In emergency medical services, information can be gathered on patients, doctors, nurses, and ambulances and uploaded to a cloud computing platform via IoT notes. The UDA-IoT idea integrates diverse IoT data into resources with a single URI. The UDA-IoT is crucial for aiding decision-making in emergency medical care. This study focuses on the unified data model and semantic data explanation for data access and storage based on ontologies. Long supply chains in some industrial regions may provide new challenges. Since there are numerous businesses engaged and the industry ecology is complex, applying a consistent data model is difficult. The proposed UDA-IoT method is ideal for information-intensive organizations, such as the healthcare industry.

References

1. Khan, A.A., Keung, J.W., Abdullah-Al-Wadud, M.: SPIIMM: toward a model for software process improvement implementation and management in global software development. IEEE Access **5**, 13720–13741 (2017)
2. Humayun, M., Jhanjhi, N.Z.: Exploring the relationship between GSD, knowledge management, trust and collaboration. J. Eng. Sci. Technol. **14**(2), 820–843 (2019)
3. Khan, S.U., Niazi, M., Ahmad, R.: Empirical investigation of success factors for offshore software development outsourcing vendors. IET Softw. **6**(1), 1–15 (2012)
4. Ilyas, M., Khan, S.U.: An empirical investigation of the software integration success factors in GSD environment. In: Proceedings of the 15th IEEE International Conference on Software Engineering Research, Management and Applications, pp. 255–262 (2017)
5. Akbar, R., Hassan, M.F., Safdar, S., Qureshi, M.A.: Client's perspective: realization as a new generation process for software project development and management. In: Proceedings of the Second International Conference on Communication Software and Networks, pp. 191–195 (2010)
6. Hamid, M.A., Hafeez, Y., Hamid, B., Humayun, M., Jhanjhi, N.Z.: Towards an effective approach for architectural knowledge management considering global software development. Int. J. Grid Util. Comput. **11**(6), 780–791 (2020)
7. Ramasubbu, N.: Governing software process improvementsin globally distributed product development. IEEE Trans. Softw. Eng. **40**(3), 235–250 (2013)
8. Chua, C.E.H., Lim, W.K., Soh, C., Sia, S.K.: Client strategies in vendor transition: a threat balancing perspective. J. Strat. Inf. Syst. **21**(1), 72–83 (2012)
9. De Farias, I., Júnior, N.L., De Moura, H.P.: An evaluation of motivational factors for distributed development teams. In: Proceedings of the IEEE/ACM Joint 5th International Workshop on Software Engineering for Systems-of-Systems, pp. 78–79 (2017)
10. Simpao, A.F., Ahumada, L.M., Gálvez, J.A., Rehman, M.A.: A review of analytics and clinical informatics in health care. J. Med. Syst. **38**, 1–7 (2014)

Association Rule Mining for Healthcare Data Analysis

Punyaban Patel, Borra Sivaiah, Riyam Patel, and Ruplal Choudhary

Abstract In the domain of healthcare, massive amounts of data both unstructured and structured are created. As a result, a significant amount of money and time are required for storing and analysing it. The health industry has seen significant changes in recent years, with a growth in the number of physicians, patients, diseases, and technology. Doctors can analyse patient symptoms using data and information technology. Data mining is widely used for analysing these data. Association rule mining is one of the most significant tasks in data mining. Techniques such as apriori and FP-growth may be used to analyse data for illness diagnosis. The prime objective is to uncover the hidden associations between symptoms as well as statistically confirming those that are already known. These connections can aid in a better knowledge of illnesses and their causes, which will aid in their prevention. Association rules connect different diseases and treatments, as well as provide important information to doctors and health institutions in society. These rules are useful in healthcare research and development in areas such as potential complications, preventive medicine, disease diagnosis, and prevention. This chapter provides the complete information about association rule mining algorithms used in the healthcare domain. It also analyses critically existing association rules, relationships among various diseases, and discovers strong association rules from the healthcare data.

P. Patel (✉) · B. Sivaiah
Department of Computer Science and Engineering, CMR College of Engineering & Technology, Hyderabad, India
e-mail: punyaban@gmail.com

B. Sivaiah
e-mail: sivabetld@gmail.com

R. Patel
Department of Computer Science and Engineering, SRM University, Chennai, India
e-mail: riyampatel2001@gmail.com

R. Choudhary
Department of Plant, Soil and Agriculture Systems, Southern Illinois University, Carbondale, USA
e-mail: choudhry@siu.edu

D. P. Acharjya and K. Ma (eds.), *Computational Intelligence in Healthcare Informatics*, Studies in Computational Intelligence 1132,
https://doi.org/10.1007/978-981-99-8853-2_8

1 Introduction

Data mining is presently employed in a variety of fields. It is very significant in clinical practice. Thousands of patients visit hospitals every day for various treatments. Every department in the hospital is seeing an increase in the number of patient records. Data mining methods are employed in the medical industry to uncover hidden knowledge in medical datasets [1]. The patterns uncovered might help decision-making and save lives. Different data mining techniques, such as classification, clustering, association rule mining, statistical learning, and link mining, are all useful in research and development in the respective field [2]. The most effective approach for extracting frequent itemsets from large datasets is Association Rule Mining (ARM). The minimal support value is utilised to discover the most common itemsets. Frequent itemsets have a support value greater than or equal to the minimum support value. If an itemset is common, all of its subgroups must be frequent as well [3]. Heart disease is one of the primary causes of death in humans. Heart disease is the leading cause of mortality for both men and women in the United States. It is an equal opportunity killer that takes the lives of around 1 million people each year. In 2011, the illness claimed the lives of over 787,000 individuals, with 380,000 people dying each year from heart disease. Every 30 s, someone suffers a heart attack, and every 60 s, someone dies from a heart-related condition [4].

Likewise, diabetes, cancer, renal disease, high blood pressure, hepatitis, TB, musculoskeletal disorders, and stroke all have a significant influence on human health [5–7]. Chronic diseases are long-term illnesses that usually worsen, and they can be caused by a variety of circumstances. A chronic illness is one that lasts more than six months in a person. Chronic illnesses are currently the leading cause of early adult mortality and disability worldwide. According to World Health Organization (WHO) statistics, adults under the age of 70 account for over half of all non-communicable disease fatalities worldwide [2]. Cardiovascular disorders, chronic respiratory illnesses, cancer, and diabetes are the principal non-communicable diseases of concern. A patient with any of these disorders will need to undergo a series of tests in order to fully understand their condition. Other factors, such as lifestyle changes, are also taken into consideration. As a result, each patient has a variety of such characteristics. Relationships between these traits can be uncovered by analysing patient-specific data, which may reveal some type of link between these attributes and other life-saving information. This information can aid in comprehending the nature of diseases, the relationships between various characteristics, the regions of the body that the disease affects, and the necessary tests [8].

Association Rule Mining (ARM) is one of the most significant tasks in data mining. ARM techniques such as apriori and FP-growth may be used to analyse data generated from doctors, patients, and illnesses. It is a strong approach for uncovering hidden associations between symptoms as well as statistically confirming those that are already known. These associations can aid in a better knowledge of illnesses and their causes, which will aid in their prevention. Besides, it connects different diseases and treatments, as well as provides important information to doctors and health institutions in society.

This chapter provides clear understanding of the ARM algorithms in the healthcare field. It also looks at current association rules, links between illnesses, and discovers strong association rules using healthcare data. The rest of the chapter is organised as follows. Section 2 describes the related works followed by ARM in Sect. 3. Section 4 describes various measures used in ARM followed by experimental analysis and results in Sect. 5. Finally Sect. 6 concludes the chapter and shows its future direction.

2 Related Works

Many literature are available on the applications of association rule mining to liver disease, heart diseases, and kidney diseases in the web and online repository. A few of them have been discussed briefly in this section.

2.1 *Liver Diseases*

The liver plays an important part in human bodily activities, from protein manufacturing to toxin removal, and it is necessary for survival. Failure of the liver to operate properly might result in major health problems. Two types of testing such as imaging and liver function tests are used to assess the liver's function and aid in the diagnosis of liver illnesses. Many factors contribute to liver disease, including stress, eating habits, alcohol usage, and drug use. It has recently been shown that it is extremely difficult to diagnose at an early stage since symptoms are difficult to define. The physician frequently fails to recognise liver illness, resulting in ineffective medical treatment. Various data mining techniques may be used to anticipate various illness stages, even early stages, to aid physicians in providing appropriate therapy [9].

Many individuals nowadays suffer from liver disease as a result of their eating habits, and a variety of other uncommon activities. Early detection of liver illness may increase the chances of cure. But, if it is not treated appropriately at an early stage, it might lead to major health problems. Although previous algorithms are effective at forecasting, they become inefficient as data expands [10]. Because clinical test reports provide a large amount of data, predicting any specific disease is quite challenging. To address such challenges, the medical field frequently partners with automation technology. Machine learning, classification, data analytics, and other computer techniques are applied. To address the concerns with liver disease prediction, a comprehensive research of prediction algorithms is conducted, followed by a comparison analysis to determine the most accurate method. Despite the fact that existing solutions are good, their accuracy, execution speed, specificity, and sensitivity must be targeted in order to create an effective system [11, 12]. The effectiveness of existing approaches is addressed as a consequence of comparing the results of various algorithms.

To forecast liver illness, initially decision tree algorithms J48, LMT, random tree, random forest, REPTree, decision stump, and Hoeffding tree were introduced [4]. A comparison of these algorithms has also been carried out in the literature. The accuracy, precision, recall, mean absolute error, F-measure, kappa statistic, and run time of each algorithm are all measured by the system [13]. According to the findings, the decision stump algorithm performs well when compared to other algorithms, with a 70.67% of accuracy rate. Another strategy for distinguishing different types of data and predicting accuracy is classification [10, 11]. Likewise, clustering is the process of dividing a collection of abstract objects into classes. The process of discovering rules that control relationships and causal objects between a group of elements is known as ARM.

Further, a survey is conducted on several categorisation algorithms for predicting liver disorders [14]. The C4.5, naive Bayes, decision tree, support vector machine (SVM), back propagation neural network, and classification and regression tree algorithms were compared and assessed using speed, accuracy, performance, and cost as criteria. When compared against other algorithms, C4.5 was good. Similarly, to detect the illness and the algorithm performance, MATLAB 2013 was used to develop classification techniques such as naive Bayes and SVM. When comparing the accuracy and execution time of the SVM and naive Bayes algorithms, it was discovered that the SVM method performs better [10].

To predict fatty liver disease, machine learning methods such as random forest, naive Bayes, artificial neural networks (ANN), and logistic regression are used [15]. When compared to other classification models, the random forest model performed better, which should aid clinicians in classifying fatty liver patients for early treatment. Similarly, the WEKA tool is used to create a model in which liver function test attributes like age, gender, total bilirubin, direct bilirubin, alkaline phosphotase, total proteins, albumin asparatamino transferase, ratio albumin, and globulin were combined with classification algorithms like decision tree, navies Bayes, and NBTree considered. In addition, the χ^2 ranking algorithm was utilised to assess the influence of various qualities [16, 17].

One of the most difficult aspects of medical data mining is automated illness prediction and diagnosis. To build a classification model logistic, linear logistic regression, Gaussian processes, logistic model trees, multilayer perceptron (MLP), K-star, ANN, rule induction, SVM, and classification and regression trees are used [16]. A comparison research was conducted utilising these algorithms, and their performance was evaluated. To obtain the optimum algorithm, the info-gain feature selection approach in classification algorithms such as C4.5, random forest, CART, random tree, and REP was utilised [18]. To improve accuracy, the datasets were partitioned into two sets of training-testing ratios such as 70–30 per cent and 80–20 per cent. It is determined that utilising an 80–20 per cent training-testing data split with 6 features, random forest achieves an accuracy of 79.22%.

Further, a hybrid model using several algorithms such as J48, MLP, SVM, random forest, and Bayes net is proposed [19]. The model contains three phases. The first phase involves applying a classification algorithm to the original dataset; the second phase involves selecting characteristics that influence liver disease; and the third phase involves comparing the results of the original dataset without and with features. The accuracy of the algorithms was measured to evaluate the performance based on the experiments. As a consequence, before performing feature selection, the SVM method is rated the best. After feature selection, the random forest method is thought to perform better than other algorithms. Likewise, a real liver disease patient data is used to develop models employing various classification algorithms to detect liver disorders. The liver function test attributes included age, gender, total bilirubin, alkphos, DB, SgptTP, A/G Ratio, albumin, sgot, and selector field, and were considered with classification algorithms such as naive Bayes, random forest, K-means, C5.0, and K-nearest neighbour (KNN). As a result, before the adaptive boosting method, the random forest approach provided great accuracy. However, after adopting the C5.0 algorithm, the accuracy has improved [20].

2.2 *Heart Diseases*

Cardiovascular Disease (CVD) is one of the world's leading causes of death, according to the WHO and the Global Burden of Disease (GBD) research. According to the WHO, CVD is estimated to impact over 23.6 million individuals by 2030. In other developed nations, such as the United States of America, the mortality rate is about 1 in 4. The Middle East and North Africa region has an even higher fatality rate, accounting for 39.2% of the total [21]. As a result, lowering the number of deaths caused by CVD requires early and precise diagnosis as well as adequate treatment. For people who are at high risk of acquiring heart disease, such services must be available [22].

Many factors influence the likelihood of developing heart disease. In the past, researchers were more concerned with selecting important traits to include in their heart disease prediction models [23]. The relevance of understanding the links between these features and deciding their priority inside the prediction model was downplayed [24]. Many data mining-related research have already been undertaken to address the challenges that impede early and accurate diagnosis [25–27]. ARM is also used over the UCI dataset to predict CVD [28, 29]. Despite the high scores achieved from these datasets, the studies have a reproduction issue because the datasets are not accessible for access [30].

2.3 *Kidney Diseases*

Kidney problems occur when the kidneys are unable to filter blood as effectively as they should. Chronic Kidney Disease (CKD) occurs when the kidneys' function deteriorates over time. Diabetes and high blood pressure are the two most common causes of CKD, accounting for up to two-thirds of cases. Diabetes affects several organs in the body, including the kidneys and heart, as well as blood vessels, nerves, and eyes, when blood sugar levels are too high. Researchers sought to use information system techniques in health care 15 years ago in order to reduce the expense and burden of illnesses on individuals, hence saving their lives [31, 32]. Knowledge discovery and data mining is one of these methods for predicting and discovering illness indications [33]. It is a critical method for extracting information from large amounts of data. It employs a variety of approaches and strategies to extract relevant information that may be utilised to aid decision-making. Association rules, classification, and clustering are among the methodologies and strategies used [34]. It involves many iterative steps to extract the significant knowledge, which is used to make a right decision in an efficient manner [35].

On the other hand, limited study and research used an integrated strategy to collecting insight from medical data by merging different methodologies. To close this gap in information, classification and ARM techniques were combined and employed in this study to create and construct a classification system for predicting CKD using the Weka tool. To predict and diagnose CKD, the classification algorithms naive Bayes (NB), decision tree (J48), SVM, KNN, and classier based on association rule (JRip) were utilised. The apriori technique may also be used to uncover strong correlation rules between characteristics. The findings are more remarkable and valuable for patients, clinicians, governments, and decision-makers in the medical and health informatics industry. After using all of these algorithms, it is found that using an integrated method that combines classification algorithms with ARM enhances prediction accuracy, particularly for medical data.

3 Association Rule Mining

Every healthcare facility organisation has a vast database of patient data. It's tough to physically break down each of these records. Data mining methods are used to extract meaningful information from a dataset with a significant volume of data. It is used in the medical industry to analyse patient data in order to inform patients who are more likely to be impacted by the ailment and to aid doctors in detecting the condition. Humans are afflicted with ailments such as dengue fever, liver disease, and kidney disease, among others. This chapter uses two algorithms such as Frequent Pattern (FP) and apriori to determine which disease belongs to which group based on clinical data.

Algorithm 6 (FP-Growth Algorithm)

1. The first step is to search the database for itemsets. Support count, also known as frequency of 1-itemset, is the number of 1-itemsets in the database.
2. The FP tree must be built in the second stage. Create the tree's root as a starting point. Null is the symbol for the root.

3. The next step is to re-scan the database and review the transactions. Examine the first transaction and identify the itemset contained in it. The highest-counting itemset is placed first, followed by the next-lowest-counting itemset, and so on. It signifies that the tree's branch is made up of transaction itemsets in descending order of count.
4. The database's next transaction is investigated. The itemsets are listed in order of decreasing count. If any of the transaction's itemsets are already present in another branch, this transaction's branch will have a common root prefix.
5. This signifies that the common itemset in this transaction is linked to the new node of another itemset.
6. In addition, as transactions occur, the count of the itemset is increased. As nodes are formed and linked according to transactions, the count of both the common node and new node increases by 1.
7. The constructed FP tree must now be mined. The lowest node, as well as the links between the lowest nodes, is inspected first in this process. The frequency pattern length is represented by the lowest node. Then, in the FP tree, follow the path. A conditional pattern base is a path or set of paths.
8. The conditional pattern base is a database of prefix pathways in the FP tree that start with the lowest node (suffix).
9. Construct a conditional FP tree from the path's count of itemsets. In the conditional FP tree, itemsets that meet the threshold support are examined.
10. The conditional FP tree generates FPs.

Algorithm 7 (Apriori Algorithm)

1. Initially, scan the database to get frequent 1 itemsets.
2. Generate $(k + 1)$ candidate itemsets from length k frequent itemsets.
3. Test the candidates against the database.
4. Terminate when no frequent or candidate set can be generated.

4 Measures Used in Association Rule Mining

The support-confidence framework is commonly used to capture a certain form of dependency among objects recorded in a database. This approach uses five parameters to assess the uncertainty of an association rule. These are support, confidence, lift, leverage, and conviction. Equation 1 defines the support whereas the confidence is defined in Eq. 2, where X, Y are the itemsets, $|XY|$ is the number of transactions of itemset that contain both X and Y, and $|D|$ represents the total number of transactions of itemsets in the database. Similarly, the other measure leverage, conviction, and lift are defined in Eqs. 3, 4, and 5, respectively:

$$Support(X \rightarrow Y) = P(X \cap Y) = \frac{|X \cap Y|}{|D|} \tag{1}$$

$$Confidence(X \rightarrow Y) = \frac{Support(X \rightarrow Y)}{Support(X)} = \frac{|X \cap Y|}{|X|} \quad (2)$$

$$Leverage(X \rightarrow Y) = Support(X \rightarrow Y) - Support(X) \times Support(Y) \quad (3)$$

$$Conviction(X \rightarrow Y) = \frac{1 - Support(Y)}{1 - Confidence(X \rightarrow Y)} \quad (4)$$

$$Lift(X \rightarrow Y) = \frac{Confidence(X \rightarrow Y)}{Support(Y)} = \frac{Support(X \rightarrow Y)}{Support(X) \times Support(Y)} \quad (5)$$

5 Experimental Analysis and Results

In this section, implementation details and dataset used for evaluation are described. Practical result analysis of the proposed approach is also presented. Publicly available datasets such as heart diseases and breast cancer datasets are used to evaluate the proposed approach. The heart disease dataset has totally 76 different attributes with 303 patients in its record; 14 attributes that are linked with the heart disease are used. The details of these 14 attributes and their specification are presented in Table 1.

On employing the apriori algorithm, the rules are generated. The generated association rules presented in Table 2 are used by doctors for analysing the relationships among the attributes in heart disease.

Table 1 Attributes of heart disease and data specification

Attribute	Data specification
Age	Real
Sex	Male, Female
Chest pain (cp)	Asymptomatic, Non-anginal, Typical, Atypical angina
Resting blood pressure (trestbps)	Real
Cholesterol (chol)	Real
First blood sugar (fbs)	True, False
Rest ECG (restecg)	Left vent hyper, Normal, St t wave abnormality
Maximum heart rate (thalach)	Real
Exercise induced angina (exang)	No, Yes
Old peak (oldpeak)	Real
Peak exercise ST segment (slope)	Up, Flat, Down
Number of major vessels (ca)	Real
Thallium scam (thal)	Fixed defect, Normal, Reversible defect
Predicted attribute (num)	Absent 0, Present (1, 2, 3, 4)

Table 2 The best 10 rules generated by apriori algorithm for the heart disease

Sl. No.	Association rule	Confidence	Lift	Leverage	Conviction
1	sex = male, cp = asympt, fbs = f, ca = (0.5-inf) ⇒ num = 1	0.98	2.15	0.09	14.43
2	cp = asympt, exang = yes, ca = (0.5-inf) ⇒ num = 1	0.98	2.15	0.08	12.8
3	sex = male, cp = asympt, ca = (0.5-inf) ⇒ num = 1	0.97	2.12	0.10	11.25
4	cp = asympt, thal = normal ⇒ fbs = f	0.96	1.13	0.02	2.57
5	cp = asympt, slope = flat, ca = (0.5-inf) ⇒ num = 1	0.96	2.11	0.09	9.26
6	cp = asympt, ca = (0.5-inf), thal = reversable defect ⇒ num = 1	0.96	2.11	0.08	9.08
7	cp = asympt, slope = flat, thal = reversable defect ⇒ num = 1	0.96	2.10	0.08	8.71
8	sex = male, slope = flat, ca = (0.5-inf) ⇒ num = 1	0.95	2.08	0.09	7.49
9	cp = asympt, exang = yes, thal = reversable defect ⇒ num = 1	0.94	2.07	0.08	7.08
10	restecg = normal, thalach = (147.5-inf), thal = normal ⇒ fbs=f	0.94	1.11	0.02	2.02

The breast cancer dataset includes 201 instances of one class and 85 instances of another class. The instances are described by 9 attributes, some of which are linear and some are nominal. The attribute information for breast cancer is data presented in Table 3. The frequent itemsets generated by apriori algorithm are shown in Table 4.

Apriori algorithm is implemented using the WEKA tool. It is observed that, on increasing the minimum support and keeping confidence the same, the number of all frequent itemsets are decreased due to the minimum support change. Similarly, on keeping the same minimum support and decreasing the confidence, it is not affecting the number of frequent itemsets generated. On taking minimum support of 20% and minimum confidence of 90%, the best 10 association rules generated from breast cancer data is shown in Table 5.

Table 3 Attributes of breast cancer data specification

Attribute	Data specification
Class	No recurrence events, Recurrence events
Age	10–19, 20–29, 30–39, 40–49, 50–59, 60–69, 70–79, 80–89, 90–99
Menopause	lt40, ge40, premeno
Tumour size	0–4, 5–9, 10–14, 15–19, 20–24, 25–29, 30–34, 35–39, 40–44, 45–49, 50–54, 55–59
Inv nodes	0–2, 3–5, 6–8, 9–11, 12–14, 15–17, 18–20, 21–23, 24–26, 27–29, 30–32, 33–35, 36–39
Node caps	Yes, No
Deg malig	1, 2, 3
Breast	Left, Right
Breast quad	Left up, Left low, Right up, Right low, Central
Irradiat	Yes, No

Table 4 Aprior algorithm for breast cancer data for generating frequent itemsets

Minimum support	Confidence	L1	L2	L3	L4	L5	L6
0.1	1.0	26	138	230	205	87	13
0.2	1.0	19	58	69	29	5	
0.3	1.0	13	26	20	4		
0.4	1.0	9	10	4	1		
0.5	1.0	6	6	4	1		
0.5	0.9	6	6	4	1		
0.6	0.8	4	3	1			
0.6	0.7	4	3	1			
0.6	0.6	4	3	1			

6 Conclusion and Future Direction

Healthcare data is generated from hospitals and diagnostic centres. It is very essential to generate most frequently occurring symptoms from health data. This chapter explains apriori algorithm for heart disease and breast cancer data to discover frequently occurring symptoms and generated strong association rules from the frequent occurring symptoms. The association rules can be used by healthcare professionals or physicians to find the strong associations among symptoms. This research can be extended by considering more risk factors to extract more useful and significant rules not only for breast cancer and heart disease but also other diseases using the association rule mining algorithm. Furthermore, it is planned to build a predictive model using machine learning techniques for all the diseases.

Table 5 The best 10 rules generated by apriori algorithm for the breast cancer disease

Sl. No.	Association rule	Confidence	Lift	Leverage	Conviction
1	Inv nodes = 0-2, Irradiat = no, class = No recurrence events ⇒ Node caps = no	0.99	1.27	0.11	10.97
2	Inv nodes = 0-2, Irradiat = no ⇒ Node caps = no	0.97	1.25	0.12	5.85
3	Node caps = no, Irradiat = no, Class = No recurrence events ⇒ Inv nodes = 0-2	0.96	1.29	0.11	5.51
4	Inv nodes = 0-2, Class = No recurrence events ⇒ Node caps = No	0.96	1.23	0.11	4.67
5	Inv nodes = 0-2, ⇒ Node caps = no	0.94	1.22	0.12	3.67
6	Node caps = no, Irradiat = no ⇒ Inv nodes = 0-2	0.94	1.26	0.13	4
7	Node caps = no, Class = No recurrence events ⇒ Inv nodes = 0-2	0.94	1.26	0.11	3.64
8	Irradiat = no, Class = No recurrence events ⇒ Node caps = no	0.92	1.19	0.08	2.62
9	Inv nodes = 0-2, Node caps = No, Class = No recurrence events ⇒ Irradiat = no	0.91	1.19	0.08	2.38
10	Node caps = no, ⇒ Inv nodes = 0-2	0.91	1.22	0.12	2.58

References

1. Han, J., Kamber, M., Pei, J.: Data Mining Concepts and Techniques, 3rd edn. Morgan Kaufmann Publishers (2012)
2. Dunham, M.H.: Data Mining: Introductory and Advanced Topics. Pearson Education India (2006)
3. Kabir, M.F., Ludwig, S.A., Abdullah, A.S.: Rule discovery from breast cancer risk factors using association rule mining. In: Proceedings of the IEEE International Conference on Big Data, pp. 2433–2441 (2018)
4. Nahar, N., Ara, F.: Liver disease prediction by using different decision tree techniques. Int. J. Data Mining Knowl. Manag. Process **8**(2), 01–09 (2018)
5. Kumari, N., Acharjya, D.P.: A hybrid rough set shuffled frog leaping knowledge inference system for diagnosis of lung cancer disease. Comput. Biol. Med. **155**(3), 106662 (2023)
6. Acharjya, D.P., Ahmed, P.K.: Knowledge inferencing using artificial bee colony and rough set for diagnosis of hepatitis disease. Int. J. Healthc. Inf. Syst. Inf. **16**(2), 49–72 (2021)
7. Kumari, N., Acharjya, D.P.: A decision support system for diagnosis of hepatitis disease using an integrated rough set and fish swarm algorithm. Concurrency Comput. Pract. Experience **34**(21), e7107 (2022)

8. Kumari, N., Acharjya, D.P.: Data classification using rough set and bioinspired computing in healthcare applications-an extensive review. Multimedia Tools Appl. **82**(9), 13479–13505 (2023)
9. Acharjya, D.P., Ahmed, P.K.: A hybridized rough set and bat-inspired algorithm for knowledge inferencing in the diagnosis of chronic liver disease. Multimedia Tools Appl. **81**(10), 13489–13512 (2022)
10. Vijayarani, S., Dhayanand, S.: Liver disease prediction using SVM and Naïve Bayes algorithms. Int. J. Sci. Eng. Technol. Res. **4**(4), 816–820 (2015)
11. Hassoon, M., Kouhi, M.S., Zomorodi-Moghadam, M., Abdar, M.: Rule optimization of boosted c5. 0 classification using genetic algorithm for liver disease prediction. In: Proceedings of the IEEE International Conference on Computer and Applications, pp. 299–305 (2017)
12. Dinesh, S., Metin, K.Ö.K.: A review on different parameters affecting the vehicle emission gases of different fuel mode operations. Res. J. Sci. Eng. Syst. **3**(4), 146–164 (2018)
13. Sasikala, B.S., Biju, V.G., Prashanth, C.M.: Kappa and accuracy evaluations of machine learning classifiers. In: Proceedings of the 2nd IEEE International Conference on Recent Trends in Electronics, Information & Communication Technology, pp. 20–23 (2017)
14. Sindhuja, D.R.J.P., Priyadarsini, R.J.: A survey on classification techniques in data mining for analyzing liver disease disorder. Int. J. Comput. Sci. Mobile Comput. **5**(5), 483–488 (2016)
15. Wu, C.C., Yeh, W.C., Hsu, W.D., Islam, M.M., Nguyen, P.A.A., Poly, T.N., Wang, Y.C., Yang, H.C., Li, Y.C.J.: Prediction of fatty liver disease using machine learning algorithms. Comput. Methods Programs Biomed. **170**, 23–29 (2019)
16. Bahramirad, S., Mustapha, A., Eshraghi, M.: Classification of liver disease diagnosis: a comparative study. In: Proceedings of the Second IEEE International Conference on Informatics & Applications, pp. 42–46 (2013)
17. Auxilia, L.A.: Accuracy prediction using machine learning techniques for Indian patient liver disease. In: Proceedings of the 2nd IEEE International Conference on Trends in Electronics and Informatics, pp. 45–50 (2018)
18. Kumar, A., Sahu, N.: Categorization of liver disease using classification techniques. Int. J. Res. Appl. Sci. Eng. Technol. **5**(5), 826–828 (2017)
19. Gulia, A., Vohra, R., Rani, P.: Liver patient classification using intelligent techniques. Int. J. Comput. Sci. Inf. Technol. **5**(4), 5110–5115 (2014)
20. Kumar, S., Katyal, S.: Effective analysis and diagnosis of liver disorder by data mining. In: Proceedings of the IEEE International Conference on Inventive Research in Computing Applications, pp. 1047–1051 (2018)
21. James, S.L., Abate, D., Abate, K.H., Abay, S.M., Abbafati, C., Abbasi, N., Briggs, A.M.: Global, regional, and national incidence, prevalence, and years lived with disability for 354 diseases and injuries for 195 countries and territories, 1990–2017: a systematic analysis for the Global Burden of Disease Study 2017. The Lancet **392**(10159), 1789–1858 (2018)
22. Maji, S., Arora, S.: Decision tree algorithms for prediction of heart disease. In: Proceedings of the Third International Conference on Information and Communication Technology for Competitive Strategies, pp. 447–454 (2017)
23. Amin, M.S., Chiam, Y.K., Varathan, K.D.: Identification of significant features and data mining techniques in predicting heart disease. Telemat. Inform. **36**, 82–93 (2019)
24. Mohammed, K.I., Zaidan, A.A., Zaidan, B.B., Albahri, O.S., Albahri, A.S., Alsalem, M.A., Mohsin, A.H.: Novel technique for reorganisation of opinion order to interval levels for solving several instances representing prioritisation in patients with multiple chronic diseases. Comput. Methods Programs Biomed. **185**, 105151 (2020)
25. Tripathy, B.K., Acharjya, D.P., Cynthya, V.: A framework for intelligent medical diagnosis using rough set with formal concept analysis. Int. J. Artif. Intell. Appl. **2**(2), 45–66 (2011)
26. Acharjya, D.P., Ahmed, K.P.: A hybrid scheme for heart disease diagnosis using rough set and cuckoo search technique. J. Med. Syst. **44**(1), 1–16 (2020)
27. Fitriyani, N.L., Syafrudin, M., Alfian, G., Rhee, J.: HDPM: an effective heart disease prediction model for a clinical decision support system. IEEE Access **8**, 133034–133050 (2020)

28. Shuriyaa, B., Rajendranb, A.: Cardio vascular disease diagnosis using data mining techniques and ANFIS approach. Int. J. Appl. Eng. Res. **13**(21), 15356–15361 (2018)
29. Srinivas, K., Reddy, B.R., Rani, B.K., Mogili, R.: Hybrid approach for prediction of cardiovascular disease using class association rules and MLP. Int. J. Electr. Comput. Eng. **6**(4), 1800–1810 (2016)
30. Thanigaivel, R., Kumar, K.R.: Boosted apriori: an effective data mining association rules for heart disease prediction system. Middle-East J. Sci. Res. **24**(1), 192–200 (2016)
31. Patel, P., Sivaiah, B., Patel, R.: Relevance of frequent pattern (FP)-growth-based association rules on liver diseases. In: Intelligent Systems: Proceedings of International Conference on Machine Learning, Internet of Things and Big Data, pp. 665–676. Springer (2022)
32. Patel, P., Sivaiah, B., Patel, R.: Approaches for finding optimal number of clusters using k-means and agglomerative hierarchical clustering techniques. In: Proceedings of the IEEE International Conference on Intelligent Controller and Computing for Smart Power, pp. 1–6 (2022)
33. Patel, P., Palakurthy, M., Ramakrishna, P., Choudhary, R.: A comprehensive classification framework for chronic kidney disease prediction. Int. J. Adv. Res. Sci. Technol. **12**(3), 929–931 (2023)
34. Patel, P., Sivaiah, B., Patel, R.: Relevance of frequent pattern (FP)-growth based association rules on liver diseases. Lect. Notes Netw. Syst. **431**, 665–676 (2022)
35. Fayyad, U., Piatetsky-Shapiro, G., Smyth, P.: From data mining to knowledge discovery in databases. AI Mag. **17**(3), 37–37 (1996)

Feature Selection and Classification of Microarray Cancer Information System: Review and Challenges

Bichitrananda Patra, Santosini Bhutia, and Mitrabinda Ray

Abstract The sickness of cancer, a fatal condition, is caused by the abnormal proliferation of cells in the body. Microarray technology has become famous for diagnosing such serious diseases. Developing a swift and precise method for detecting cancer and discovering drugs that can help eliminate the disease are crucial. The microarray data presents a significant challenge for accurate classification due to its high number of attributes and a relatively small sample size. A noisy, irrelevant, and redundant gene is also present in these microarray genes, resulting in poor diagnosis and categorization. Researchers used machine learning techniques to extract the most significant aspects of gene expression data to achieve this goal. This study explores microarray data, encompassing feature selection, cancer-specific classification algorithms, and future scopes in this field.

1 Introduction

DNA (Deoxyribonucleic acid), found in every cell, represents the genetic information of the organs. Genes are the coding segments of this DNA. The DNA microarray method provides an entire view of the cell, allowing for easy differentiation between normal and malignant cells [1]. Cancer is a fatal disease divided into two types: malignant and benign. This malignant tumor develops quickly due to abnormal cell proliferation in one organ and then spreads to all other human body regions, potentially resulting in death [2]. As per the World Health Organization (WHO), cancer is the leading cause of mortality [3]. Among men, the most frequent types are lung, prostate, central nervous system (CNS), brain, leukemia, small round blue

B. Patra · S. Bhutia (✉) · M. Ray
Department of Computer Science and Engineering, SOA University, Bhubaneswar, Odisha, India
e-mail: santosinibhutia@gmail.com

B. Patra
e-mail: bichitranandapatra@soa.ac.in

M. Ray
e-mail: mitrabindaray@soa.ac.in

D. P. Acharjya and K. Ma (eds.), *Computational Intelligence in Healthcare Informatics*, Studies in Computational Intelligence 1132,
https://doi.org/10.1007/978-981-99-8853-2_9

cell tumors (SRBCT), and bladder cancers. At the same time, women commonly experience breast, endometrial, lung, cervical, and ovarian cancers [4]. In 2012, approximately 14 million new cancer cases were reported, which is anticipated to increase by 70% over the next two decades [5]. The fatality rate has grown when cancer spreads from its primary site to other body sections. Early detection and treatment, on the other hand, can lower cancer mortality rates. Cancer diagnosis typically relies on conventional methods, often slow, ineffective, and sometimes yield inaccurate results. Recently, gene expression profiles obtained from microarray datasets have been employed to address characteristics and make more reliable predictions [6]. This microarray technique analyzes gene expression data, facilitating cancer detection.

Microarray technology proves to be a powerful technique in detecting diseases like cancer. It has the maximum throughput, allowing for simultaneous experimentation with as many as ten thousand DNA microarrays. These instruments produce raw data, subsequently subjected to preprocessing to eliminate redundant and noisy information. Finally, data mining algorithms are applied to extract biological information from the data [7]. The "curse of dimensionality" is a phenomenon that occurs when microarray data has thousands upon thousands of genes in proportion to the number of samples. However, none of these genes or features are linked to cancer classification. Some of these characteristics are highly connected with improved categorization accuracy. As a result, before cancer classification, efficient feature selection algorithms are utilized to discover positively linked features [8].

Because there are so few samples and many features for each sample, feature selection becomes extremely difficult [9]. Most genes need help distinguishing between class labels effectively, rendering them unhelpful [10]. Feature selection involves the choice of a subset of pertinent features from a broader range of features, guided by specific criteria or criteria-based selection methods. Deleting redundant and useless genes improves the compression of data processing processes. Learning accuracy improves, and learning time decreases due to better feature selection [11]. Four types of feature selection approaches are commonly used [12]. To begin, there is the filter technique, which entails analyzing each attribute separately using statistical features. Second, using machine learning (ML) approaches, the wrapper strategy finds the best feature subset. The accuracy of the supplied classifier determines the wrapper technique's quality. Third, an embedded technology incorporated within the classifier merges the hypothesis space, seeking the optimal feature subset. Finally, a hybrid approach that includes filter and wrapper strategies is employed. Consequently, this hybrid approach combines the filter method's computational efficiency with the wrapper strategy's enhanced performance. The primary aim of this feature selection technique is to identify the most concise and informative set of attributes while minimizing time complexity, ultimately leading to improved classification accuracy when compared to utilizing the complete feature set [13].

Several research papers have predicted the discriminative prognosis based on the region. Biomarkers play an essential role in cancer illness prognosis and prediction. ML offers significant gains as compared to pathologists. ML is a method of learning that allows machines to learn without having to be explicitly programmed.

Many complicated real-world problems are solved using ML approaches, which have been proven effective in analyzing gene expression data. By a factor of 15:25, ML improves the accuracy of cancer susceptibility, mortality, and recurrence prediction [14]. ML, which decreases manual work, will be the most splendid future cancer diagnosis technology. In cancer research, ML algorithms empower doctors to provide better prevention, diagnosis, therapy, and care. In cancer prediction and prognosis modeling, machine learning algorithms have consistently demonstrated superior performance compared to traditional methods. ML-based algorithms offer several benefits, such as automating the process of hypothesis formulation and evaluation and assigning parameter weights to predictors based on their correlation with outcome prediction.

This chapter is focused on learning the use of machine learning in the classification of cancer research and medical oncology applications. It reviews significant publications since the past five years (2017–2021) with advanced machine learning models for cancer detection, classification, and prediction. The objective is chosen using ML algorithms to select and classify features from a microarray cancer dataset. The purpose of this study is furnished below.

1. Examining the utilization of microarray technology for the classification of cancer
2. Investigating the concept of feature selection within the realm of machine learning
3. Delving into diverse classifiers in the field of machine learning
4. Evaluating existing approaches for cancer classification employing machine learning
5. Exploring various metrics employed to assess performance in classification tasks
6. Scrutinizing multiple cancer datasets with varying dimensions for classification purposes.

The following paragraphs begin with an overview of the background research, which includes the microarray technique, feature selection approaches, classification algorithms, and various cancer datasets. Following that, in Sect. 2, the emphasis is on reviewing previously suggested studies on the classification of cancer using feature selection and classification methodologies. Section 3 provides an overview of related studies. Section 4 reviews several evaluation metrics and analyzes using reduced features, datasets, and classifiers. Finally, Sect. 5 summarizes the chapter.

2 Background Study

Microarray technology is extensively employed in cancer diagnosis and is recognized as a potent tool. DNA microarray technology has been harnessed to produce extensive datasets of gene expression microarray data. These datasets are leveraged for gene studies and to assist in diagnosing diseases like cancer [15]. Nonetheless, analyzing these microarray data is complicated due to constraints such as a limited sample size, high dimensionality, and class imbalance. To address this challenge, a feature

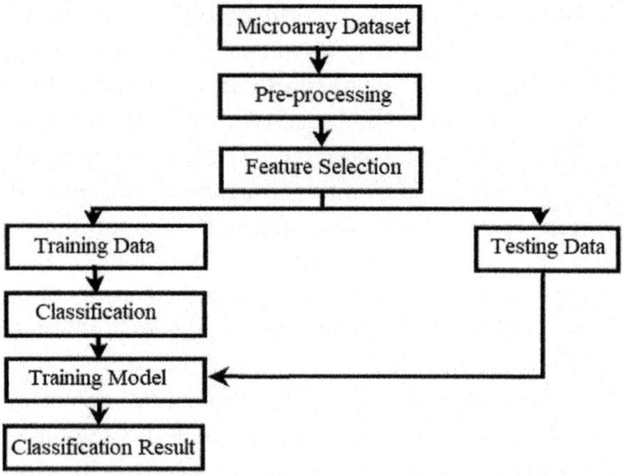

Fig. 1 Block diagram of microarray technology

selection process defines closely associated attributes that can be applied in disease diagnosis [16].

Classifying diseases is inherently challenging, and within this classification task, the feature selection process assumes paramount importance. The effectiveness of feature selection significantly influences the performance of the classification. When dealing with limited sample size and high-dimensional features, the classification model established on the training dataset often fails to yield a highly accurate model. Therefore, the process of feature selection becomes indispensable for reducing the size of microarray data [17]. Data cleansing is carried out during the preprocessing phase, followed by applying feature selection techniques to retain mainly illuminating features. Subsequently, the dataset is partitioned into two subsets: the training and testing datasets. A learning model is then trained for diagnosing cancer subtypes using the training subset, as depicted in Fig. 1.

2.1 *Fundamentals of Feature Selection*

The administration of large, multidimensional datasets needs to be revised [18]. The dataset contains noisy and duplicated data, lowering job effectiveness. However, using data mining and machine learning approaches, this problem can be solved. Many researchers have emphasized the relevance of feature selection in categorization during the last few decades [19]. Feature selection is the process of finding a collection of features from an initial feature set without changing their original meaning, and it is mainly focused on relevance and redundancy. From the powerfully relevant, weakly relevant, and irrelevant features, relevance selects a subset of relevant features. Redundancy is responsible for identifying and removing duplicate features from the final set of essential features, thereby refining the feature subset [20].

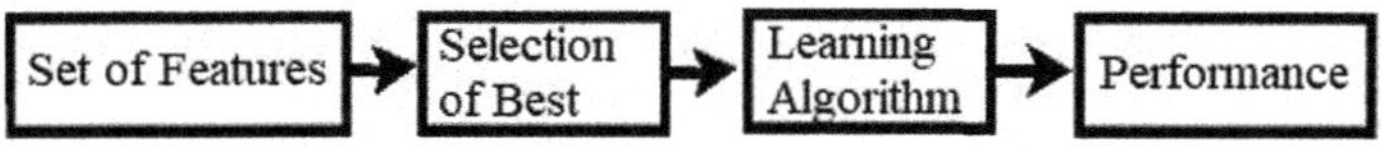

Fig. 2 Filter method

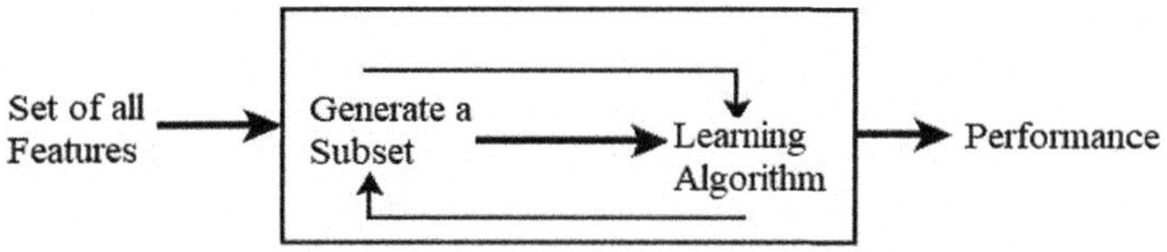

Fig. 3 Wrapper method

Although classifiers are the most significant component in microarray data processing, feature selection is more critical. Selecting key characteristics has several benefits, including memory savings, increased accuracy, lower computing costs, and eliminating overfitting issues. Different aspects are effective for other models. The two most crucial processes of a task are data processing and the feature section for building the optimal model [21]. Feature selection is a procedure for selecting the optimum subset of characteristics to improve classification and prediction accuracy [22].

Filter Method: The effectiveness of individual attributes is evaluated using evaluation metrics, including distance and dependence in the filter technique, independent of the classification algorithm. The computational cost is lower due to the separation from the categorization model. The filter approach is straightforward, effective, and completed before categorization. Features are ranked based on specific criteria, and from this ranking, the top-performing attributes are selected to create a subset that is then inputted into the classification algorithm [23]. The selection process involves grading all features, and those deemed top-ranking are assembled into a feature subset that is subsequently employed in the classification process. Figure 2 depicts the filter method of feature selection.

Wrapper Method: The wrapper method is a feature selection technique in machine learning that systematically assesses and selects the most relevant subset of features for a particular predictive model [24]. Unlike filter methods that rely on statistical measures to evaluate feature importance, the wrapper employs a specific machine learning algorithm to gauge feature subsets. It operates through an iterative process, where various combinations of features are tested by training and evaluating the model using performance metrics like accuracy or $F1$ score. While computationally intensive, this approach can yield highly effective results because it directly measures the impact of different feature subsets on the model's predictive power. However, it can also be more resource-intensive compared to other feature selection techniques due to the need for repeated model training and evaluation. Figure 3 depicts the wrapper method for feature selection.

Embedded Method: Embedded methods incorporate feature selection directly into training a machine learning model. Unlike filter methods that assess features

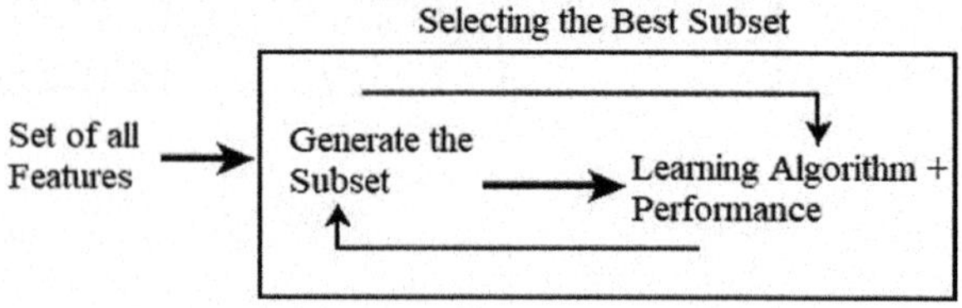

Fig. 4 Embedded method

independently or wrapper methods that use a specific algorithm for feature subset evaluation, embedded methods determine feature relevance during model training. Algorithms like decision trees, random forests, and regularized linear models such as Lasso regression have built-in mechanisms to assign importance scores to features or automatically select the most relevant ones during training. An overview of the embedded method is depicted in Fig. 4.

Hybrid Method: Hybrid feature selection methods combine elements from filter and wrapper methods to achieve a more balanced and practical approach. These techniques aim to harness the strengths of both filter methods, which are computationally efficient but may not consider the interaction between features and the learning algorithm, and wrapper methods, which incorporate the learning algorithm's performance but can be computationally expensive. Hybrid methods often begin by applying a filter to preselect a subset of promising features based on statistical measures or other criteria. Then, a wrapper method is employed to fine-tune the feature subset, considering the specific machine learning algorithm's performance. This combined approach seeks to balance computational efficiency and model performance. It is beneficial when working with large datasets and complex models where a compromise between speed and accuracy is needed for feature selection.

Filter, wrapper, embedding, and hybrid are examples of feature selection strategies, and their characteristics are listed in Table 1. Data attributes are prioritized using filter methods based on specified criteria. They have no impact on the learning algorithm. Consequently, filter methods exhibit faster computational speeds in comparison to wrapper methods. However, they have inferior classification accuracy because the learning step is disregarded. Hybrid approaches are designed to bring the best of both ways together.

2.2 *Fundamental Classification Techniques*

The two most common types of ML are supervised and unsupervised machine learning. In supervised learning, the learner possesses previous information about the data content. Key algorithms, including decision trees (DT), random forest (RF), logistic regression, Bayesian network (BN), support vector machines (SVM), k-nearest neighbors (KNN), and neural networks, have been successfully applied to classify various types of cells. In contrast, in unsupervised learning, the learner operates without previous information about the input data or the expected outcomes. Clustering,

Table 1 Characteristics of feature selection techniques

Method	Algorithm	Advantages	Disadvantages
Filter	Variance, Chi-square, Correlation, Information value, Mutual information	Independent of specific algorithm, Fast, Simple	Performance is poor No interaction with the classification model
Wrapper	Genetic algorithm Forward selection Backward elimination Exhaustive feature selection	Less error compared to other methods, Selects a nearest best subset	Overfitting, Dependent on a specific algorithm on which it has been tested
Embedded	Lasso (L1) Random forest Gradient boosted trees	Interaction with the classification model, Less computation-intensive	Overfitting
Hybrid	Recursive feature elimination Recursive feature addition	Takes the advantages from various methods	Increase in time complexity

self-organizing maps (SOM), and other unsupervised techniques were initially used in studying the relationships between distinct genes. Classification and regression are the two main types of supervised procedures. In classification, the output variable is designed to accommodate class labels, while in regression, the output variable is intended to accommodate continuous values.

There are six well-known classifiers in ML, which are briefly discussed below. The utilization of microarray data for classifying various types of cancer is increasingly prevalent. Machine learning techniques are especially well-suited for microarray gene expression databases because they can learn and construct classifiers that unveil intricate relationships within the data [25]. Below, we briefly outline six widely recognized classifiers in machine learning.

Logistic Regression: It is a regression model employed for predicting the likelihood of a specific dataset. Its functionality relies on the logistic function, commonly called the sigmoid function, a well-established model. In this model, probabilities are used to assess various outcomes of a single trial, and these probabilities are transformed using a sigmoid function. The essential advantage of this classification model is its ease of application to independent variables, although identifying the independent variable in high-dimensional data can be pretty challenging.

Naive Bayes: The Naive Bayes (NB) classifier is a popular and straightforward probabilistic machine learning algorithm for classification tasks. It is based on Bayes' theorem. It employs conditional probability to classify data. The noteworthy characteristic of this algorithm is its assumption that all features are independent. There are three variants of NB-based algorithms: Gaussian NB, multinomial NB, and Bernoulli NB. One advantage of this approach is its ability to make accurate estimations with a small training dataset for conditional parameters. However, it is essential to note that estimation time is directly proportional to the dataset size. Therefore, if the dataset dimension is exceedingly large, NB can become a less effective estimator due to the increasing estimation time and cost associated with dataset size.

K-Nearest Neighbor: KNN is a classification method that relies on neighbor-based and order-based principles. It falls under lazy learning, where instances from the training data are stored rather than constructing a global model. When classifying, the majority vote from the k-nearest neighbors of each data point is considered. One of the significant advantages of this classification algorithm is its simplicity of implementation and robustness in handling noise within the training data. However, a potential drawback lies in selecting the optimal K value, as an incorrect choice can result in suboptimal outcomes.

Support Vector Machine: The SVM classifier is a versatile and robust machine learning algorithm commonly used for classification and regression tasks. SVMs are particularly effective for binary and multi-class classification problems. It excels at finding an optimal hyperplane to separate data into different classes while maximizing the margin between them. SVMs can handle linear and non-linear data separation using various kernel functions, such as linear, polynomial, radial basis function, and sigmoid kernels. This flexibility allows SVMs to model complex decision boundaries accurately. SVMs have several advantages, including their ability to handle high-dimensional data effectively, resistance to overfitting, and firm performance even with small to medium-sized datasets.

Random Forest: The RF classifier is a powerful ensemble machine learning technique that utilizes multiple decision trees. To introduce diversity, the criteria for selecting nodes to split are randomized. This approach selects features randomly for specific purposes rather than solely focusing on the best feature. It is a bagging method combining deep trees to produce a low-variance output. When provided with an input vector x comprising various attributes, RF employs the results from multiple Decision Trees to make the final prediction.

Decision Trees: The DT classifier is a straightforward yet effective machine learning algorithm for classification and regression tasks. It builds a tree-like structure by recursively splitting the data based on the most significant features, resulting in a series of decision nodes that lead to final class or value predictions at the leaves. DTs are interpretable and suitable for both categorical and numerical data. They excel at capturing complex decision boundaries and feature interactions, making them valuable for various applications like customer churn prediction, credit scoring, and medical diagnosis. However, they can be prone to overfitting on noisy data, which can be mitigated using pruning and ensemble methods.

3 Related Research Work

Feature selection is used in many data mining and machine learning applications. The purpose of feature selection is to identify attributes that lower the error rates for predictions made by classifiers. Researchers have carried out a multitude of theories in the field of classification involving feature selection. A variety of cancer prediction models have also been presented. Some of these previously published works are discussed in this section.

The literature has discussed the classifications of 87 endometrial samples, each comprising five markers [26]. The related genes with missing data were eliminated for each indicator. Further, signal-to-noise ratios (SNRs) were computed to exclude irrelevant genes. The size of the newly obtained samples was then reduced using principal component analysis (PCA). Finally, ten random samples were chosen to serve as classification testing samples. The five cancer-related indicators collectively classify the 87 new endometrial samples as malignant or cancer-free. They proposed a model combining particle swarm optimization (PSO) algorithm, gray wolf optimizer (GWO), and Elman recurrent neural network (ERNN) to optimize the parameters. The classification accuracy obtained by the model is 88.85%. The proposed model surpasses both ERNN and ERNN optimized by PSO or GWO in accuracy.

Furthermore, the literature discusses the classification of five distinct cancer datasets: lung, central nervous system, brain, endometrial, and prostate [27]. In this research, a hybrid approach is proposed, combining the adaptive neuro-fuzzy inference system (ANFIS), fuzzy c-means clustering (FCM), and the simulated annealing (SA) algorithm. Various techniques are used to assess the performance of this proposed method against other algorithms, including the backpropagation algorithm, hybrid algorithm, genetic algorithm, and statistical procedures like the Bayesian network, support vector machine, and J48 decision tree. The findings indicate that training FCM-based ANFIS with the SA algorithm for classifying all cancer datasets leads to improvements, achieving an average accuracy rate of 96.28%.

A comprehensive framework model is established using TCGA gene expression data, which integrates linear regression, differential expression analysis, and deep learning to facilitate the precise biological interpretation of DNA methylation patterns [28]. In the context of uterine cervical cancer, this model employs linear regression to predict gene expression based on pre-filtered methylation data, removing outliers from the dataset. Differential expression analysis using Limma and empirical Bayes is then conducted to identify differentially expressed genes (DEGs). Subsequently, the deep learning method "nnet" is employed for classifying cervical cancer labels among these DEGs, utilizing a 10-fold cross-validation technique, achieving a classification accuracy of 90.69%. Additionally, a cystoscope is used in this study to investigate hub gene networks, identifying the top five genes with the highest in-degree and the top five with the highest out-degree. Furthermore, the model utilizes web gestalt to explore DEG KEGG pathway enrichment and gene ontology. This framework, which combines linear regression, differential expression analysis, and deep learning, aims to enhance the comprehension of DNA methylation and gene expression data in disease-related studies.

Additionally, an integrated two-step feature selection approach was introduced [29]. This method initially employed the correlation coefficient, T-statistics, and Kruskal-Wallis test to select relevant features. Subsequently, further optimization of the selected features was performed using techniques such as Central Force Optimization (CFO), Lighting Attachment Procedure Optimization (LAPO), Genetic Bee Colony Optimization (GBCO), and Artificial Algae Optimization (AAO). The study focused on ovarian cancer classification and employed at least five different classifiers. Notably, the Support Vector Machine (SVM) and logistic regression with GBCO achieved a high accuracy rate of 99.48% in both the Kruskal-Wallis and correlation coefficient test scenarios.

Furthermore, this study introduces the utilization of Recursive Feature Elimination (RFE) and Least Absolute Shrinkage and Selection Operator (LASSO) as feature selection techniques to identify the most significant attributes for predicting cervical cancer [30]. The dataset contains missing values and exhibits a high degree of imbalance. Therefore, a combined approach called SMOTE Tomek is employed, which integrates both under-sampling and over-sampling techniques along with a decision tree classification algorithm. Notably, the Decision Tree (DT) classifier, when coupled with RFE and SMOTE Tomek features, demonstrates improved performance, achieving an accuracy of 98.72% and a sensitivity of 100%. This enhanced performance is observed when feature reduction and addressing the significant class imbalance are considered.

Similarly, an approach called Grid Search-Based Hyperparameter Tuning (GSHPT) was introduced to classify microarray cancer data using random forest parameters [31]. Fixed parameters were utilized to ensure optimal accuracy within grid searches, employing n-fold cross-validation. The model incorporated a 10-fold cross-validation procedure. The grid search algorithm provided the most valuable parameters, including splits, the number of trees in the forest, maximum depth, and the number of leaf node samples needed to split at the leaf node. The maximum number of trees considered for ovarian and 3-class leukemia datasets was 10, 20, and 70. The highest classification accuracy was achieved with 50 trees for MLL and SRBCT. Across all datasets, a maximum depth of 2 was maintained, and node splitting was conducted using the Gini index. The performance of the proposed method was assessed using classification accuracy, precision, recall, $F1$ score, confusion matrix, and misclassification rate.

Additionally, a statistical test, analysis of variance based on MapReduce, was introduced to identify the most important features [32]. A MapReduce-based KNN classifier for microarray data classification and Hadoop was successfully used to implement these algorithms. A comparison was made between the dataset sizes and the processing times using a traditional system and a Hadoop cluster. Similarly, to categorize three microarray cancer gene expressions, an artificial intelligence-based ANFIS model was proposed [33]. The model's performance was compared to statistical techniques like Naive Bayes (NB) and Support Vector Machine (SVM). The ANFIS model parameters were optimized using the backpropagation and hybrid algorithms. The best average classification performance of ANFIS was 95.56%, whereas for statistical approaches, the performance was 87.65%.

Besides, a feature ranking framework as minimum redundancy maximum relevance (mRMR) was introduced, followed by a hybrid GA that employs the features ranked by mRMR [34]. The NB classifier selects the proper feature subsets for improved classification. Experiment results indicate that the gene subset selection in the mRMR-GA pipeline is acceptable. To develop the cervical cancer predictive model, principal component analysis (PCA) was used for dimensional reduction, and SVM and RF were used as classifiers [35]. It is observed that the RF outperforms the SVM algorithm with a precision of 94.21% when the two algorithms are compared to one another in performance.

4 Result and Analysis

Results and analysis of various models carried out by researchers are presented in this section. The study assessed performance using multiple parameters, including accuracy, sensitivity, specificity, positive predictive value, negative predictive value, and F-measure. These performance metrics are defined as follows, with TP (true positive), TN (true negative), FP (false positive), and FN (false negative) being the relevant terms:

$$Accuracy = \left(\frac{TP + TN}{TP + FP + FN + TN}\right) \times 100 \tag{1}$$

$$Sensitivity = \left(\frac{TP}{TP + FN}\right) \times 100 \tag{2}$$

$$Specificity = \left(\frac{TN}{TN + FP}\right) \times 100 \tag{3}$$

The detailed specification of datasets considered for analysis is furnished below. Gene expression microarray datasets, accessible through the Internet for free, are employed for data classification. These datasets can be categorized into two main types: binary and multi-class. Multi-class imbalance is the issue in the real-world dataset. The quantity of samples in one or more classes in the dataset is more significant than in others, causing the classification task to perform poorly. Table 2 lists several regularly used datasets, both binary and multi-class.

Table 2 Description of datasets

Dataset	Total instances	Total features	Number of classes
Brain	28	1070	2
Breast	49	7149	2
Cervical	215	27578	2
Cervical	858	36	2
Cervical	58	714	2
CNS	34	857	2
Endometrial	87	27578	2
Leukemia-2	72	7129	2
Lung	181	1626	2
Ovarian	253	15154	2
Prostate	102	339	2
Bladder	40	7129	3
Endometrial	42	8872	3
Leukemia-3	72	7129	3
MLL	72	12600	3
Endometrial	42	1771	4
SRBCT	83	2308	5

4.1 Analysis Based on Feature Selection

Several cancer datasets were employed for examination in this study. The literature reviewed in this research suggests that enhancing classification accuracy can be achieved by reducing irrelevant genes. Table 3 summarizes the number of features removed upon applying feature selection approaches. A compelling feature selection approach should yield high learning accuracy while demanding minimal processing power. As per the review findings, there is a need to further enhance the prediction accuracy of cancer classification methods, particularly for early-stage detection, to facilitate improved results for treatment and rehabilitation. The primary objective is to improve classification accuracy performance while reducing computational time.

4.2 Analysis Based on Dataset

The datasets for this study came from several different Internet repositories. Both binary and multi-class datasets were used in this study. Figure 5 depicts the classification accuracy of seventeen microarray datasets examined in this study. Except for the binary breast and CNS tumor datasets, all other datasets achieve a classification accuracy exceeding 90%.

Table 3 Original with reduced number of features

Dataset	Total features	Selected features
Brain	1070	36
Breast	7149	41
Cervical	27578	6287
Cervical	36	20
Cervical	714	–
CNS	857	30
Endometrial	27578	–
Leukemia-2	7129	31
Lung	1626	92
Ovarian	15154	50
Prostate	339	24
Bladder	7129	44
Endometrial	8872	53
Leukemia-3	7129	51
MLL	12600	26
Endometrial	1771	50
SRBCT	2308	61

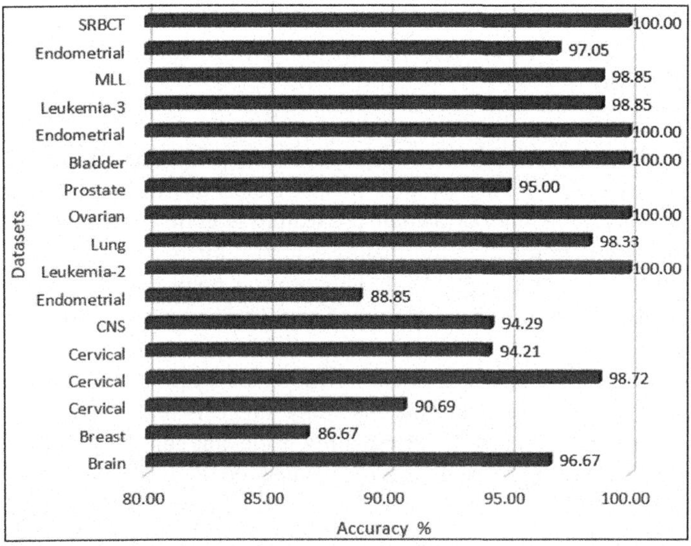

Fig. 5 Analysis based on datasets

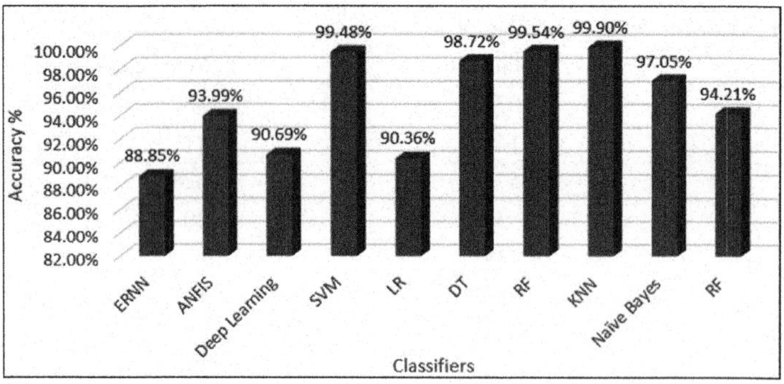

Fig. 6 Analysis based on classifiers

4.3 *Analysis Based on Classifier*

This study used binary and multi-class datasets with different classifiers such as ERNN, ANFIS, Deep Learning, SVM, LR, DT, RF, KNN, NB, and RF. The average classification accuracy of ten different classifiers analyzed in this investigation is depicted in Fig. 6. Apart from the ERNN classifier, all the others exhibit an average classification accuracy exceeding 90%.

5 Conclusion

The greatest threat to human life is complex sicknesses like cancer. Microarray technology has advanced, allowing for more precise cancer diagnosis. The most essential strategies for analyzing microarray data are feature selection and classification. Feature selection is an approach used to address high-dimensional data by eliminating irrelevant and duplicated information. This process helps conserve computational resources, enhancing learning accuracy and simplifying the understanding of learning models or data. We covered some of the available approaches for feature selection in this study, and we used additional classification techniques to assess classification accuracy. The literature reviewed in this study suggests that improving classification accuracy can be achieved by eliminating irrelevant genes. As indicated by the survey results, there is a pressing need to enhance the prediction accuracy of cancer classification, particularly in early stage, to enhance treatment and rehabilitation. In conclusion, the improvement of classification accuracy is the primary objective.

References

1. Patra, B.: Reliability analysis of classification of gene expression data using efficient gene selection techniques. Int. J. Comput. Sci. Eng. Technol. **1**(11) (2011)
2. Lopez-Rincon, A., Tonda, A., Elati, M., Schwander, O., Piwowarski, B., Gallinari, P.: Evolutionary optimization of convolutional neural networks for cancer miRNA biomarkers classification. Appl. Soft Comput. **65**, 91–100 (2018)
3. Piscaglia, F., Ogasawara, S.: Patient selection for transarterial chemoembolization in hepatocellular carcinoma: importance of benefit/risk assessment. Liver Cancer **7**(1), 104–119 (2018)
4. Bray, F., Ferlay, J., Soerjomataram, I., Siegel, R.L., Torre, L.A., Jemal, A.: Global cancer statistics 2018: GLOBOCAN estimates of incidence and mortality worldwide for 36 cancers in 185 countries. CA: A Cancer J. Clin. **68**(6), 394–424 (2018)
5. Hambali, M.A., Oladele, T.O., Adewole, K.S.: Microarray cancer feature selection: review, challenges and research directions. Int. J. Cogn. Comput. Eng. **1**, 78–97 (2020)
6. Qiu, P., Wang, Z.J., Liu, K.R.: Ensemble dependence model for classification and prediction of cancer and normal gene expression data. Bioinformatics **21**(14), 3114–3121 (2005)
7. Guzzi, P.H., Cannataro, M.: Challenges in microarray data management and analysis. In: Proceedings of 24th International Symposium on Computer-Based Medical Systems, pp. 1–6. IEEE Xplore (2011)
8. Dash, S., Patra, B., Tripathy, B.K.: A hybrid data mining technique for improving the classification accuracy of microarray data set. Int. J. Inf. Eng. Electron. Bus. **4**(1), 43–50 (2012)
9. Dash, S., Patra, B.: Feature selection algorithms for classification and clustering in bioinformatics. In: Global Trends in Intelligent Computing Research and Development, pp. 111–130. IGI Global, USA (2014)
10. Golub, T.R., Slonim, D.K., Tamayo, P., Huard, C., Gaasenbeek, M., Mesirov, J.P., Coller, H., Loh, M.L., Downing, J.R., Caligiuri, M.A., Bloomfield, C.D., Lander, E.S.: Molecular classification of cancer: class discovery and class prediction by gene expression monitoring. Science **286**(5439), 531–537 (1999)
11. Zhao, Z., Morstatter, F., Sharma, S., Alelyani, S., Anand, A., Liu, H.: Advancing feature selection research. ASU Feature Selection Repository, pp. 1–28 (2010)
12. Almugren, N., Alshamlan, H.: A survey on hybrid feature selection methods in microarray gene expression data for cancer classification. IEEE Access **7**, 78533–78548 (2019)
13. Guyon, I., Elisseeff, A.: An introduction to feature extraction. In: Feature Extraction, pp. 1–25. Springer, Berlin, Heidelberg (2006)
14. Kourou, K., Exarchos, T.P., Exarchos, K.P., Karamouzis, M.V., Fotiadis, D.I.: Machine learning applications in cancer prognosis and prediction. Comput. Struct. Biotechnol. J. **13**, 8–17 (2015)
15. Li, Z., Xie, W., Liu, T.: Efficient feature selection and classification for microarray data. PloS One **13**(8), e0202167 (2018)
16. Patra, B., Bisoyi, S.S.: CFSES optimization feature selection with neural network classification for microarray data analysis. In: Proceedings of 2nd International Conference on Data Science and Business Analytics, pp. 45–50. IEEE Xplore (2018)
17. Elkhani, N., Muniyandi, R.C.: Review of the effect of feature selection for microarray data on the classification accuracy for cancer data sets. Int. J. Soft Comput. **11**(5), 334–342 (2016)
18. Dash, S., Patra, B.: Genetic diagnosis of cancer by evolutionary fuzzy-rough based neural-network ensemble. In: Data Analytics in Medicine: Concepts, Methodologies, Tools, and Applications, pp. 645–662. IGI Global, USA (2020)
19. Patra, B., Bhutia, S., Panda, N.: Machine learning techniques for cancer risk prediction. Test Eng. Manag. **83**, 7414–7420 (2020)
20. Sahu, B., Dehuri, S., Jagadev, A.: A study on the relevance of feature selection methods in microarray data. Open Bioinf. J. **11**(1), 117–139 (2018)
21. Bączkiewicz, A., Wątróbski, J., Sałabun, W., Kołodziejczyk, J.: An ANN model trained on regional data in the prediction of particular weather conditions. Appl. Sci. **11**(11), 4757 (2021)

22. Tai, S.K., Dewi, C., Chen, R.C., Liu, Y.T., Jiang, X., Yu, H.: Deep learning for traffic sign recognition based on spatial pyramid pooling with scale analysis. Appl. Sci. **10**(19), 6997 (2020)
23. Patra, B., Jena, L., Bhutia, S., Nayak, S.: Evolutionary hybrid feature selection for cancer diagnosis. In: Intelligent and Cloud Computing, pp. 279–287. Springer, Singapore (2021)
24. Jain, I., Jain, V.K., Jain, R.: Correlation feature selection based improved-binary particle swarm optimization for gene selection and cancer classification. Appl. Soft Comput. **62**, 203–215 (2018)
25. Tan, A.C., Gilbert, D.: Ensemble machine learning on gene expression data for cancer classification. In: Proceedings of New Zealand Bioinformatics Conference, pp. 1–10. Te Papa, Wellington, New Zealand (2003)
26. Hu, H., Wang, H., Bai, Y., Liu, M.: Determination of endometrial carcinoma with gene expression based on optimized Elman neural network. Appl. Math. Comput. **341**(1), 204–214 (2019)
27. Haznedar, B., Arslan, M.T., Kalinli, A.: Optimizing ANFIS using simulated annealing algorithm for classification of microarray gene expression cancer data. Med. Biol. Eng. Comput. **59**(3), 497–509 (2021)
28. Mallik, S., Seth, S., Bhadra, T., Zhao, Z.: A linear regression and deep learning approach for detecting reliable genetic alterations in cancer using DNA methylation and gene expression data. Genes **11**(8), 931 (2020)
29. Prabhakar, S.K., Lee, S.W.: An integrated approach for ovarian cancer classification with the application of stochastic optimization. IEEE Access **8**, 127866–127882 (2020)
30. Tanimu, J.J., Hamada, M., Hassan, M., Kakudi, H., Abiodun, J.O.: A machine learning method for classification of cervical cancer. Electronics **11**(3), 463 (2022)
31. Shekar, B.H., Dagnew, G.: Grid search-based hyperparameter tuning and classification of microarray cancer data. In: Proceedings of Second International Conference on Advanced Computational and Communication Paradigms, pp. 1–8. IEEE Xplore (2019)
32. Kumar, M., Rath, N.K., Swain, A., Rath, S.K.: Feature selection and classification of microarray data using MapReduce based ANOVA and K-nearest neighbor. Procedia Comput. Sci. **54**, 301–310 (2015)
33. Haznedar, B., Arslan, M.T., Kalınlı, A.: Using adaptive neuro-fuzzy inference system for classification of microarray gene expression cancer profiles. Tamap J. Eng. **2018**, 39 (2018)
34. Thangavelu, S., Akshaya, S., Naetra, K.C., AC, K.S., Lasya, V.: Feature selection in cancer genetics using hybrid soft computing. In: Proceedings of Third International Conference on I-SMAC, pp. 734–739. IEEE Xplore (2019)
35. Abdullah, A.A., Sabri, N.A., Khairunizam, W., Zunaidi, I., Razlan, Z.M., Shahriman, A.B.: Development of predictive models for cervical cancer based on gene expression profiling data. In: IOP Conference Series: Materials Science and Engineering, vol. 557(1), pp. 012003 (2019)

Early Detection of Osteoporosis and Osteopenia Disease Using Computational Intelligence Techniques

T. Ramesh and V. Santhi

Abstract People are now more frequently affected by the disorders of osteoporosis and osteopenia. Food habits and genetics have been recognized as the main contributing factors; hence earlier disease prediction is necessary for the conditions mentioned above. Osteoporosis and osteopenia generally affect older adults and women who have passed menopause to a greater extent. Early identification of osteoporosis and osteopenia is crucial to prevent bone fractures and fragility. Low bone mineral density is the cause of bone fragility and fracture. Clinical information such as bone mineral density and radiographic images are used to perform a predictive study of the disorders mentioned above. Bone mineral density values are beneficial when determining T-Score and Z-Score values, which can be used to categorize osteoporosis and osteopenia. Several computational intelligence strategies forecast and categorize osteoporosis and osteopenia disorders. The various computational methodologies utilized to predict the same have been thoroughly explored in this chapter. Further, the chapter has been concluded with the appropriate metrics.

1 Introduction

New diseases that pose a significant threat to human life are discovered daily. Some disorders can be fatal if they are not appropriately diagnosed, making them exceedingly risky. Most people have developed eating habits that result in the degeneration of bone density, which causes diseases like osteoporosis and osteopenia. These eating habits have been discovered to be critical for various ailments. There is more room for bone fractures and fragility when bone mass declines. Low bones are often replaced by new bones every time to preserve the structure of the skeleton in healthy

T. Ramesh · V. Santhi (✉)
Vellore Institute of Technology, Vellore, Tamilnadu, India
e-mail: vsanthi@vit.ac.in

T. Ramesh
e-mail: t.ramesh2014@vit.ac.in

D. P. Acharjya and K. Ma (eds.), *Computational Intelligence in Healthcare Informatics*, Studies in Computational Intelligence 1132,
https://doi.org/10.1007/978-981-99-8853-2_10

individuals. Due to poor bone mineral density (BMD), new bones may replace old ones in the process of replacing bones. Osteoporosis, brought on by low bone mineral density, causes further bone mineral loss. The most reliable method for detecting osteoporosis is BMD testing [1].

A prior definition of osteoporosis said that it is a condition characterized by reduced bone mass and microarchitectural deterioration of bone cells resulting in heightened bone fragility and a subsequent rise in fracture incidence. A fracture in this criteria is required for the diagnosis of osteoporosis. According to the World Health Organization, the new classification of the condition based on BMD, osteoporosis, may be identified and treated before a fracture occurs. Osteoporosis is determined if a woman's BMD drops by 2.5 standard deviations or below the young adult criterion (a T score of 2.5). Women with healthy bone mass (less than 1) and osteopenia (less than 2.5 but greater than 1) may both have detectable bone mass. A person's bone mass changes throughout their lifespan. Women's bone mass increases rapidly until their mid-20s to mid-30s, reaching their ideal bone mass. After that, women lose bone mass gradually and long before menopause, following a few years of relative stability after achieving their peak bone density. Estrogen deficiency results in bone loss at a rate of up to 7% per year. Besides, it extends up to seven years following menopause. After that, the rate of bone loss slows down over time to 1–2% annually, while certain older women may experience a more significant loss in bone density. According to a study, halting bone loss at any time can lower the risk of fractures. It is anticipated that a 14% increase in bone density in 80-year-old women will cut the likelihood of hip fracture in half.

Osteoporosis and osteopenia can be predicted using tests of BMD such as the T-Score and Z-Score [2]. Numerous computational methods are available to categorize disorders like osteopenia and osteoporosis. Extreme learning machine (ELM) is a method that is quite effective in predicting osteoporosis. It requires less time for processing because it doesn't need to be adjusted, and more hidden nodes are necessary to obtain more outstanding performance, which increases network complexity. Categorization accuracy is quite good, even using just ELM. While ELM is being trained, the output matrix it generates could occasionally be unconditioned due to the random values. Therefore, optimization is crucial for ELM to offer the best possible solution. Optimization is maximizing or minimizing an objective function, or fitness function is given a set of constraints to produce an accurate or fitness value and ensure the system operates as efficiently as feasible.

In addition to the ELM, one of the most important computational methods, artificial neural networks (ANN), has been modified to identify osteoporosis and osteopenia. Numerous meta-heuristic techniques are utilized for training ANN for classification. For example, a monarch butterfly optimization (MBO) effectively classifies osteoporotic data with healthy data [3]. This algorithm, which is simple and dependable, can balance the two competing tactics of exploration and exploitation to produce the best possible result. However, most meta-heuristic algorithms cannot have optimal values when dealing with data with a significant dimension, which lowers the convergence rate.

By utilizing several computational algorithms and the values of BMD, osteoporosis and osteopenia can be predicted extremely successfully. Some of the computational techniques, such as ANN and ELM, for the classification of osteoporosis, will be presented in Sect. 2 to predict the early diagnosis of osteoporosis. In Sect. 3, the scheme of evaluation for the prediction of osteoporosis is presented with a block diagram. Finally, in Sect. 4, results are analyzed for the prediction of osteoporosis and osteopenia. Section 5 concludes the chapter.

2 Methods of Computational Intelligence

Heuristic optimization, learning, and adaptive features are included in a group of computer models and techniques referred to as computational intelligence. It is used to support research into problems that are difficult to tackle using conventional computer techniques. Fuzzy systems, evolutionary computation, and neural networks are the three fundamental pillars of computational intelligence. Computational intelligence is a powerful tool for various pattern detection and data analysis problems, including financial forecasting and industrial, scientific, and social media applications. The two applications where these techniques have been most successful are pattern recognition and data analytics. It is believed that computational intelligence methods and techniques can be successfully used in pattern recognition, given their effectiveness in big data analytic applications. In addition, recent developments in artificial intelligence, big data, and machine learning have enhanced the importance of research in biological signal and image processing.

2.1 Artificial Neural Networks for Osteoporosis Classification

The input layer, a hidden layer in the middle, and the output layer are the three layers that comprise the essential ANN [4]. This network structure is extensive, parallel, and intricately interwoven. Multi-layer perceptron (MLP) is often trained using the backpropagation network (BPN) approach [5, 6]. To lessen training-related errors, it uses a gradient descent method. An ANN's three-tiered architecture is depicted in Fig. 1. A summary of ANN's training procedures is furnished below

1. Use the synaptic weights and biases as the model's initial starting point.
2. Confirm that the training data have been loaded.
3. For the jth sample, compute the net input at the hidden layer using Eq. 1:

$$net_h^j = \sum_{n=1}^{N} x_n w_{h,n}^1 + b_h^1; \quad h = 1, 2, \ldots, H \tag{1}$$

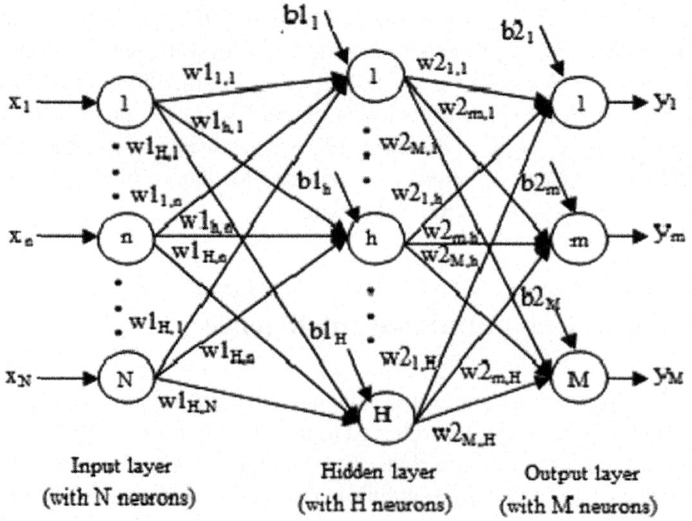

Fig. 1 Three-layered ANN architecture

where N and H are the sizes of the input and hidden layers, respectively. The notion $w^1_{h,n}$ is the weight between input neuron n and hidden neuron h. Similarly, notion b^1_h is the bias value of hidden neuron h. The output of the hidden layer is calculated using Eq. 2:

$$O^j_h = sigmoid(net^j_h) = \frac{1}{1 + e^{-net^j_h}} \tag{2}$$

4. At the output layer, computation of the input value is calculated using (3):

$$net^j_h = \sum_{h=1}^{h} O^j_h W^2_{m,j} + b^2_m; \quad m = 1, 2, \ldots, M \tag{3}$$

where $W^2_{m,j}$ stands for the weight between the hidden neuron h and the output neuron m. The notion b^2_m stands for the bias of the output neuron m, and M stands for the number of output nodes. The network output is determined using Eq. 4:

$$y^j_m = f(net^j_m) \tag{4}$$

5. The error value is calculated by comparing the network output to the goal output using Eq. 5, where T^j_m refers to goal output:

$$E^j = \sum_{m=1}^{M} (T_m^j - y_m^j)^2 \tag{5}$$

6. Changes have been made to the weight and bias parameters.
7. Continue carrying out the processes for each training example until the error is minimized when the training is finished.

Any data not noticed during the training phase is used to classify the system. Ideal weights and biases are the most crucial elements when training ANN. A three-layer ANN is typically chosen since it is seen to be sufficient for the majority of tasks. When there are fewer hidden neurons, the function approximation is decreased, and when there are more hidden neurons, the function approximation is increased [7, 8]. As a result, the number of hidden neurons must be carefully considered. This quantity is found using the MBO algorithm [9, 10]. In this method, the weights and biases are optimized before training an ANN. The performance of the ANN classifier is assessed in terms of how well it learns and generalizes the MBO. The mapping of outputs to inputs during the learning phase yields the most precise outcomes.

2.2 *Extreme Learning Machine in Osteoporosis Classification*

The ELM has consistently been more successful than gradient algorithms like ANNs, even after several repetitions with random probability. Since it does not require tuning, it performs better with fewer computations, and more hidden nodes [10, 11]. In addition, ELM offers fairly accurate classification services. However, because random variables are introduced into the equation, the output matrix produced by ELM during training may not be well-conditioned. As a result, ELM needs optimization to make the optimum solution. It is widely known in the literature that ELMs optimized using meta-heuristic approaches are quite effective even when there are fewer hidden neurons [12, 13]. An objective function is maximized or minimized under given constraints to provide an objective value or successfully manage the system via optimization.

To increase the effectiveness of this approach even further, two meta-heuristic algorithms—the migration operator of the MBO [14], and the evolution phase of an artificial algae algorithm with numerous light sources are merged. Applying the hybrid monarch butterfly optimization (HMBA) is necessary to optimize ELM to distinguish between osteoporotic and normal data. The ELM is a tuning-free method to train single hidden-layer feed-forward neural networks (SHFNs) [15]. Figure 2 shows the internal organization. The input layer is home to n input neurons, L neurons are found on a hidden layer, and M neurons are found on the output layer. The weights are adjusted with just one repetition. Numerous activation functions are used to excite neurons in different ways.

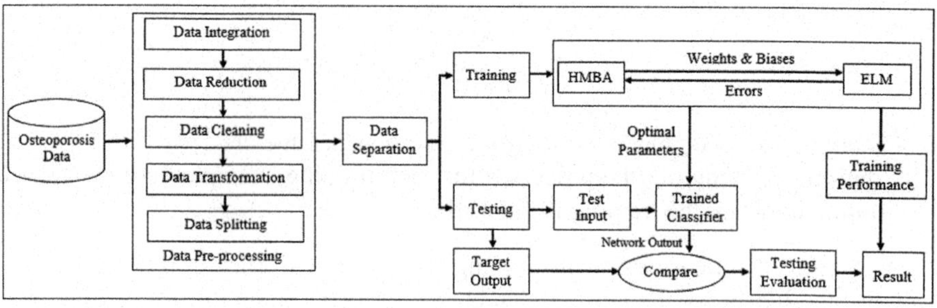

Fig. 2 HMBA-ELA classifier structure

Using Eq. 6 the output hidden layer matrix, H, is computed as the first step in ELM's training of single hidden-layer feed-forward neural networks:

$$H = \begin{bmatrix} h(X_1) \\ h(X_N) \end{bmatrix} = \begin{bmatrix} g(w_1).x_1 + b_1 + \cdots + g(w_L).x_1 + b_L \\ g(w_1).x_N + b_1 + \cdots + g(w_L).x_N + b_L \end{bmatrix}_{N \times L} \tag{6}$$

In other words, the input is converted to the ELM feature space described by the equation $h(X) = g(W, X, b)$, where W and b are the random weight, and bias values of the hidden neurons' inputs [16, 17]. Any activation functions, such as the sigmoidal, hyperbolic tangent, and Gaussian, may be utilized for the computation. The next step is calculating the output weight, which is done using Eq. 7 to connect the neurons in the hidden layer to the output layer:

$$\beta = H^t T = \begin{bmatrix} \beta_1 \\ \ldots \\ \beta_L \end{bmatrix}_{L \times M} \tag{7}$$

where H^t is the Moore-Penrose generalized inverse of matrix H, the equation would represent (8). The notion T is the training sample target matrix. The solution to the problem β is to find a value that minimizes the approximation error, which would be the objective function defined in Eq. 9. The last step is to use Eq. 10 to determine outputs y for the training data input:

$$H^t = (H^T H)^{-1} H^T \tag{8}$$

$$\beta = Min_{\beta \in R^{L \times M}} ||H\beta - T||^2 \tag{9}$$

$$y = f_L(X) = \sum_{i=1}^{L} \beta_i h_i(X) = h(X)\beta = H\beta \tag{10}$$

ELM requires an input layer, a hidden layer, and an output layer, each of which must appropriately have sizes n, h, and m because it is a network with only one hidden layer [18, 19]. The input and output properties of the datasets, respectively, determine the values of n and m. It is generally known that ELM generates precise diagnostics even when it reaches a significant number of hidden units. The steps used while classifying the osteoporosis sample using HMBA-ELM [13, 20] are listed below

1. First, load the training osteoporosis data and initialize all parameters.
2. Just normalize the data for training purposes.
3. As a starting point, choose some random numbers for the algal colony.
4. Determine whether each algal colony is viable.

    a. Calculate the hidden neuron output matrix using Eq. 6.
    b. Calculate the output weight using Eq. 7.
    c. Apply Eq. 9 to determine the training data output.
    d. Based on the training data, ascertain your level of fitness.

5. Based on the health status, algal colonies need to be arranged.
6. Calculate the sizes of algal colonies.
7. Identify the frictional surfaces and energy levels of the algal colonies.
8. Complete the helical movement phase.
9. Ascertain the best alternatives to the movement of algae in adaptation phase.
10. Repeat Steps 4 through 9 as necessary to reach the maximum number of generations.
11. Identify the best potential feasible solution.
12. Input the results of osteoporosis testing into the HMBA-ELM classifier through its paces.
13. Based on the obtained results through evaluation, the recommendation would be given.

3 A General Evaluation Scheme with a Block Diagram

A general evaluation scheme is discussed in this section. The model's performance is analyzed with data not utilized in training. This statistic makes it possible to assess the model's performance with previously unexplored data. It is meant to show how the model might operate in real-world situations. An 80/20 or 70/30 split between training and testing evaluation is a general rule. Accuracy, precision, and recall are considered the metrics of assessment. The source dataset's starting size has a significant impact on this. Accuracy is determined by the proportion of predictions that match the test data correctly [21]. By dividing the total number of guesses by the number of correct predictions, it is easy to calculate.

A low standard deviation suggests that data are grouped around the mean, whereas a significant standard deviation shows that data are more spread out. A standard deviation close to zero indicates that data points are relative to the mean. Similarly,

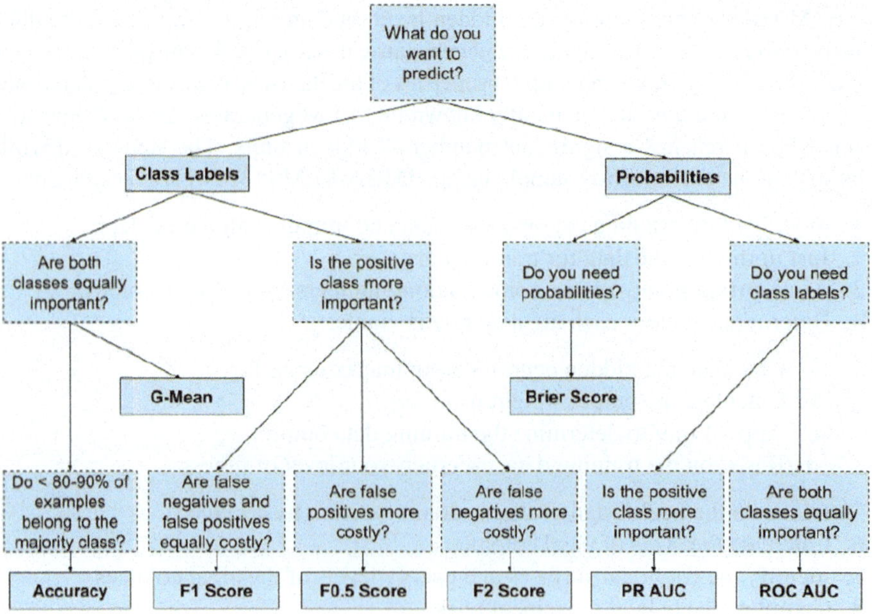

Fig. 3 General evaluation procedure

a high or low standard deviation indicates that data points are above or below the mean. The minimum possible value for the standard deviation is zero, which occurs only in contrived conditions in which every single number in the dataset in identical outliers have an impact on the standard deviation [22, 23]. It is because the standard deviation is calculated based on the distance from the mean. Therefore, outliers affect the mean as well. The standard deviation is measured in the same units as the original data. The general evaluation procedure is depicted in Fig. 3.

4 Findings and Evaluation

The MBO-ANN classifier's ability to generalize across the two datasets is evaluated using 10-fold cross-validation [4, 24]. Each dataset is divided into ten equal parts, and Table 1 lists the specifications of the input attributes. The last subset is used for testing after the classifier has been trained using nine subsets. After repeating this procedure ten times and giving each subgroup a chance to be assessed, the classifier's performance is documented. Ten relevant values from the classifier's output are averaged to estimate the performance parameters. Finally, experiments are conducted for all approaches using the 10-fold cross-validation with ten different trials to reduce the influence of chance.

Table 1 Details of the input attributes in an osteoporotic dataset

Serial number	Attribute name	Attribute description
1	Age	Age of the patient
2	Height	Height of the patient
3	Weight	Weight of the patient
4	BMI	Body mass index
5	C.Width	Cortical bone width or thickness
6	C.FD	Cortical bone fractal dimension
7	Tr.thick	Trabecular bone thickness
8	Tr.FD	Trabecular bone fractal dimension
9	Tr.Numb	Trabecular number
10	Tr.separa	Trabecular separation

Table 2 Performance comparison for osteoporotic datasets on 10-fold cross-validation

Approach	Data	Mean	SD	Best	Cc	InC	Ct
MBO-ANN	LS	0.00132	0.00021	0.00073	97.9	2.1	36.09
	FN	0.00105	0.00019	0.00068	99.3	0.7	36.13
ACO-ANN	LS	0.01438	0.0.0073	0.01218	92.2	7.8	36.68
	FN	0.01791	0.00574	0.00975	92.9	7.1	36.80
ABC-ANN	LS	0.14112	0.01297	0.02752	90.8	9.2	37.38
	FN	0.08602	0.01002	0.02147	87.23	12.77	36.17
BBO-ANN	LS	0.00280	0.00126	0.00183	95		40.15
	FN	0.00254	0.00092	0.00217	96.5	3.5	41.04
DE-ANN	LS	0.10189	0.14021	0.07231	90.1	9.9	42.20
	FN	0.09271	0.09120	0.09221	88.65	11.35	42.91
SGA-ANN	LS	0.04183	0.01496	0.11918	88.7	11.3	32.79
	FN	0.03872	0.19002	0.11149	85.8	14.2	32.71

All of the performance parameters for the trials, including the mean fitness value (Mean), a standard deviation (SD) of fitness value, best fitness value (Best), and the average of correctly classified data, are combined to yield the average of correctly classified (CC) data in percentage, an average of incorrectly classified (InC) data in percentage, and average computation time (Ct) in seconds. Table 2 compares the performance of MBO-ANN to that of other classifiers using these parameters. The results demonstrate that the MBO-ANN classifier obtained good classification accuracy of 97.9 and 99.3% for the LS and FN datasets, respectively. It is superior to alternative approaches. The accuracy obtained for the BBO-ANN is 95%, whereas for LS and FN datasets it is 96.5%. Thus, MBO-ANN has better accuracy. Figures 4 and 5 refer to the convergence curves of algorithms for lumbar spine (LS) and femoral neck (FN) datasets, respectively.

Fig. 4 Convergence curves of algorithms for LS dataset

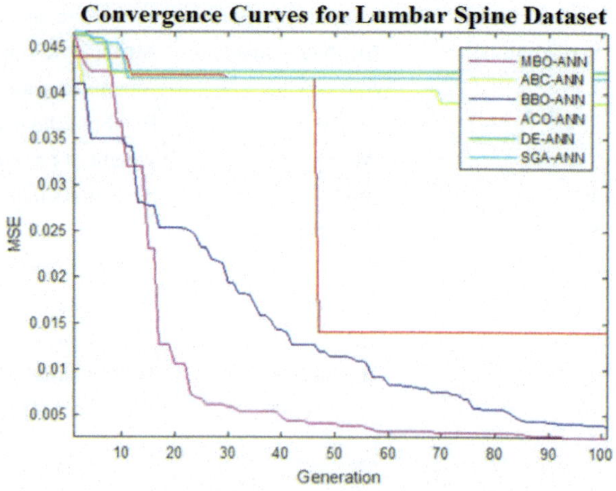

Fig. 5 Convergence curves of algorithms for FN dataset

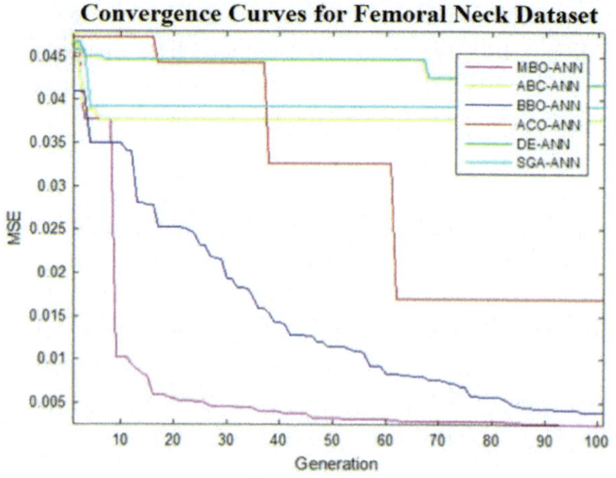

Among all the approaches, it is noted that MBO-computation ANN's time for the LS and FN datasets is 36.09 and 36.13, respectively. Similarly, SGA-ANN's time for LS and FN datasets is 32.79 and 32.71, respectively. However, it is discovered that SGA-ANN accuracy is only 88.7 and 85.8% for the LS and FN datasets, respectively. Medical field diagnostics rely on good accuracy rather than computing time. Similar computation has also been carried out for LS, FN, and femoral spine (FS) datasets using ELM-based classifiers. The results are presented in Table 3. The convergence curves of hybrid ELM classifiers for the FN, LS, and FS datasets are presented in Fig. 6.

As seen in Fig. 6, the HMBA-ELM curve steadily steepens toward the global optimum. It demonstrates that HMBA-ELM has received a quality education [25]. The HMBA-ELM computation for FN, LS, and FS dataset has produced a better mean fitness and SD. It is the HMBA-ELM classifier's lowest RMSE value among

Table 3 Learning ability of ELM-based classifiers for all datasets

Dataset	Approach	Mean	SD	Best	Tr.ct (s)
FN	HMBA-ELM	2.81E-06	0.00004	1.30E-07	115.28
	MBO-ELM	0.00089	0.00031	0.00061	128.63
	AAAML-ELM	0.00018	0.00022	0.00052	113.17
	GA-ELM	0.00332	0.00098	0.00125	172.33
	PSO-ELM	0.00981	0.01758	0.00235	178.37
	DE-ELM	0.01893	0.00871	0.01131	183.31
LS	HMBA-ELM	2.92E-06	0.00007	1.11E-07	116.23
	MBO-ELM	0.00041	0.00001	0.00036	129.13
	AAAML-ELM	0.00085	0.00031	0.00053	113.72
	GA-ELM	0.00345	0.00262	0.00313	173.23
	PSO-ELM	0.00827	0.02151	0.00821	177.24
	DE-ELM	0.00919	0.00525	0.00901	181.41
FS	HMBA-ELM	3.11E-06	0.00012	2.71E-07	120.13
	MBO-ELM	0.00038	0.0002	0.00054	132.44
	AAAML-ELM	0.00026	0.00011	0.00032	116.38
	GA-ELM	0.00344	0.00027	0.00298	177.25
	PSO-ELM	0.00991	0.00281	0.00913	181.26
	DE-ELM	0.01131	0.01412	0.01114	185.13

all others. In comparison of HMBA-ELM to AAAML-ELM and MBO-ELM, the computation time is slightly more significant for HMBA-ELM [13, 20]. Additionally, for the FN, LS, and FS dataset, HMBA-ELM has attained the best RMSE of 1.30E-07, 1.11E-07, and 2.71-07, respectively.

5 Conclusion

The prediction of osteoporosis is necessary, especially during the early stages, so that necessary treatment is taken to avoid further loss. Computational techniques such as ANN and ELM have been discussed in this chapter with the help of meta-heuristic optimization techniques like MBO, artificial algae algorithm with the multi-light source to improve the accuracy using text-based clinical data to predict osteoporosis at its early stages. Regarding text-based data, ANN and ELM techniques are more suitable for predicting osteoporosis and osteopenia at the early stages. The LS and FN datasets, two real-world osteoporotic datasets, are separated from healthy ones using the MBO-ANN classifier. The 10-fold cross-validation approach is used to conduct the studies. The MBO-ANN algorithm provides an accuracy of 97.9%, specificity of 98.33%, and sensitivity of 95.24% for the dataset LS. Similarly, the MBO-ANN algorithm provides an accuracy of 99.3%, specificity of 99.2%, and sensitivity of

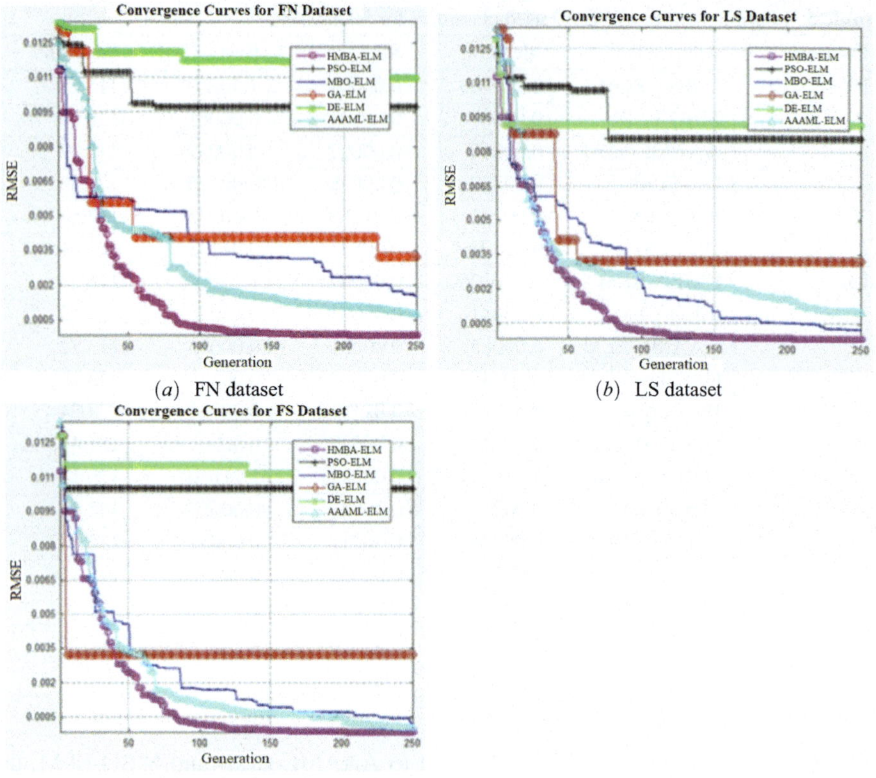

(*a*) FN dataset (*b*) LS dataset

(*c*) FS dataset

Fig. 6 Convergence curves of hybrid ELM classifiers

100% for the dataset FN. Three datasets on osteoporosis included in the study are trained using the HMBA algorithm to enable ELM to classify them. It is noticed that, except for calculation time, HMBA-ELM has surpassed all other strategies. It has a classification accuracy of 99.7% for datasets with osteoporotic conditions. Observing these data, one can conclude that ELM offers greater accuracy in the prediction of osteoporosis when compared to ANN.

References

1. Klibanski, A., Adams-Campbell, L., Bassford, T., Blair, S.N., Boden, S.D., Dickersin, K., Gifford, D.R., Glasse, L., Goldring, S.R., Hruska, K., Johnson, S.R., McCauley, L.K., Russell, W.E.: Osteoporosis prevention, diagnosis, and therapy. J. Am. Med. Assoc. **285**(6), 785–795 (2001)
2. Law, A.N., Bollen, A.M., Chen, S.K.: Detecting osteoporosis using dental radiographs: a comparison of four methods. J. Am. Dental Assoc. **127**(12), 1734–1742 (1996)

3. Devlin, H., Karayianni, K., Mitsea, A., Jacobs, R., Lindh, C., Stelt, P.V.D., Marjanovic, E., Adams, J., Pavitt, S., Horner, K.: Diagnosing osteoporosis by using dental panoramic radiographs: the OSTEODENT project. Oral Surg. Oral Med. Oral Pathol. Oral Radiol. Endodontology **104**(6), 821–828 (2007)
4. Devikanniga, D., Raj, R.J.S.: Classification of osteoporosis by artificial neural network based on monarch butterfly optimisation algorithm. Healthc. Technol. Lett. **5**(2), 70–75 (2018)
5. Kavitha, M.S., Ganesh Kumar, P., Park, S.Y., Huh, K.H., Heo, M.S., Kurita, T., Asano, A., An, S.Y., Chien, S.I.: Automatic detection of osteoporosis based on hybrid genetic swarm fuzzy classifier approaches. Dentomaxillofacial Radiol. **45**(7), 20160076 (2016)
6. Ordonez, C., Matias, J.M., de Cos Juez, J.F., García, P.J.: Machine learning techniques applied to the determination of osteoporosis incidence in post-menopausal women. Math. Comput. Model. **50**(5–6), 673–679 (2009)
7. Tafraouti, A., El Hassouni, M., Toumi, H., Lespessailles, E., Jennane, R.: Osteoporosis diagnosis using fractal analysis and support vector machine. In: Proceedings of Tenth International Conference on Signal-Image Technology and Internet-Based Systems, pp. 73–77. IEEE Xplore (2014)
8. Chang, H.W., Chiu, Y.H., Kao, H.Y., Yang, C.H., Ho, W.H.: Comparison of classification algorithms with wrapper-based feature selection for predicting osteoporosis outcome based on genetic factors in a Taiwanese women population. Int. J. Endocrinol. **2013** (2013)
9. Karaboga, D., Basturk, B.: A powerful and efficient algorithm for numerical function optimization: artificial bee colony algorithm. J. Global Optim. **39**(3), 459–471 (2007)
10. Simon, D.: Biogeography-based optimization. IEEE Trans. Evol. Comput. **12**(6), 702–713 (2008)
11. Storn, R., Price, K.: Differential evolution-a simple and efficient heuristic for global optimization over continuous spaces. J. Global Optim. **11**(4), 341–359 (1997)
12. Hegazy, T., Fazio, P., Moselhi, O.: Developing practical neural network applications using back-propagation. Comput. Aided Civil Infrastruct. Eng. **9**(2), 145–159 (1994)
13. Devikanniga, D.: Diagnosis of osteoporosis using intelligence of optimized extreme learning machine with improved artificial algae algorithm. Int. J. Intell. Netw. **1**, 43–51 (2020)
14. Khatib, W., Fleming, P.J.: The stud GA: a mini revolution?. In: Proceedings of International Conference on Parallel Problem Solving from Nature, pp. 683–691. Springer, Berlin, Heidelberg (1998)
15. Zhu, Q.Y., Qin, A.K., Suganthan, P.N., Huang, G.B.: Evolutionary extreme learning machine. Pattern Recognit. **38**(10), 1759–1763 (2005)
16. Xu, Y., Shu, Y.: Evolutionary extreme learning machine-based on particle swarm optimization. In: International Symposium on Neural Networks, pp. 644–652. Springer, Berlin, Heidelberg (2006)
17. Ma, C.: An efficient optimization method for extreme learning machine using artificial bee colony. J. Digital Inf. Manag. **15**(3), 135–147 (2017)
18. Yu, X., Ye, C., Xiang, L.: Application of artificial neural network in the diagnostic system of osteoporosis. Neurocomputing **214**, 376–381 (2016)
19. Harrar, K., Hamami, L., Akkoul, S., Lespessailles, E., Jennane, R.: Osteoporosis assessment using multilayer perceptron neural networks. In: Proceedings of 3rd International Conference on Image Processing Theory, Tools and Applications, pp. 217–221. IEEE Xplore (2012)
20. Huang, G.B., Zhu, Q.Y., Siew, C.K.: Extreme learning machine: a new learning scheme of feedforward neural networks. In: Proceedings of IEEE International Joint Conference on Neural Networks, vol. 2, pp. 985–990 (2004)
21. LeBoff, M.S., Greenspan, S.L., Insogna, K.L., Lewiecki, E.M., Saag, K.G., Singer, A.J., Siris, E.S.: The clinician's guide to prevention and treatment of osteoporosis. Osteoporos. Int. **33**(10), 2049–2102 (2022)
22. Mohapatra, P., Chakravarty, S., Dash, P.K.: An improved cuckoo search based extreme learning machine for medical data classification. Swarm Evol. Comput. **24**(10), 25–49 (2015)
23. Iliou, T., Anagnostopoulos, C.N., Anastassopoulos, G.: Osteoporosis detection using machine learning techniques and feature selection. Int. J. Artif. Intell. Tools **23**(05), 1450014 (2014)

24. Devikanniga, D., Vetrivel, K., Badrinath, N.: Review of meta-heuristic optimization based artificial neural networks and its applications. J. Phys. Conf. Ser. **1362**(1), 012074 (2019)
25. Matias, T., Araújo, R., Antunes, C.H., Gabriel, D.: Genetically optimized extreme learning machine. In: Proceedings of 18th Conference on Emerging Technologies & Factory Automation, pp. 1–8 (2013)

Pathway to Detect Cancer Tumor by Genetic Mutation

Aniruddha Mohanty, Alok Ranjan Prusty, and Daniel Dasig

Abstract Cancer detection is one of the challenging tasks due to the unavailability of proper medical facilities. The survival of cancer patients depends upon early detection and medication. The main cause of the disease is due to several genetic mutations which form cancer tumors. Identification of genetic mutation is a time-consuming task. This creates a lot of difficulties for the molecular pathologist. A molecular pathologist selects a list of gene variations to analyze manually. The clinical evidence strips belong to nine classes, but the classification principle is still unknown. This implementation proposes a multi-class classifier to classify genetic mutations based on clinical evidence. Natural language processing analyzes the clinical text of evidence of gene mutations. Machine learning algorithms like K-nearest neighbor, linear support vector machine, and stacking models are applied to the collected text dataset, which contains information about the genetic mutations and other clinical pieces of evidence that pathology uses to classify the gene mutations. In this implementation, nine genetic variations have been taken, considered a multi-class classification problem. Here, each data point is classified among the nine classes of gene mutation. The performance of the machine learning models is analyzed on the gene, variance, and text features. The gene, variance, and text features are analyzed individually with univariate analysis. Then K-nearest neighbor, linear support vector machine, and stacking model are applied to the combined features of a gene, variance, and text. In the experiment, support vector machine gives better results as compared to other models because this model provides fewer misclassification points. Based

A. Mohanty
Computer Science and Engineering, CHRIST (Deemed to be) University, Bengaluru, India
e-mail: aniruddha.mohanty@gmail.com

A. R. Prusty (✉)
DGT, RDSDE, NSTI(W), Kolkata, West Bengal, India
e-mail: dralokrprusty@gmail.com

D. Dasig
Graduate Studies College of Science and Computer Studies, De La Salle University, Dasmarinas, Philippines
e-mail: dddasig@dlsud.edu.oh

D. P. Acharjya and K. Ma (eds.), *Computational Intelligence in Healthcare Informatics*, Studies in Computational Intelligence 1132,
https://doi.org/10.1007/978-981-99-8853-2_11

on the variants of gene mutation, the risk of cancer can be detected, and medications can be given. This chapter will motivate the readers, researchers, and scholars of this field for future investigations.

1 Introduction

A few human body cells grow uncontrolled and spread to other parts of the body, called cancer. Generally, human body cells multiply, which is called cell division. This process generates new cells, and the body needs them. When cells become older, the new cell substitutes them. Sometimes this orderly process breaks down and multiplies the cells where they should not be. These cells may form tumors which are nothing but lumps of tissues. Cancerous tumors spread into nearby tissues and apply to inner places in the body. Also, it can develop new tumors through metastasis. Thousands of genetic mutations cause cancer tumors.

Gene is the basic physical and functional unit of heredity, a sequence of nucleotides in DNA or RNA that encodes the synthesis of a gene product, either RNA or protein. Nucleotides are composed of three subunit molecules a nucleobase, a five-carbon sugar (ribose or deoxyribose), and a phosphate. The variation in the DNA sequence in each of our genomes is known as genetic variation. Genetic variations caused in multiple ways, such as genetic recombination and mutation, cause cancer [1]. Cancer is a multistage process that progresses over several years and results from chromosomal changes. When it got detected, some cancers gathered as many as 50 chromosomal reshuffles and most likely an even greater number of changes in the nucleotide sequence of the tumor DNA [2]. Interpretation of genetic mutations is time-consuming when it has done manually.

The exact prediction of survival in patients with cancer is a tedious task. This is because of the heterogeneity and complexity of cancer, treatment options, patient populations, health care providers and health systems, lack of satisfactory authentication, and labor force training before disseminating new skills into clinical practice. In current practice, clinicians use the patient's data in consultation with the doctors. A doctor may feel areas of the patient's body for lumps that may indicate cancer such as changes in skin color or expansion of an organ. Besides, it may specify the occurrence of cancer, based on data collected from the previous medical records like laboratory tests [3], such as urine and blood tests. It helps the doctor to identify abnormalities that can be caused by cancer. Even some of the image tests like a computerized tomography (CT) scan, bone scan, magnetic resonance imaging (MRI), positron emission tomography (PET) scan, ultrasound, and X-ray also help with prediction and decision making [4]. If the prediction is perfectly alright, then adapt the care and the treatment.

In order to avoid manual collection of data from various sources for diagnosis, the aim is to develop a machine learning algorithm which can be used by the consultants for the prediction of cancer. The designed algorithm is used by an expert interpreted knowledge base where the researchers and oncologists have manually verified a number of gene mutations as a baseline. The aim of this initiative is to reduce the manual effort in diagnosing the cancer disease and prescribe the proper medicine for survival. Recently, gene sequencing has moved from the research domain to a clinical setting. In the past couple of years, researchers focused more on genetically understanding the disease and selecting the treatments most suited to the patients. Physicians use genetic testing as one of the innovative areas to treat cancer disease. The main hurdle is in gene classification, which requires considerable manual work [5]. The process starts with the molecular pathology laboratory of "reportable" and "not reportable" variants to generate a clinical report. Variant identification can be made using bioinformatics pipelines. This is the process of detecting variants and overturning irrelevant ones. Cancer tumors have thousands of genetic mutations. This experiment only focuses on those Mutations that contribute to tumor growth from neutral mutations [6].

Genetic mutation can be classified through textual pieces of evidence, which helps to detect cancer tumors in a more efficient and faster manner as compared to manual tactics in clinical investigation. Natural Language Processing (NLP) concepts have processed the collected texts. The applications of machine learning classification techniques like logistic regression classifier, random forest classifier, and Extreme Gradient Boosting (XGB) classifier, along with deep learning Recurrent Neural Network (RNN) classifier, have been used [7].

The rest of the chapter is organized as follows. Section 2 will explain the contextual details of the experimentation. Section 3 explores the done related to gene mutations and discusses the system design, including exploratory data analysis, pre-processing, and the training and testing datasets segregation. Section 4 explains methodologies that describe the various machine learning techniques, text transformation models, and classification models employed in this research. Section 5 deals with the experimental results and analysis. Section 6 concludes the entire exploration and proposes future areas of study.

2 Background Study

The basic unit of the human body is body cells. The body requires the cells to grow and divide, which creates new cells. Even cells die when they become older. Then the new cells will replace the older cells once this orderly process deviates, then cancer causes. Cancer is the uncontrolled development of abnormal cells in the human body. Everything that develops normal body cells unusually potentially causes cancer. This spreads to other parts of the body, which is in Fig. 1. However, the tumor is not spread to other parts of the body.

Figure 1 shows the overrun and malfunction of the body's cells [8]. The over production and the malfunction of the body's cells because of chemical or toxic compound exposures, ionizing radiation, some pathogens, and human genetics. The indications of cancer depend upon the specific type of cancer. It can be detected by fatigue, weight loss, pain, skin changes, change in bladder function, unusual bleeding,

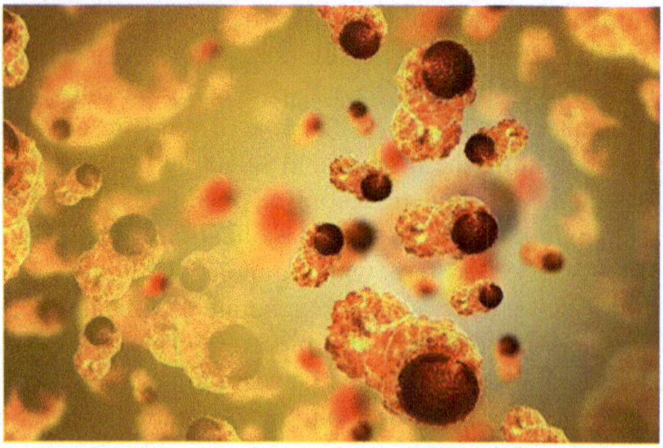

Fig. 1 Cancer tumor cells

persistent cough or voice change, fever, lumps, or tissue masses, unexplained muscle or joint pain, persistent, unexplained fevers or night sweats, unexplained bleeding.

Genetic and epigenetic variations in the human body are called cancer [9]. The uncommon gene function and changed patterns of gene expression are the cause of cancer [10]. Cancer cells develop because of multiple changes in the genes. Lifestyle habits, genes those get from your parents, and exposure to cancer-causing agents in the environment can all play a significant role in creating cancer. At some point in time, excess body growth also causes cancer [11]. A deficiency of any of the micronutrients folic acid, vitamin B12, vitamin B6, niacin, vitamin C, vitamin E, zinc mimic radiation in damaging DNA. A level of folate deficiency causes chromosome breaks which cause an increase in cancer risk [12]. As per World Health Organization (WHO), the cause of cancer is due to the three categories of external agents such as physical carcinogens, chemical carcinogens, and biological carcinogens.

Tobacco use, alcohol use, unhealthy diet, physical inactivity, and air pollution are risk factors for cancer [13]. Cancers are mainly two categories such as hematologic (blood) cancers and solid tumor cancers. Most cancers that involve tumor stages are in five broad categories. Other kinds, like blood cancers, lymphoma, and brain cancer, have their staging systems. But The stages of cancer do not change over time [14, 15].

- Stage 0: No cancer
- Stage I: The cancer is small and only in one area. This is also called primary-stage cancer.
- Stage II: The cancer is bigger and has grown up deep into nearby tissues or lymph nodes.
- Stage III: The cancer is bigger and has grown up deep into nearby tissues or lymph nodes (Same as Stage II).
- Stage IV: The cancer has spread to other parts of the body. It's also called advanced stage cancer.

A single test cannot confirm accurately diagnosed cancer. A thorough history of the patient, physical examination, diagnostic testing the complete evaluation of cancer can be performed. Effective diagnostic testing benefits observing the disease process and planning and evaluating the effective treatment. Cancer diagnosis methods are laboratory examinations, investigative imaging, endoscopic exams, genetic examinations, and tumor surgeries [16, 17].

Cancer is a generic disease which affects any parts of the body. It also creates abnormal cells rapidly. One needs to go through various laboratory tests with different methods to detect cancer. Early detection helps to prevent cancer, especially in stage 0. Detection of Cancer is quite a tedious task. Various research and tons of literature are available to detect and prevent cancer. This motivates us to work on cancer detection, as cancer tumors are thousands of genetic mutations. The primary motivation is to differentiate the mutations which back tumor growth from natural mutations.

2.1 Literature Survey

A recent study throws light on deep learning models such as the self-attention mechanism to understand the maps of tumor tiles with the help of Whole Slide Images (WSI) [18]. The model can predict six essential genes (AUC 0.680.85) and a copy of another six genes (AUC 0.690.79), an alteration of the original genes. This implementation also validates lung and liver cancer related to metastatic breast cancer. Similarly, machine learning and deep learning techniques are explained to analyze the complex biology of cancer [19]. In this review paper, cancer diagnosis, prognosis, and treatment management are highlighted with the help of genomic, methylation, and transcriptomic data by using deep learning techniques such as Multilayer Perceptron (MLP), Recurrent Neural Network (RNN), and Convolutional Neural Network (CNN). This paper also discussed Graph Convolutional Neural Networks (GCNNs) to analyze cancer.

In another study, a deep learning model to detect cancer dependencies which is nothing but to reveal genes is proposed [20]. DeepDEP, a conventional machine learning method verified on 8000 tumor samples with clinical relevance, helps identify cancer dependency with increasing genomic resources using transfer-learning techniques. Further, a method for classifying genetic variations/mutations based on the evidence from the text-based clinical literature is proposed [21]. With natural language processing, the text data can be processed to reduce the multifaceted nature of the information like missing quality, trait esteems, containing commotion or exceptions, and copying or wrong information. Then apply the univariate analysis to get more data about the quality of information. Then use random forest and logistic regression to classify the words into potential classes.

Likewise, to classify a large number of genetic mutations based on the cells that promote cell proliferation (drivers) is also proposed [22]. But most of them have not been classified correctly. This paper proposes a reliable method to automatically organize gene mutations by using deep learning concepts. Initially pre-processed the collected data provided by Memorial Sloan Kettering Cancer Center (MSKCC). Word2Vec is a tool for embedding text. Skip the Gram used for semantic searches. Then the sequence obtained from word2vec model was inserted into Long Short Term Memory (LSTM). The output of the model contains nine neurons which classify into nine classes. This LSTM model shows better log loss compared to other traditional models.

3 System Modeling

Identifying gene mutations is a tedious task that significantly affects millions of lives. The problem can be solved with the help of machine learning techniques. There are nine different classes of genetic mutations, which is considered a multi-class classification problem. Estimate the probability of individual data point fitting each of the nine classes. Comma separated data is present in the dataset, containing descriptions of gene mutations in the format below. The format contains four fields such as ID, gene, variation, and class. The field ID refers to detect clinical evidence whereas gene refers to locate genetic mutations. The field variation refers to change of amino acids and class refers to 1–9 classes of gene mutations.

For implementation, the data pre-processing step tokenizes the data, removes the stop words unnecessary spaces, and converts the text to lower cases. After segregating the data into training, testing, and cross-validation, the below modeling is applied to classify gene mutations and compare the performances as discussed in [23]. The classification algorithms considered are K-nearest neighbor, linear support vector machines, and stacking models. To measure the performance of each model, the confusion matrix, the correctly classified points, and the incorrectly classified points have been determined. Also, hyperparameter tuning is performed to get the best hyperparameters. This optimal hyperparameter will improve the performance of the models.

4 Materials and Methods

Memorial Sloan Kettering Cancer Center (MSKCC) data is used in this implementation which automatically classifies the genetic variations. Each class type in the classification is detected by unique characteristics of each data point with the relations of the gene and its mutations. All possible information of collected data is visualized, and extracted information to implement the machine learning algorithm. To visualize the collected data, the Exploratory Data Analysis (EDA) approach is used to summarize the main features of the data [24]. The EDA is implemented by graphical EDA, univariate analysis, and multivariate analysis.

Graphical EDA analyzes and summarizes the statistical characteristics of data, whereas in univariate analysis a dependable variable is derived and patterns of data are

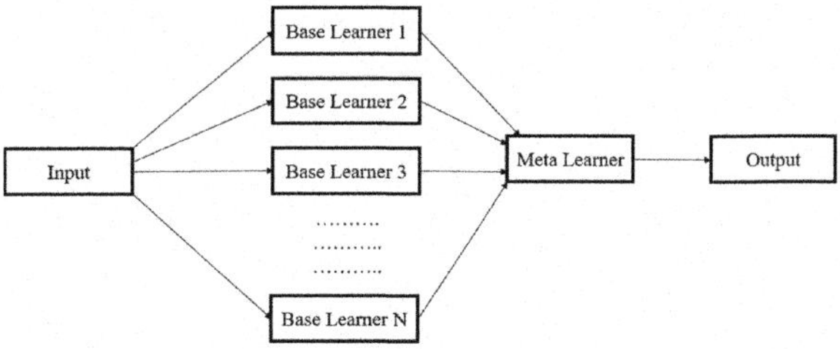

Fig. 2 Overview of stacking model

analyzed. The multivariate analysis analyzes more than two variables simultaneously. The various machine learning techniques used are briefly discussed below.

4.1 Stacking Model

Stacking model is an ensemble machine learning set of rules. It is used to get the estimates from two or more base learning models shown in Fig. 2. It is also used both in classification and regression tasks. The stacking model involves a base model (Level 0 Model) to fit on the training data and meta model (Level 1 Model) that learns how to best combine the predictions of the base models. The meta model uses base model output as input for classification. Stacking gives better performance compared to any of the single models [25].

4.2 K-Nearest Neighbor

K-Nearest Neighbor (KNN) focuses on similar things and keeping them close to each other and works on class labels and feature vectors. In KNN, a spatial vector is used to represent text which is denoted by Sentence (Text 1, Word 1; Text 2, Word 2, …, Text n, Word n). For any text, the highest similarity is selected from the similarity of training and testing data similarity value. Finally, the classes are computed based on K-neighbors [26]. The steps of KNN involved in the classification of text are furnished below.

1. Feature vectors are computed for both training and test data
2. Then feature vector of incoming text and each training text comparison is performed by using Eq. 1, where p_i and p_j are the feature vectors of the incoming and training text, respectively. The notion M being the dimension of the feature vector. $Wordi_k$ and $Wordj_k$ are the kth elements of vectors p_i and p_j, respectively.

$$similarity(p_i, p_j) = \frac{\sum_{k=1}^{M} wordi_k wordj_k}{\sqrt{\sum_{k=1}^{M}(wordi_k)^2}\sqrt{\sum_{k=1}^{M}(wordj_k)^2}} \tag{1}$$

3. Finally, the K-nearest neighbors are computed based on the similarity of texts, $similarity(p_i, p_j)$ using Eq. 2.

$$\delta(p_i, C_m) = \begin{cases} 0, & \text{if } a = 1 \\ 1, & \text{otherwise} \end{cases} \quad (2)$$

This algorithm is implemented here because of its quick calculation time, simplicity of interpretation, and usefulness for regression and classification. Also, the accuracy rate is high.

4.3 Linear Support Vector Machines

Support Vector Machine (SVM) is used to solve two-class pattern recognition problem introduced in 1995. The technique helps to find a decision surface which separates the data into two classes. The decision surface is a hyperplane can be written as $WX + b$, where X is an arbitrary object; the vector W and constant b are learned from a training set [27]. SVM works comparatively well to perform separation between classes more efficiently in high-dimensional space. So, the model is considered in this implementation.

5 Experiment and Result Analysis

To classify the genetic variations, stacking models, K-nearest neighbor, and linear support vector machines are used. Scikit-learn and Natural Language Toolkit (NLTK) is the standard Python library used. Matplotlib and seaborn are used to visualize the data. Other Python libraries like pandas, regular expression, collections, and math Python libraries are used in this implementation. The classification concept is implemented with the help of Python codes to visualize live principles.

5.1 Dataset

The dataset is collected from the Kaggle competition [28] and Memorial Sloan Kettering Cancer Center (MSKCC) [29]. Genes, variations, and clinical text are the three parameters present in the dataset. In the training dataset, there are nine classes of current mutations. The dataset (training/test variants) offers evidence about the genetic mutations, whereas (training/test text) offers the clinical indication (text) that our human experts used to classify the genetic mutations by inking via ID fields. This dataset is used to categorize the nine categories of gene variants which are considered a multi-class classification.

5.2 Performance Analysis

The problem statement is "classify the given genetic variations or mutations based on evidence from text-based clinical literature." So, a machine learning algorithm helps to classify genetic variations automatically. The objective of the implementation is to forecast the probability of each data point belonging to each of the nine classes. In order to verify the performance of class probabilities, multi-class log loss, and confusion matrices are used. In this experiment, the dataset has been split randomly into three parts, with 60%, 20%, and 20% of data as train, cross-validation, and test, respectively.

Reading the Data: Data points need to be read from the files as the variant and text data. A variant file is a comma-separated file used for training with fields like ID, gene, variation, and class.

Performance Metrics: In order to measure the performance of the planned models below performance metrics are used. Namely, multi-class log loss and confusion matrix. Multi-class log loss is also called categorical cross-entropy. It is defined using Eq. 3, where N is the number of data points, P is the number of classes [30].

$$logloss = -\frac{1}{n}\sum_{i=1}^{N}\sum_{j=1}^{P} y_{ij} log(p_{ij}) \tag{3}$$

Confusion Matrix: A confusion matrix is a $(Z \times Z)$ matrix for evaluating a classification model, where Z is the number of target classes. The matrix compares actual target values with those predicted by the machine learning model [31]. The matrix having positive and negative values, targeted variables represented by columns, rows represent predicted variables. The vital part of the confusion matrix is true positive, true negative, false positive, and false negative.

Featurization: Gene and variation need to be converted to numeral vectors as both are categorical features. For vectorization, one hot encoding and response coding approach has been used. One hot encoding creates a new binary variable for each categorical feature containing either 0 or 1 [32]. Here, 0 signifies the absence, and 1 signifies the presence of that category. Response coding is the probability of the data points represent the category of a class that belongs to a particular class. So, for an N-class classification problem, N new features of the data point belonging to each class are created. Mathematically, the formulation is defined in Eq. 4.

$$P(class = X|category = A) = \frac{P(category = A \cap class = X)}{P(category = A)} \tag{4}$$

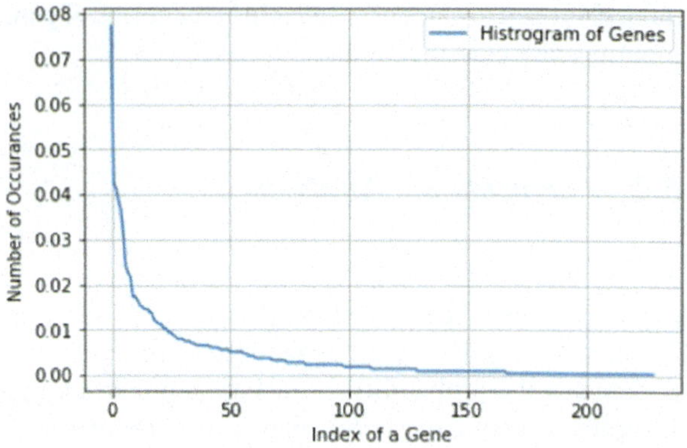

Fig. 3 Frequency visualization of genes

In this concept, probabilities have been used. In some scenarios, the numerator value is zero or extremely small. So, it is difficult to predict the class probabilities. To avoid this, Laplace smoothing is used [33]. In this implementation, below formulation is applied for Laplace smoothing. Here, 90 is selected because there exist 9 classes for classifications and alpha is the smoothing parameter. One hot encoding works better for linear and logistic regression, and SVM. For Naïve Bayes and random forest, response coding works better.

$$\frac{(Numerator + 10 * \alpha)}{Denominator + 90 * \alpha}$$

Univariate Analysis: Univariate analysis is to derive the data, define, summarize, and analyze the pattern in the data. In this implementation, univariate analysis can be performed for gene, variance, and text features. In the data set, there exist 235 distinct genes, which are represented in Fig. 3. This Fig. 3 represents the shorted form of the occurrences of genes. It represents a skewed distribution. This presentation shows very few genes occur a large number of times, and a lot of genes occur very few times.

Logistic regression is used to design the model with the help of gene feature only. In logistic regression, log loss is used to evaluate the loss of the model. In this implementation, an SDG classifier is used, which helps to implement logistic regression, and α is used to regularize the hyperparameter. This model is evaluated with various values of α like $1e-05, 0.0001, 0.001, 0.01, 0.1$, and 1. It is represented in Fig. 4. Figure 4 shows the log loss increase initially, then decreases, and again increases. For values of best α are 0.0001, and the train, cross-validation, and test log loss are 1.0425604300119806, 1.2325868001617826, and 1.200905436534172, respectively. It is also observed that there is no significant difference between train loss, cross-validation loss, and test loss values. So, the model is not overfitting. The gene features in a train, cross-validation, and test data overlap quite a lot. So, it is a stable model.

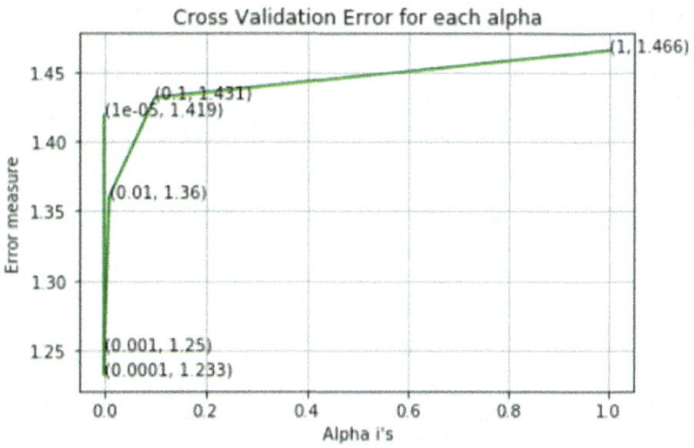

Fig. 4 Logloss visualization of gene features only

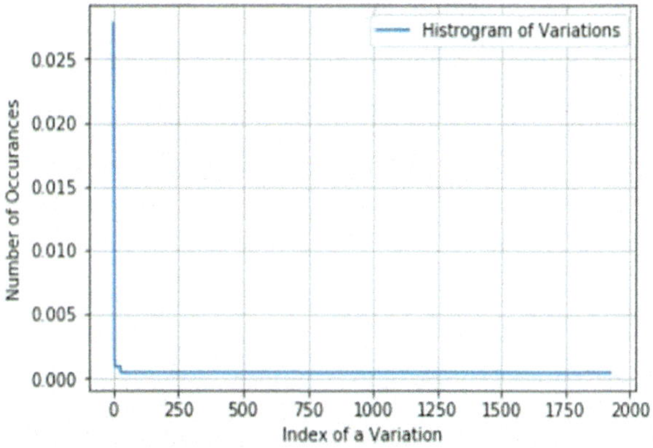

Fig. 5 Variation count of the data

Univariate Analysis using Variance Feature: Variation is a categorical feature, just like gene features. Training data contains a total of 2124 data points, and from these, 1927 are the unique variations. The histogram representation in Fig. 5 shows the descending order of the number of occurrences of the various features. A very smaller number of points occur a greater number of times.

To feature variance features, both hot encoding and response coding are used. After featurization, implementing logistic regression with calibration verified the variance features stability. The hyperparameter used in this model to measure the errors is α. Various alpha values like $1e-05$, 0.0001, 0.001, 0.01, 0.1, and 1 are observed in this model. Initially, the error is more, and then it is reduced again; it increases as the alpha value increases shown in Fig. 6. Also, it observed that the best hyperparameter value is 0.0001. The train, cross-validation, and test log loss are 0.8255455900343496, 1.689842322110077, and 1.7362385768757977, respectively. The difference between the cross-validation and test loss is very close. So, the variance feature is useful.

Univariate Analysis using Text Feature: In the text feature, every data point is considered as "TEXT" and splits each row into words by space and creates a dictionary

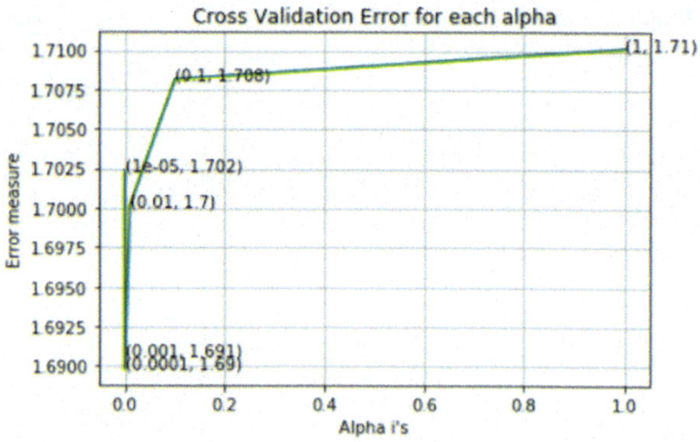

Fig. 6 Cross-validation error for each hyperparameter α

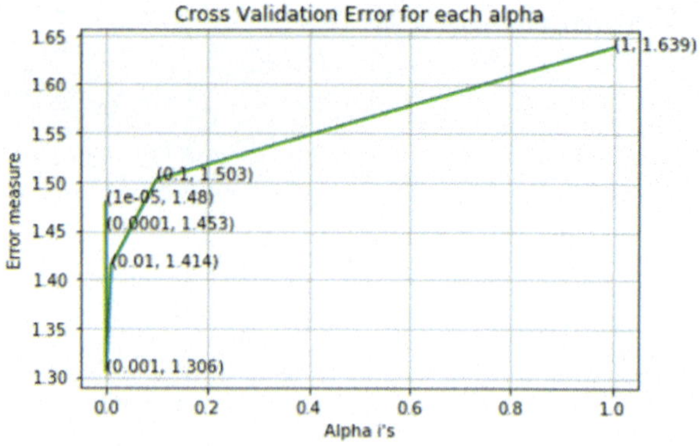

Fig. 7 Cross-validation log loss value analysis

of words. In this implementation, the words considered whose frequencies are more than three for train, test, and cross-validate data. Now train the model with logistic regression and calibration. Here one hot encode is used for featurization. To measure the model's performance, various hyperparameter values like $1e-05$, 0.0001, 0.001, 0.01, 0.1, and 1 used to verify the log loss. The best hyperparameter value is 0.001. For the best hyperparameter value of 0.001, the train, cross-validation, and test log loss are 0.7614133457365512. 1.3062465612808916, 1.1902669954542044 respectively. Cross-validation and test errors are less, so the text feature is a stable and useful feature. It is depicted in Fig. 7.

5.3 *Machine Learning Model Implementations*

This section discusses the results pertaining to machine learning implementation. The machine learning models considered for analysis are KNN, SVM, and stacking classifiers.

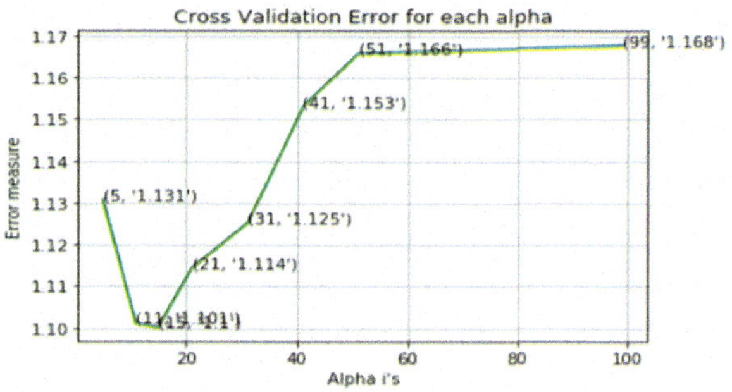

Fig. 8 Hyperparameter verification for KNN model

K-Nearest Neighbor: K-Nearest Neighbor is a simple, supervised machine learning algorithm that can solve both classification and regression problems. This algorithm does not work very well in high-dimension data. Hence, one hot Encoding may not work very well in KNN. So, response coding is used in this implementation. The number of neighbors (K) is used as a hyperparameter, and different K values like 5, 11, 15, 21, 31, 41, 51, and 99 are verified in this implementation. In the experiment, K value 15 gives the lowest error. So, $K = 15$ is the best hyperparameter. It is depicted in Fig. 8. Training, cross-validation, test log loss is 0.7056, 1.1002, 1.0911.

KNN model performance can be observed with the help of the confusion matrix as depicted in Fig. 9. The confusion matrix shows the misclassified points are 39.47%. There is also confusion between classes 1 and 4 and classes 2 and 7.

Support Vector Machine: Linear support vector machine performs better on high-dimensional data. It is interpretable and very similar to logistic regression. The only difference is linear SVM used hinge loss to validate the loss in the model. To train linear SVM, balancing of data is required. Further for evaluation various hyperparameter values like $1e - 05$, 0.0001, 0.001, 0.01, 0.1, 1, 10, 100 also required. The hyperparameter analysis is depicted in Fig. 10. From the Fig. 10, the best hyperparameter identified is 0.01.

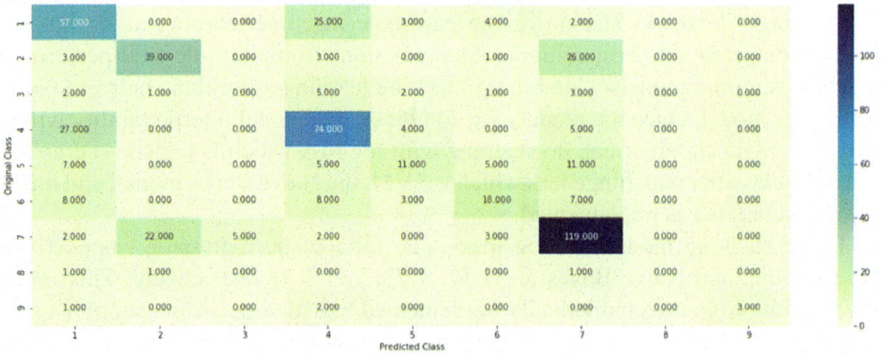

Fig. 9 Confusion matrix representation of KNN

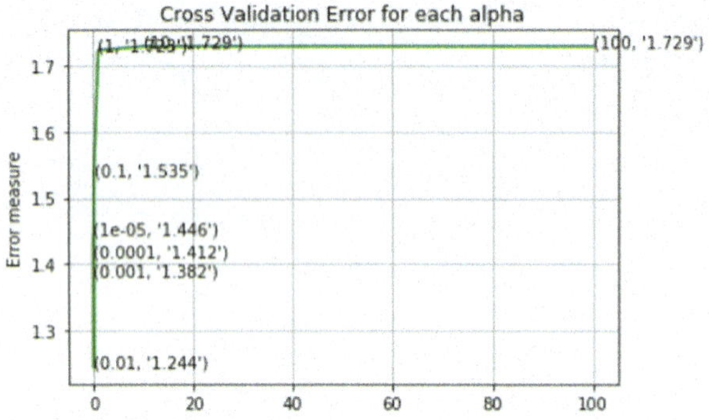

Fig. 10 Hyperparameter verification for SVM model

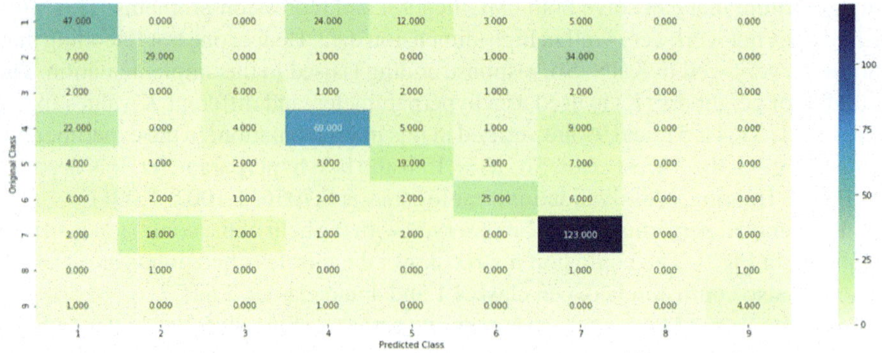

Fig. 11 Confusion matrix representation of SVM algorithm

Further, train cross-validation, and test losses obtained are 0.7628309867716067, 1.2442964974823838, and 1.1541891969863685 respectively. It is observed that 36.46% of data was misclassified in the SVM model implementation. In the confusion matrix, the diagonal values are showing protuberant in Fig. 11.

Stacking Classifier: Stacking is an ensemble machine learning algorithm that helps to detect how to best combine the predictions from multiple well-performing machine learning models. Combining machine learning algorithms helps to obtain a stronger model. Once the models are combined, the model interpretability will be lost. The SGD classifier can do stacking with log loss which is logistic regression, the SGD classifier with hinge loss which is SVM, the Naïve Bayes model, and finally stacking classifier is implemented.

In the stacking model, the log loss value for logistic regression, support vector machine, and Naïve Bayes are 1.17, 1.73, and 1.31 respectively. This model gives more errors than individually implemented logistic regression, support vector machine, and Naïve Bayes because the data is shared between the base learner and the meta learner. So, stacking is not a good idea. From this model, train log loss,

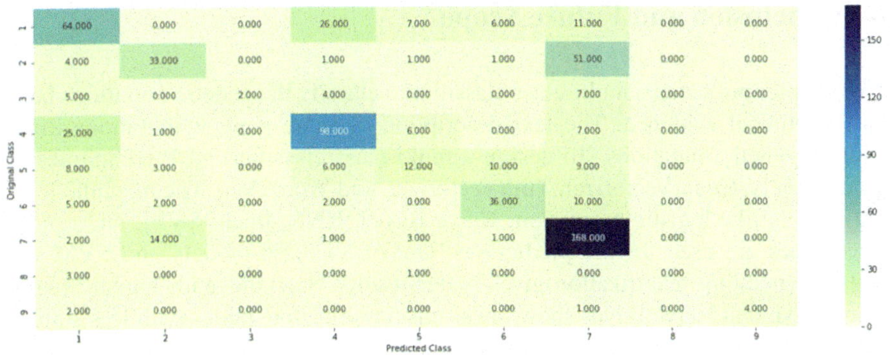

Fig. 12 Confusion matrix representation of stacking algorithm

cross-validation log loss, and test log loss are 0.662816676788, 1.1769826219, and 0.362406015034555, respectively, shown in Fig. 12.

5.4 Comparison of Machine Learning Models

The comparison of Naïve Bayes, logistic regression with class balancing and unbalancing, and random forest models are explained in [23]. Apart from those models, the performance of the KNN, SVM, and stacking models is explained in Table 1. KNN model used with response coding as featurization in which the misclassified points are 39.66%. The gap between training and cross-validation losses is more. So, this model is not performing well. Similarly, in stacking, the misclassified points are 36.24%, and the gap between the training and cross-validation loss is more. So, stacking is not considered a better-performing model. Compared to another model, misclassified points have a higher value. So, the model cannot be considered.

Table 1 Comparison of implemented models

Model	Train loss	Cross-validation loss	Test loss	Misclassified points
KNN	0.7056	1.1002	1.0911	39.66% (Response coding)
SVM	0.762	1.244	1.154	36.46% (One hot)
Stacking	0.663	1.176	1.081	36.24% (One hot)

6 Conclusion and Future Scope

This chapter projected a multi-class classifier to classify the genetic mutations based on the clinical evidence. The text description of these genetic mutations differentiates genetic mutations. To design a multi-class classifier, NLP techniques are used, namely, CountVectorizer, TfidfVectorizer, and Word2Vec. The machine learning models like logistic regression, Naïve Bayes, KNN, linear SVM, and random forest trees are used for the predictions. Out of all these models, the SVM with one hot encoding featurization gives better results. Stacking with logistic regression, SVM, and Naïve Bayes show lower misclassification values with less stability because the difference between train loss and cross-validation loss is higher. So, it creates overfitting. As cancer detection is a time-consuming process, this proposed implementation helps to diagnose the disease more easily and quickly. These models show consistent results in the prediction of cancer. As part of future work, the datasets can be evaluated with different deep learning models and compared to the performances.

References

1. Kukovacec, M., Kukurin, T., Vernier, M.: Personalized medicine: redefining cancer treatment. Text Analysis Retrieval, pp. 69–71 (2018)
2. Jackson, A.L., Loeb, L.A.: The mutation rate and cancer. Genetics **148**(4), 1483–1490 (1998)
3. Tagliabue, G., Maghini, A., Fabiano, S., Tittarelli, A., Frassoldi, E., Costa, E., Nobile, S., Codazzi, T., Crosignani, P., Tessandori, R., Contiero, P.: Consistency and accuracy of diagnostic cancer codes generated by automated registration: comparison with manual registration. Popul. Health Metrics **4**(1), 1–8 (2006)
4. Iın, A., Direkoğlu, C., ah, M.: Review of MRI-based brain tumor image segmentation using deep learning methods. Proc. Comput. Sci. **102**, 317–324 (2016)
5. Dienstmann, R., Dong, F., Borger, D., Dias-Santagata, D., Ellisen, L.W., Le, L.P., Iafrate, A.J.: Standardized decision support in next generation sequencing reports of somatic cancer variants. Mol. Oncol. **8**(5), 859–873 (2014)
6. Tenaillon, O., Matic, I.: The impact of neutral mutations on genome evolvability. Curr. Biol. **30**(10), R527–R534 (2020)
7. Gupta, M., Wu, H., Arora, S., Gupta, A., Chaudhary, G., Hua, Q.: Gene mutation classification through text evidence facilitating cancer tumour detection. J. Healthcare Eng. **2021**, 1–16 (2021)
8. Dunn, G.P., Old, L.J., Schreiber, R.D.: The immunobiology of cancer immunosurveillance and immunoediting. Immunity **21**(2), 137–148 (2004)
9. Prendergast, G.C., Metz, R., Muller, A.J.: Towards a genetic definition of cancer-associated inflammation: role of the IDO pathway. Am. J. Pathol. **176**(6), 2082–2087 (2010)
10. Jones, P.A., Baylin, S.B.: The epigenomics of cancer. Am. J. Pathol. **128**(4), 683–692 (2007)
11. Bergstrom, A., Pisani, P., Tenet, V., Wolk, A., Adami, H.O.: Overweight as an avoidable cause of cancer in Europe. Int. J. Cancer **91**(3), 421–430 (2001)
12. Ames, B.N.: DNA damage from micronutrient deficiencies is likely to be a major cause of cancer. Mut. Res./Fundam. Mol. Mech. Mutagen. **475**(1–2), 7–20 (2001)
13. Ferlay, J., Ervik, M., Lam, F., Colombet, M., Mery, L., Pineros, M., Znaor, A., Soerjomataram, I., Bray, F.: Global cancer observatory: cancer today. Lyon, France: Int. Agency Res. Cancer **3(20)**, (2019)

14. Bookstein, R., MacGrogan, D., Hilsenbeck, S.G., Sharkey, F., Allred, D.C.: p53 is mutated in a subset of advanced-stage prostate cancers. Cancer Res. **53(14)**, 3369–3373 (1993)
15. Tokunaga, R., Sakamoto, Y., Nakagawa, S., Miyamoto, Y., Yoshida, N., Oki, E., Watanabe, M., Baba, H.: Prognostic nutritional index predicts severe complications, recurrence, and poor prognosis in patients with colorectal cancer undergoing primary tumor resection. Dis. Colon Rectum **58**(11), 1048–1057 (2015)
16. Ahdoot, M., Wilbur, A.R., Reese, S.E., Lebastchi, A.H., Mehralivand, S., Gomella, P.T., Bloom, J., Gurram, S., Siddiqui, M., Pinsky, P., Pinto, P.A.: MRI-targeted, systematic, and combined biopsy for prostate cancer diagnosis. New Engl. J. Med. **382**(10), 917–928 (2020)
17. Dehan, P., Kustermans, G., Guenin, S., Horion, J., Boniver, J., Delvenne, P.: DNA methylation and cancer diagnosis: new methods and applications. Expert Rev. Mol. Diagnost. **9**(7), 651–657 (2009)
18. Li, G., Yao, B.: Classification of Genetic mutations for cancer treatment with machine learning approaches. Int. J. Desig. Anal. Tools Integrat. Circuits Syst. **7**(1), 63–66 (2018)
19. Tran, K.A., Kondrashova, O., Bradley, A., Williams, E.D., Pearson, J.V., Waddell, N.: Deep learning in cancer diagnosis, prognosis and treatment selection. Genome Med. **13**(1), 1–17 (2021)
20. Chiu, Y-C., Zheng, S., Wang, L-J., Iskra, B.S., Rao, M.K., Houghton, P.J., Huang, Y., Chen, Y.: Predicting and characterizing a cancer dependency map of tumors with deep learning. Genome Med. **7(34)**, eabh1275 (2021)
21. Bosnjak, M., Kovac, G., Sesto, F.: Personalized Medicine: Redefining Cancer Treatment classification using Bidirectional Recurrent Convolutions. Text Analysis and Retrieval 2018 Course Project Reports, 28–32 (2018)
22. Zhang, S., Bamakan, S.M.H., Qu, Q., Li, S.: Learning for personalized medicine: a comprehensive review from a deep learning perspective. IEEE Rev. Biomed. Eng. **12**, 194–208 (2018)
23. Mohanty, A., Prusty, A.R., Cherukuri, R.C.: Cancer tumor detection using genetic mutated data and machine learning models. In: Proceedings of the IEEE International Conference on Intelligent Controller and Computing for Smart Power, pp. 1–6 (2022)
24. Sahoo, K., Samal, A.K., Pramanik, J., Pani, S.K.: Exploratory data analysis using Python. Int. J. Innov. Technol. Explor. Eng. **8**(12), 4727–4735 (2019)
25. Dou, J., Yunus, A.P., Bui, D.T., Merghadi, A., Sahana, M., Zhu, Z., Chen, C.-W., Han, Z., Pham, B.T.: Improved landslide assessment using support vector machine with bagging, boosting, and stacking ensemble machine learning framework in a mountainous watershed. Jpn. Landslides **17**(3), 641–658 (2020)
26. Shah, K., Patel, H., Sanghvi, D., Shah, M.: A comparative analysis of logistic regression, random forest and KNN models for the text classification. Jpn. Augment. Hum. Res. **5**(1), 1–16 (2020)
27. Zhang, W., Yoshida, T., Tang, X.: Text classification based on multi-word with support vector machine. Knowl.-Based Syst. **21**(8), 879–886 (2008)
28. Personalized Medicine: Redefining Cancer Treatment. https://www.kaggle.com/c/msk-redefining-cancer-treatment/data
29. Ortiz, M.V., Kobos, R., Walsh, M., Slotkin, E.K., Roberts, S., Berger, M.F., Hameed, M., Solit, D., Ladanyi, M., Shukla, N., Kentsis, A.: Integrating genomics into clinical pediatric oncology using the molecular tumor board at the Memorial Sloan Kettering Cancer Center. Pediatr. Blood Cancer **63**(8), 1368–1374 (2016)
30. Kabani, A., El-Sakka, M.R.: Object detection and localization using deep convolutional networks with softmax activation and multi-class log loss. In: Proceedings of the International Conference on Image Analysis and Recognition, pp. 358–366 (2016)
31. Karabatak, M.: A new classifier for breast cancer detection based on Naive Bayesian. Measurement **72**, 32–36 (2015)
32. Dahouda, M.K., Joe, I.: A deep-learned embedding technique for categorical features encoding. IEEE Access **9**, 114381–114391 (2021)
33. Kikuchi, M., Yoshida, M., Okabe, M., Umemura, K.: Confidence interval of probability estimator of Laplace smoothing. In: Proceedings of the 2nd International Conference on Advanced Informatics: Concepts, Theory and Applications, pp. 1–6 (2015)

A Knowledge Perception: Physician and Patient Toward Telehealth in COVID-19

Ritu Chauhan and Aparajita Sengupta

Abstract Telehealth is the delivery, management, and monitoring of healthcare services using data and communication technologies. In the current scenario of the pandemic, usage of telemedicine has become relevant, as it has significant potential to safeguard both medical professionals and patients. Additionally, it limits the social activity of patients, therefore minimizing the sprsead of the virus during the COVID-19 pandemic. Furthermore, the role of telemedicine is discussed at the global level where the World Health Organization has announced it to be the need for the delivery of healthcare services at a distance. Hence, utilizing electronic methods for diagnosis and treatment, research, evaluation, illness, injury prevention, and education of healthcare workers. The proposed study of approach focuses on the concept of telemedicine while generalizing the perception of patients and physicians toward telehealth. The results discuss the adaptability of both patients and healthcare practitioners toward telemedicine. To support the current study of approach, a survey was conducted where 105 respondents participated, in which 80 respondents were people who availed of the telehealth service and 25 respondents were physicians who provided the service. Further, their level of perception and satisfaction was observed using varied statistical inferences using SPSS. Lastly, a significant relationship between the satisfaction level of patients and physicians is measured and whether they will continue with the telehealth service in the future or not.

1 Introduction

A progressive stipulated growth has been acknowledged in the informatics community over the last two decades to build a systematic approach in health care. Hence,

R. Chauhan (✉)
Center for Computational Biology and Bioinformatics, Amity University, Noida, India
e-mail: rituchauha@gmail.com

A. Sengupta
Center for Medical Biotechnology, Amity University, Noida, India
e-mail: aparajitasengupta96@gmail.com

D. P. Acharjya and K. Ma (eds.), *Computational Intelligence in Healthcare Informatics*, Studies in Computational Intelligence 1132,
https://doi.org/10.1007/978-981-99-8853-2_12

a new level of technology has been adopted such as machine-level communication infrastructures like HL7 messaging, SNOMED, and so on, which have paved the way toward the future. This technology is supported by the back-end development of electronic records and clinical decision support systems which are growing at a faster pace. These junctures are critical in the process of telemedicine application standardization and expansion. It can be said that telehealth is the use of digital technology to link numerous people in different places to give medical treatment, health education, and public health services. Similarly, it can be discussed as a wide term that refers to technology-enabled healthcare services.

In addition, telehealth services have been utilized in several arenas to measure the prognosis of disease which may include evaluation, monitoring, communication systems, prevention, and training, as well as diagnosis and treatment of sickness or injury. Moreover, keeping in consideration the economic factors, the government is taking steps to develop a research center focusing on Telehealth which is known as the National Consortium of Telehealth Resource Centers (NCTRC). NCRTC has released a Telehealth definition framework to assist regulators, physicians, insurers, and the general public in properly defining "telehealth" and its many components. Telehealth clinical services are currently delivered in three ways:

1. Real-time consultations, physician-to-patient discussions, and language translation services are all possible with video conferencing.
2. Electronic gadgets convey patient health information to healthcare providers via remote patient monitoring.
3. Store and forward technologies allow medical professionals to electronically communicate pre-recorded footage and digital pictures such as X-rays, video clips, and photographs.

Telehealth technology can assist in improved patient outcomes while providing limited access to treatment, as well as cost efficacy in the health service system. Likewise, it also offers useful technologies to improve health outcomes and access to care while making healthcare delivery systems more efficient and cost-effective. Time, distance, and provider scarcity may all be overcome with telehealth, allowing vital medical treatments to be delivered where they are most needed. This encompasses medically undeserved metropolitan regions as well as isolated, rural places. Advances in telehealth delivery were aided by the introduction of high-speed Internet and the rising usage of Information Communication Technology (ICT) in traditional ways of treatment. Telehealth became increasingly viable when more portable devices, such as laptops and cellphones, became available; the market then expanded into health promotion, prevention, and education. To speed up telehealth delivery, healthcare companies are increasingly utilizing self-tracking and cloud-based technology, as well as novel data analysis methodologies [1–3].

In the current study of approach, telehealth services are integrated with the experience of physicians and other healthcare practitioners to anticipate and discover an assisted technology that can benefit the end users with quick findings and better healthcare services. The chapter discusses the vision of telemedicine with a focus on the shift to a new diagnosis-based system. The questionnaire is designed to understand the relationship between healthcare practitioners and patients regarding adaptability and whether they will choose telehealth for further health issues or not. Analysis of the data is conducted on SPSS using chi-square, and cross-tabulation is run to correlate them with the variables. The result suggested that the majority of the patients and practitioners are satisfied with the usage of telehealth services.

The chapter highlights the major concern of telehealth in Sect. 1 followed by the literature review in Sect. 2. Section 3 discusses the overall methodology applied with patterns, and results are deliberated in Sect. 4. The manuscript is concluded in Sect. 5.

2 Review of Telehealth and Telemedicine Services

The use of electronic information and communication technology to offer healthcare treatments around the globe when participants are spread throughout the country is termed telemedicine [4, 5]. It is frequently used to refer to a larger use of technologies for remote education, consumer engagement, and other purposes in which communications and information systems are employed to assist healthcare services. Telemedicine and telehealth include the transmission of still pictures, telemonitoring of vital signs, e-health, videoconferencing, continuous medical training, and nurse service centers [6–8].

Remote health service delivery is utilized for several purposes. Expert referral services are a solution in which an expert assists a general practitioner in making a diagnosis. This might include a patient "visiting" a specialist via a live, remote consult or transmitting diagnostic images and video, together with patient data, to a professional for review. The exchange of audio, video, and health information between a patient and a health practitioner to provide an evaluation, treatment regimen, and prescription is referred to as direct patient care. Patients in a distant clinic, a doctor's office, and own home may be included. Devices that collect and transmit data to a monitoring station for processing are used in remote patient monitoring. Telemedicine is sometimes best described in terms of the services offered, and the methods utilized to deliver those services are listed below

1. A primary care or allied health provider may be involved in patient health care and specialty referral services.
2. Medical education offers health educational certificates to physicians and other health professionals. Special medical education seminars are held in remote areas for certain communities.
3. A practitioner who consults with a patient assists with primary care in making a diagnosis. This might include utilizing live interactive video or utilizing save and send diagnostic pictures, vital signs, and clips, as well as patient information saved for subsequent review.
4. These applications might incorporate a specific vital indication, such as blood glucose or heart rate. For homebound patients, several indications may be used. These services can be utilized for a variety of purposes.

5. The use of the Internet and wireless devices for consumer medical and health information to supply customers with professional health information and online discussion groups mentor assistance.
6. Remote patient surveillance, including personal telehealth, employs equipment to gather and transmit data remotely and transfer information to a health center or a Remote Diagnostic Testing Facility (RDTF) for analysis such as interpretation [9, 10].

Establishing and maintaining records is another important task in telehealth and telemedicine. Over 30 years of creating telemedicine networks and providing remote medical services have produced insights that are vital to grasp and expand national Health Information Technology (HIT) targets. The existing telemedicine networks include a wealth of information, with 200 multilateral institutions and numerous home health and remote monitoring programs relying on sophisticated HIT [2, 3]. Many telehealth networks have necessitated the collaboration of numerous different businesses with the shared purpose of providing health care.

Telemedicine network developers have had to discuss inter-organizational meetings to discuss problems such as governance, responsibility, and supervision in an environment with multiple partners and health information exchange organizations. These networks have had to focus on developing trusting connections among organizations, practitioners, and clients for a specialist at another institution to provide care [11, 12]. The physician-patient interaction frequently exists between remote geographic areas when the practitioner and patient are placed in distinct institutions and have never met face to face. Over time, the creators of these platforms and programs have improved their technical design and functional protocols to effectively adapt to actual conditions within their service region and among the collaborating institutions [13].

Several effective telemedicine networks have established a grid of all main health consumers like clinical medical services provided in-person remotely, services offered in-office, medical services from a distance, in-classroom distance learning health and medical education, and information management for electronic records. Many hospitals and medical groups have launched early telehealth projects during the last 10 years. Due to the necessity for social separation, telemedicine swiftly became the primary way of physician-patient connection with the commencement of the COVID-19 pandemic. As a result of the epidemic, many physicians and other competent care providers have had their first experience administering care through video links. During the COVID-19 epidemic, even going to the doctor's office was too dangerous for a while. Many patients and clinicians had their first encounter with telehealth, which used a variety of telecommunications and software techniques to interact [14]. Some employed complex video systems were linked with email and online messaging, while others depended on the smartphone for audio and video communication. To allow for the use of telehealth during the pandemic, federal and state authorities, as well as health insurance companies, loosened several telemedicine laws and payment constraints [15–17].

3 Methodology

The experience and attitudes of physicians and other frontline clinicians during the COVID-19 epidemic through the physician survey are discussed in this section. It is anticipated that these findings will assist medical practices, funders, and government authorities in the coming months to establish a new standard for clinical care. The study focuses on defining patients' experiences and views through the patient survey toward the COVID-19 epidemic. It, in turn, assists medical practitioners, funders, and government authorities in mapping the way for the future "virtual healthcare" system.

This analysis includes all responses from physicians and other professionals. Structured and free-text answers were used to collect clinical subspecialties. Individual questions might be left unanswered by respondents. As a result, answer volumes differ from question to question. Twenty-five physicians and other certified healthcare professionals participated in the survey. The patient survey was sent via a digital form to those dealing with health issues and impairments. Patients responded to questions about their most recent telehealth interaction. Questions focused on a variety of major topics, such as clinical problems addressed, satisfaction, and if they intend to use telehealth services shortly. Patients were permitted to choose multiple responses to some questions and to skip any question. Respondents to the survey included 50 people who received telehealth during the epidemic. Respondents were permitted to skip questions, resulting in variation between questions.

A survey was done where 105 respondents participated. 80 respondents were people who availed the telehealth service and 25 respondents were physicians who provided the service. Their level of perception and satisfaction was observed. A chi-square test was run to prove the hypothesis. The data collected was analyzed in SPSS to prove that there is a significant relationship between the satisfaction level of patients and physicians and they will continue with the service:

1. In patient's survey, a test was run to show a significant association between both the variables which are I will continue to use telehealth services in the future and I was confident that my health concern could be addressed during the visit.
2. In physician's survey, a test was run to show a significant association between both the variables which are I am personally motivated to increase the use of telehealth in my practice and telehealth has improved the financial health of my practice.

4 Results Analysis

Due to the quick shift from face-to-face health consulting to online health consulting, many people of every generation faced some difficulty. From doctors, nurses, healthcare practitioners, and compounders to patients of varied age groups had to adapt to the new change. Many healthcare providers had to learn how to deal with patients online. They had to update their technical skills. Patients who are of age 50 and above have been seen to struggle a bit more than the younger generation. Therefore, they also had to brush their technical ability. The result shows the patient's and practitioner's perception of the shift. The questions have varied responses from the respondents. The results are shown in the form of a bar graph for each question.

4.1 Patient's Perception of Telehealth

The results of patient's perception toward telehealth is discussed using a bar graph. Figure 1 interprets that 63.7% agree and 26% strongly agree that they had a sense of access and continuity of care. Only 5% are neutral and 5% disagree with this. The reason the 8 respondents did not get a sense of access and continuity because of a lack of proper network connection and they could not accustom to the technology. Similarly, Fig. 2 interprets that 66.3% of the patients agree and 23.8% strongly agree that they will continue using telehealth services in the future. Besides, 10% of the patients are neutral with this statement as they are the 8 respondents who faced technological difficulty while using telehealth services.

Figure 3 interprets that 61.3% of the patients agree and 28.7% of the respondents strongly agree with the statement that their physician joined within 15 min of their appointment. Likewise, 10.1% of the patients are neutral or disagree with this. They are the 8 respondents who faced technical issues while using the telehealth service. Additionally, Fig. 4 states that 61.3% of patients agree and 28.7% of the patients strongly agree that the way the physicians explained things was easy to understand. This shows development in the artificial intelligence used in telehealth services. Physicians used pictures and videos to explain the patient their disease which made it much clear to them and they got well aware of how to take care of themselves with the medication provided by the physician.

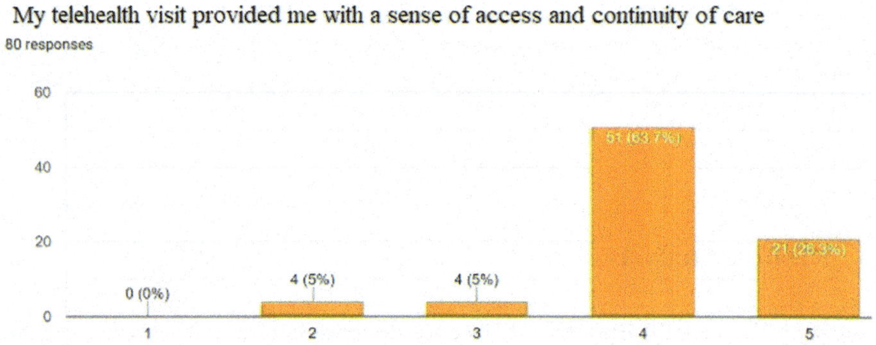

Fig. 1 Patient's sense of access and continuity of care

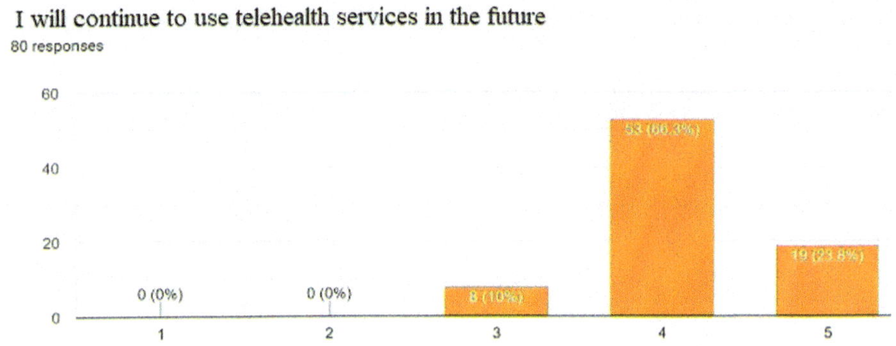

Fig. 2 Patient's perception toward using telehealth in the future

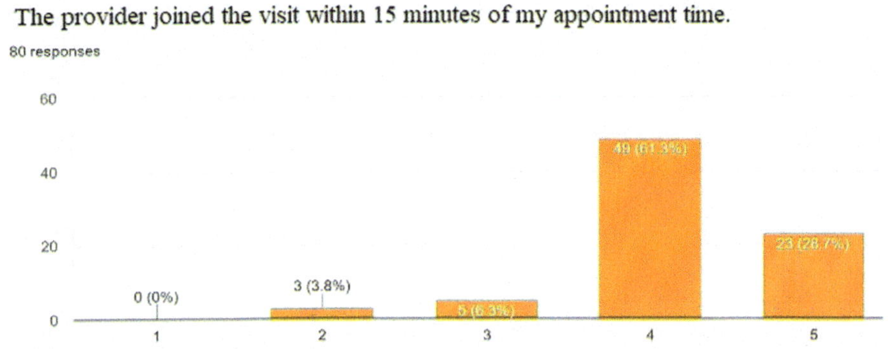

Fig. 3 Patient's sense of satisfaction with service

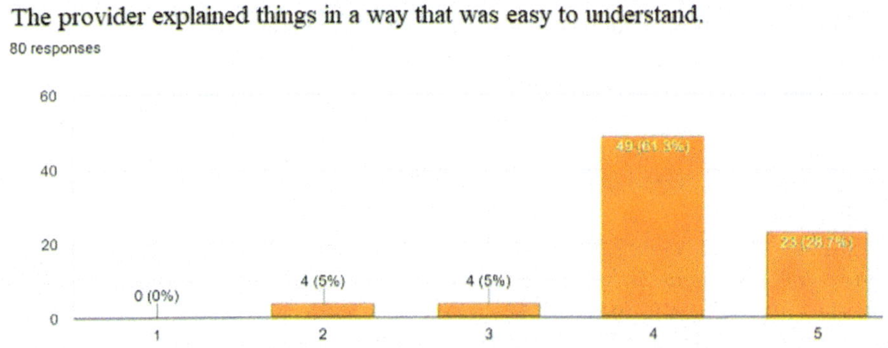

Fig. 4 Patient's level of understanding

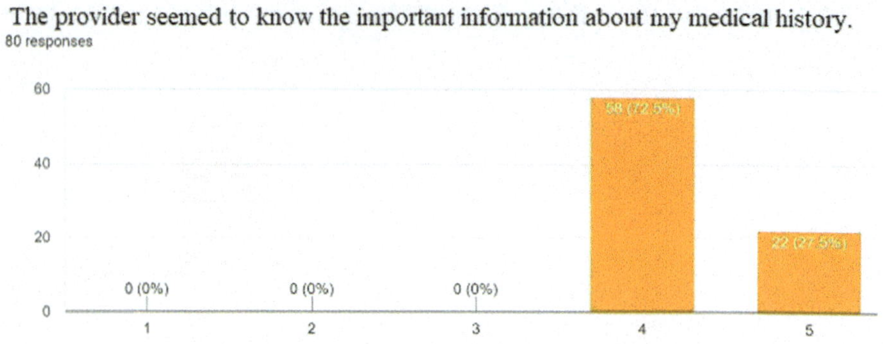

Fig. 5 Artificial intelligence role in telehealth

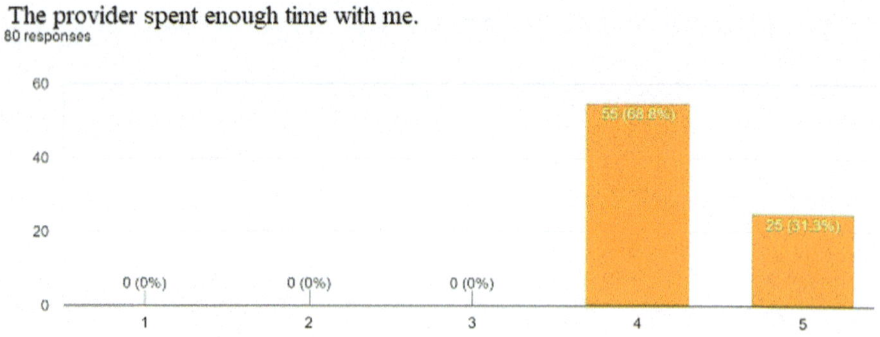

Fig. 6 Patient's satisfaction with service

Figure 5 interprets that 100% of the patients agree and strongly agree with the statement that their physician seemed to know information about their medical history. This shows advancement in artificial intelligence in the health sector and in medical records. Besides, Fig. 6 depicts that 100% of the patients agree and strongly agree with the statement that the physician spent enough time with them. This shows a positive response of patients toward telehealth services. It means they were satisfied with the service.

Figure 7 depicts that 67.5% of the patients agree, and 6.3% of the patients strongly agree with this statement that they were able to show the provider the physical concern they had like rash, burn, and mole. 13% of the patients disagree and 10% are neutral to this as their concern was different. Likewise, Fig. 8 depicts that 100% of the patients agree and strongly agree with the statement that they were confident about their health concern as it was properly addressed during the visit. This shows a positive response toward the telehealth service.

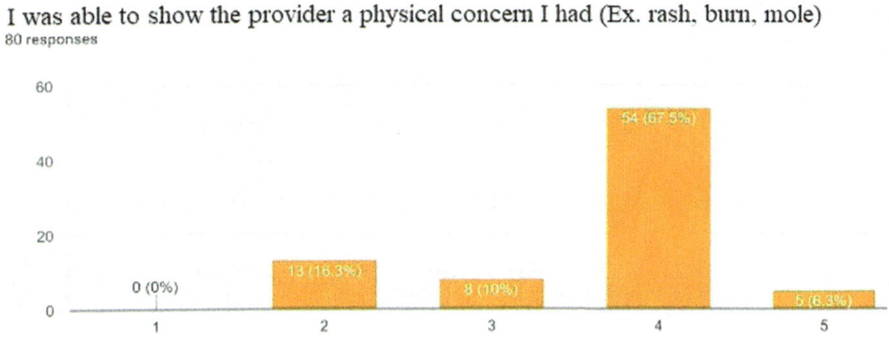

Fig. 7 Patient's being able to discuss physical concerns

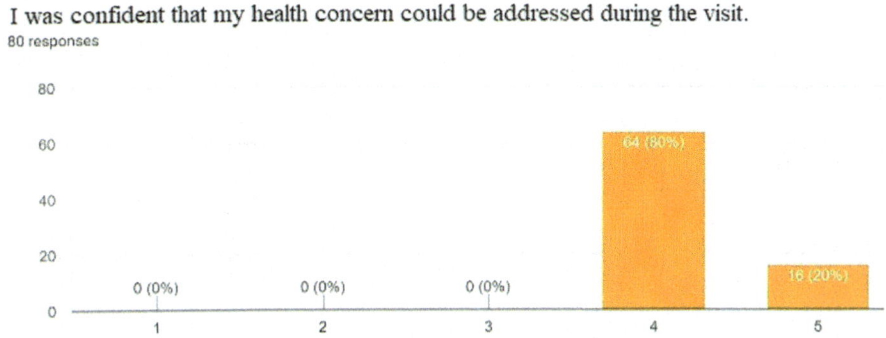

Fig. 8 Patient's level of confidence during the service

4.2 *Physician's Perception of Telehealth*

Figure 9 depicts that 14 health physicians agreed and 5 health physicians strongly agreed to the fact that telehealth enables them to deliver quality care for COVID-19-related care. In case of serious conditions for instance patients with critical COVID-19, cases needed immediate care from the physician and telehealth could not suffice in those cases. Similarly, Fig. 10 interprets that 52% of physicians agree and 16% of physicians strongly agree to the fact that their patients have better access to care since the telehealth practice began. Besides, 68% of the respondents responded positively to the advent of telehealth. Likewise, 20% and 12% disagree and are neutral about this, respectively. These 32% of physicians believe in face-to-face interactions as they majorly faced issues with technology.

Figure 11 depicts that 32% of physicians strongly disagree, 52% of the physicians disagree, and 8% of the physicians are neutral to the fact that they were able to deliver quality care for chronic disease management. It can be said that telehealth is useful as long as patients know about the disease and minimum care is needed. In the case of chronic disease, telehealth services could not give proper care to the patient.

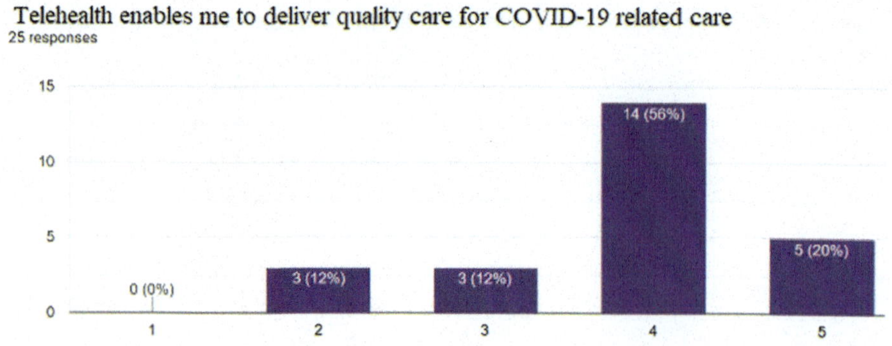

Fig. 9 Physician's satisfaction level in providing the service

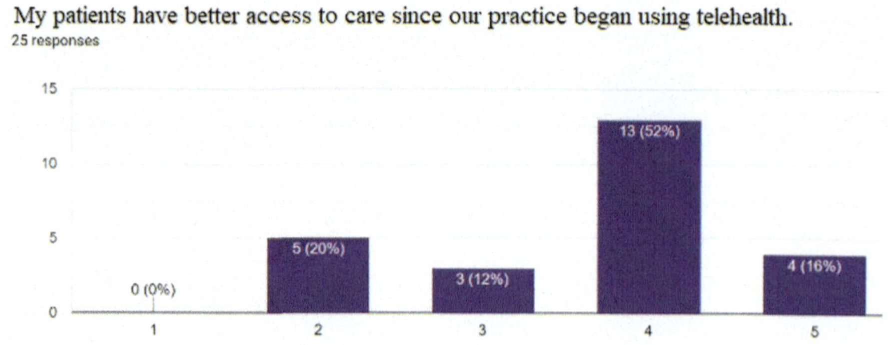

Fig. 10 Physician's perception of their patient's experience

Only 8% of the physicians agree to this, and they were able to deliver proper care to their patients as the patient's family was able to set up all that was needed for the patient in their home. This kind of situation is seen rarely in a developing country. To provide care to a larger population suffering from chronic disease, face-to-face interaction of physician and patient is needed. Similarly, Fig. 12 depicts that the shift from face-to-face interaction to online interaction was smooth for the physicians. This shows growth in the use of artificial intelligence during the pandemic. 44% of the physicians agree and 56% of physicians strongly agree with this.

From Fig. 13, it can be interpreted that the majority of the physicians' financial conditions improved by using telehealth as they were able to reach patients in different states through the online mode of remedy. Likewise, Fig. 14 interprets that according to the physicians, their patients well received the transition of the mode of treatment from face-to-face to the virtual mode of remedy. This shows that patients also learned and got acquainted with remote health services.

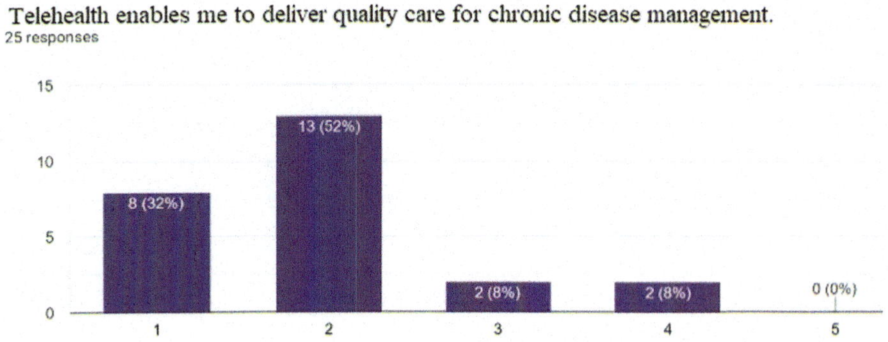

Fig. 11 Physician's satisfaction level in chronic disease

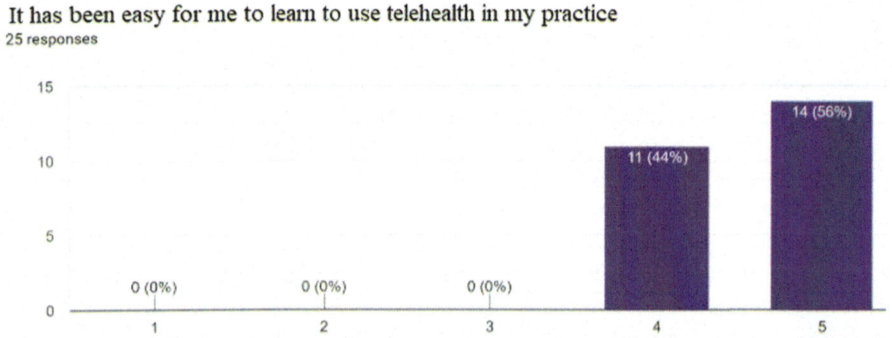

Fig. 12 Ease of understanding of telehealth in physician's practice

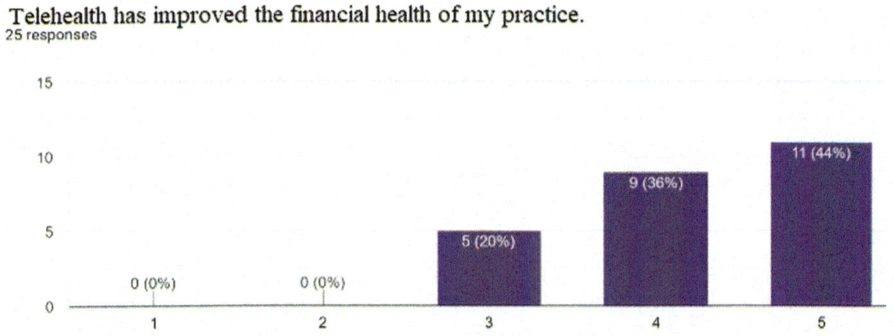

Fig. 13 Financial improvement of physicians after using telehealth

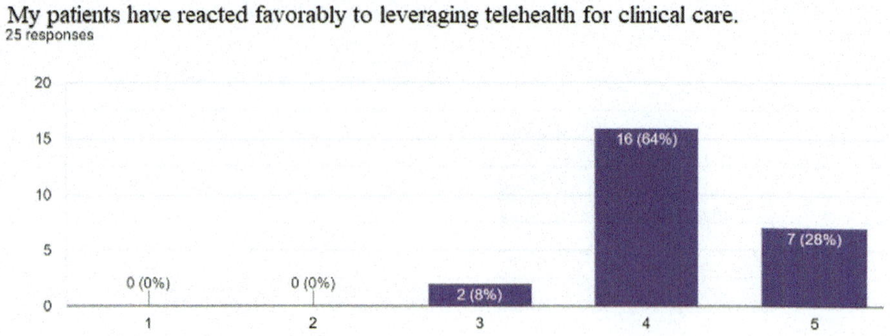

Fig. 14 Favor in leveraging telehealth

4.3 Data Interpretation on Patient's Response

The case processing summary that appears in front and tells about the cases is discussed in this section. In this case, the total number of valid cases is $N = 80$, whereas missing cases is $N = 2$. In total, it is $N = 82$. The valid cases comprise 97.6%, whereas missing cases comprise 2.4%. Only those cases which have non-missing values for both variables, that is, "I will continue to use telehealth services in the future" and "I was confident that my health concern could be addressed during the visit" are considered as valid cases.

Further, chi-square test is carried out and is presented in Table 1. In this case, the value of the chi-square statistic is 30.965. The "p-value" is taken to be significant if it is either less than or equal to the assumed alpha value which in this case is 0.05 ($\alpha = 0.05$, level of significance = 95%). It can be visualized that the p-value, which is visible in the first row of the asymptotic significance (2-sided) column, is 0.000. Since the p-value is lesser than the α value, the null hypothesis is rejected. It indicates that both the variables, "I will continue to use telehealth services in the future" and "I was confident that my health concern could be addressed during the visit", are associated with each other thereby making the result significant. So,

Table 1 Chi-square test for patient's response

	Value	Degree of freedom	Asymptotic significance (2-sided)
Pearson Chi-square	30.965	2	0.000
Likelihood ratio	29.027	2	0.000
Linear by linear association	14.733	1	0.000
N of valid cases	80		

Table 2 Chi-square test for physician's response

	Value	Degree of freedom	Asymptotic significance (2-sided)
Pearson Chi-square	11.883	2	0.003
Likelihood ratio	15.302	2	0.000
Linear by linear association	10.859	1	0.001
N of valid cases	25		

the null hypothesis is rejected which states that there is no significant relationship between the variables. Simultaneously, the alternate hypothesis is accepted which gives a significant association between both the variables as mentioned above.

4.4 Data Interpretation on Physician's Response

The case processing summary that appears in front and tells about the cases is discussed in this section. In this case, the total number of valid cases is $N = 25$, whereas missing cases is $N = 1$. In total, it is $N = 26$. The valid cases comprise 96.2%, whereas missing cases comprise 3.8%. Only those cases which have non-missing values for both variables that is, "I am personally motivated to increase the use of telehealth in my practice" and "telehealth has improved the financial health of my practice".

Further, chi-square test is carried out and is presented in Table 2. In this case, the value of the chi-square statistic is 11.883. The "*p*-value" is taken to be significant if it is either less than or equal to the assumed alpha value which in this case is 0.05 ($\alpha = 0.05$, level of significance = 95%). It can be visualized that the *p*-value, which is visible in the first row of the asymptotic significance (2-sided) column, is 0.003. Since the *p*-value is lesser than the α value, the null hypothesis is rejected. It indicates that both the variables, "I am personally motivated to increase the use of telehealth in my practice" and "telehealth has improved the financial health of my practice", are associated with each other thereby making the result significant. So, the null hypothesis is rejected which states that there is no significant relationship between the variables. Simultaneously, the alternate hypothesis is accepted which gives a significant association between both the variables as mentioned above.

5 Conclusion

The government has recognized telemedicine as a critical tool for providing adequate and improved medical care. The revolutionary COVID-19 epidemic has significantly altered how doctors treat patients. Furthermore, as a result of the aim to fatter the transmission curve, the focus is now an emphasis on infection prevention via quarantine and socialization distancing. Currently, health professionals who engage in patient care are utilizing telemedicine and e-health applications. However, telemedicine has not always been accepted as a viable treatment option for patients since e-health platforms do not give essential information from virtual consultations and diagnostics. Recent legislative developments throughout the world have enabled the use of popular social video chat programs like Microsoft Teams, Zoom, Skype, Google Duo, FaceTime, and others for patient consultations via mobile devices or tablets. A stable network connection as well as reliable IT assistance are also necessary. A digital examination can be performed by physicians with the addition of supplementary add-ons such as a digital stethoscope. However, to make use of the capabilities of telemedicine and e-health platforms in creating a seamless patient experience, dedicated staff training and education are necessary. Many countries have recognized telemedicine as a critical tool for providing adequate and equitable medical care. As a result, it is incorporated into national healthcare programs defined by several countries. To attain the same operational standards, all relevant stakeholders must be involved and reach agreements. Telemedicine is made up of numerous systems that are linked together by pre-defined data formats and protocols. Telemedicine has grown in popularity and has great potential for improving access to health care, boosting patient disease control, and facilitating healthcare monitoring. Although the future seems promising, more study is needed to find the best ways to incorporate telemedicine, particularly remote monitoring, into normal clinical treatment. A strong political advocacy message is that policy reforms are required to address regulatory and reimbursement problems.

References

1. Russi, C.S., Heaton, H.A., Demaerschalk, B.M.: Emergency medicine telehealth for COVID-19: minimize front-line provider exposure and conserve personal protective equipment. Mayo Clin. Proc. **95**(10), 2065–2068 (2020)
2. Moazzami, B., Razavi-Khorasani, N., Moghadam, A.D., Farokhi, E., Rezaei, N.: COVID-19 and telemedicine: immediate action required for maintaining healthcare providers well-being. J. Clin. Virol. **126**, 104345 (2020)
3. Mahnke, C.B., Jordan, C.P., Bergvall, E., Person, D.A., Pinsker, J.E.: The Pacific Asynchronous TeleHealth (PATH) system: review of 1,000 pediatric teleconsultations. Telemed. e-Health **17**(1), 35–39 (2011)
4. Khemapech, I., Sansrimahachai, W., Toachoodee, M.: Telemedicine-meaning, challenges and opportunities. Siriraj Med. J. **71**(3), 246–252 (2019)
5. Craig, J., Petterson, V.: Introduction to the practice of telemedicine. J. Telemed. Telecare **11**(1), 3–9 (2005)
6. Leite, H., Gruber, T., Hodgkinson, I.R.: Flattening the infection curve-understanding the role of telehealth in managing COVID-19. Leadersh. Health Serv. **33**(2), 221–226 (2020)
7. Leite, H., Hodgkinson, I.R., Gruber, T.: New development: 'healing at a distance'-telemedicine and COVID-19. Public Money . Manag. **40**(6), 483–485 (2020)
8. Serper, M., Volk, M.L.: Current and future applications of telemedicine to optimize the delivery of care in chronic liver disease. Clin. Gastroenterol. Hepatol. **16**(2), 157–161 (2018)

9. Perednia, D.A., Allen, A.: Telemedicine technology and clinical applications. J. Am. Med. Assoc. **273**(6), 483–488 (1995)
10. AlDossary, S., Martin-Khan, M.G., Bradford, N.K., Armfield, N.R., Smith, A.C.: The development of a telemedicine planning framework based on needs assessment. J. Med. Syst. **41**, 1–9 (2017)
11. Smith, A.C., Bensink, M., Armfield, N., Stillman, J., Caffery, L.: Telemedicine and rural health care applications. J. Postgrad. Med. **51**(4), 286–293 (2005)
12. Polisena, J., Coyle, D., Coyle, K., McGill, S.: Home telehealth for chronic disease management: a systematic review and an analysis of economic evaluations. Int. J. Technol. Assess. Health Care **25**(3), 339–349 (2009)
13. Vidal-Alaball, J., Acosta-Roja, R., Hernández, N.P., Luque, U.S., Morrison, D., Pérez, S.N., Perez-Llano, J., Verges, A., S.,& Seguí, F. L.: Telemedicine in the face of the COVID-19 pandemic. Prim. Care **52**(6), 418–422 (2020)
14. Shokri, T., Lighthall, J.G.: Telemedicine in the era of the COVID-19 pandemic: implications in facial plastic surgery. Facial Plastic Surg. Aesthet. Med. **22**(3), 155–156 (2020)
15. Wright, J., Purdy, B., McGonigle, S.: E-care: a viable option for remote ambulatory oncology nursing care. Oncol. Nurs. Forum **33**(2), 402–403 (2006)
16. Pecchia, L., Piaggio, D., Maccaro, A., Formisano, C., Iadanza, E.: The inadequacy of regulatory frameworks in time of crisis and in low-resource settings: personal protective equipment and COVID-19. Heal. Technol. **10**(6), 1375–1383 (2020)
17. Roine, R., Ohinmaa, A., Hailey, D.: Assessing telemedicine: a systematic review of the literature. Can. Med. Assoc. J. **165**(6), 765–771 (2001)

Computational Intelligence in Electronic Health Record

Classification of Cardiovascular Disease Information System Using Machine Learning Approaches

Subham Kumar Padhy, Anjali Mohapatra, and Sabyasachi Patra

Abstract Recent advancements in computational approaches have facilitated the storage and collection of medical data for accurate medical diagnosis. Various computational techniques are used to improve the accuracy of disease diagnosis and reduce the diagnosis time, and the mortality rate. Advanced learning methods need to be used to improve efficacy and clinical significance. Machine learning methods are widely used in healthcare systems for screening, risk identification, prediction, and decision-making for different diseases. The sample sizes, features, location of data collection, performance metrics, and applied machine learning techniques play a significant role in the results of the machine learning-based cardiovascular disease data classification. This chapter discusses the performance of various machine learning algorithms relating to cardiovascular disease. The evaluation is done using different performance matrices like accuracy, precision, and recall. A comparative study of individual results of the models like support vector machine, K-nearest neighbor, Naïve Bayes, decision tree, random forest, and artificial neural network for predicting cardiovascular disease is carried out. It has been observed that random forest has a better accuracy of 0.92 when compared with other machine learning models.

1 Introduction

Cardiovascular disease has become a critical global human problem. According to the world health organization (WHO), one in every three deaths is due to Coronary Artery Disease (CAD) [1]. Health disorders may depend upon different factors such

S. K. Padhy (✉) · A. Mohapatra · S. Patra
Department of Computer Science and Engineering, IIIT, Bhubaneswar, Odisha, India
e-mail: c120006@iiit-bh.ac.in

A. Mohapatra
e-mail: anjali@iiit-bh.ac.in

S. Patra
e-mail: sabyasachi@iiit-bh.ac.in

D. P. Acharjya and K. Ma (eds.), *Computational Intelligence in Healthcare Informatics*, Studies in Computational Intelligence 1132,
https://doi.org/10.1007/978-981-99-8853-2_13

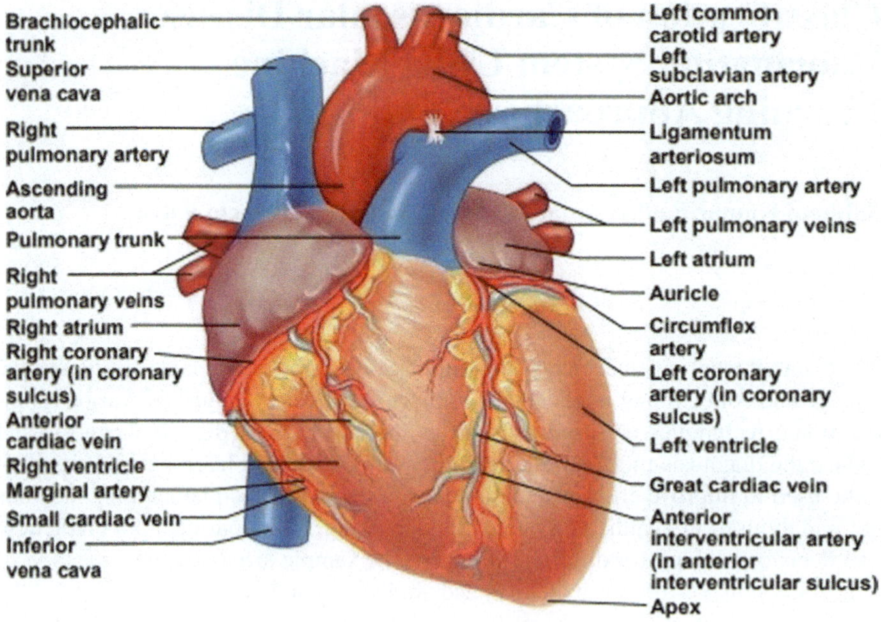

Fig. 1 Function of heart [9]

as injuries, maternal, parental, and nutritional conditions. The CAD has the highest mortality rate globally, followed by stroke and chronic obstructive pulmonary disease. The WHO estimates that the total extent of deaths due to CAD is almost the same as the totality of deaths in line with the rest of the top six diseases globally [1].

Machine learning techniques are used for decision-making and predicting cardiovascular diseases. These methods convert the raw healthcare data into informational data [2]. The proper integration of prediction models can predict cardiovascular disease more accurately. So, predicting the disease with satisfactory accuracy and encouraging results has become challenging for researchers [3]. The pre-processing of clinical data into well-organized learning models can classify the dataset quickly and accurately predict cardiovascular disease [4].

Cardiovascular disease is a condition that affects the heart and blood arteries. Low-density lipoprotein or bad cholesterol, an absence of high-density lipoprotein or good cholesterol, lack of physical exercise, fat consumption, hypertension, and obesity cause CAD, hypertension, and stroke [5]. 17.9 million public died from the cardiovascular disease per year, accounting for 31% of all fatalities worldwide [6, 7]. Stroke and heart attack are responsible for 85% of all deaths in this group. Insufficient, timely preventative measures and risk factors are causing the mortality rate of cardiovascular disease to skyrocket [8]. Figure 1 depicts the basic functionality of heart [9].

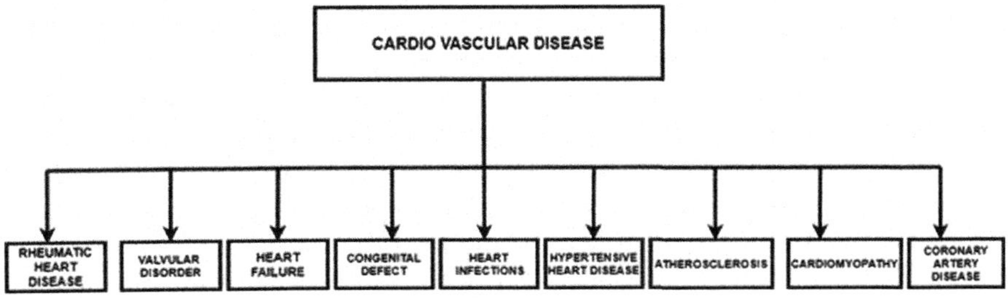

Fig. 2 Types of cardiovascular disease

The human heart contains four chambers such as the right atrium, right ventricle, left atrium, and left ventricles [10]. The de-oxygenated blood enters the right atrium by the inferior and superior vena cava and goes through the tricuspid valve from the right atrium to the right ventricle. The blood then flows to the lungs via the pulmonary artery [9]. The oxygenated blood returns to the left atrium from the lungs through the pulmonary vein and to the left ventricle by the bicuspid valve and flows by the aorta to the body from the left ventricle [11].

Various types of cardiovascular diseases are rheumatic heart disease, valvular disorder, heart failure, and congenital defect. Rheumatic heart disease is the most frequent kind of heart disease among adults under the age of 25, affecting mostly children and youths in low and middle-income nations. The narrowing of the tricuspid valve is known as tricuspid stenosis. Similarly, the narrowing of the bicuspid valve or the mitral valve is known as bicuspid stenosis or mitral stenosis. This tricuspid stenosis or mitral stenosis leads to rheumatic heart disease [12]. Similarly, any valve in the heart can be damaged or diseased, resulting in valvular heart disease. In valvular disorder, the incomplete closure of either the bicuspid or the tricuspid valve causes a defect in the heart's valves [13]. The heart consists of the sinoatrial node (S.A. node), and the atrioventricular node (A.V. node). By electrical impulses, the heart is regulated. The atrium contracts from the sinoatrial node and forces the blood to enter the ventricle. The atrioventricular node picks up the electrical signal and directs it to the purkinje fibers in the ventricular wall, which causes the ventricles to contract. Any irregularity with the electrical impulses causes a chronic condition. Therefore, the required quantity of blood does not pump inside the heart, causing heart failure [14]. At the fetus stage of a newborn infant, inadequate closure of the atrial septum or ventricular septum produces a cardiac abnormality, leading to a congenital disability. African Americans have a greater infant death rate from congenital cardiac disease. The types of cardiovascular diseases are depicted in Fig. 2.

Similarly, other heart diseases include heart infections, hypertensive heart disease, atherosclerosis, cardiomyopathy, and coronary artery disease. Heart infections are generally caused due to bacteria, viruses, and parasites that may cause infections inside the heart. Hypertensive heart disease is caused due to excessive blood pressure, often known as hypertension. In atherosclerosis, the heart's muscle hardens due to the deposit of plaques. Likewise, in cardiomyopathy, the muscle of the heart known as the myocardium does not function normally. It is a genetic or hereditary disease that causes problems inside the heart's muscles. The formation of the plaques creates a blockage, and the heart muscle does not get enough oxygen. This symptom is also known as a myocardial infarction, which causes less blood and oxygen flow to the heart's muscles. It is recognized as the ischemic effect inside the heart [15]. Due to plaque build-up in the heart's arteries, coronary artery diseases are occurring. Major blood vessels get damaged by this type of disease [16].

The rest of the chapter is organized as follows. Following an introduction to cardiovascular diseases, Sect. 2 briefly discusses the machine learning algorithms relating to cardiovascular diseases. Section 3 briefly discusses on cardiovascular disease information system. Section 4 describes the exploratory data analysis. Performance measures are discussed in Sect. 5. Finally, a conclusion is drawn in Sect. 6.

2 Machine Learning for Cardiovascular Disease Classification

It is widely accepted that no single algorithm is better than the others in building and evaluating machine learning models. In supervised learning, the dataset is labeled, whereas in unsupervised learning occurrences are left unlabelled. Various machine learning approaches used for cardiovascular disease dataset classification are described below [17, 18].

Binary Logistic Regression: Each self-determining variable with every single instance of the data is increased by weight, and the consequence is given to the sigmoid function in the binary logistic regression. Then the sigmoid function converts the actual numbers into probabilities ranging from 0 to 1. The model predicts that the dependent variable for patients with CAD will be 1 if the values are greater than or equal to 0.5. For patients with CAD, the dependent variable is expected to be 1.0. The model predicts the need for the variable to be 0 if the values are less than or equal to 0.5 [19].

Decision Tree: A regression tree, also known as a decision tree, categorizes objects based on their variable values. Every node represents a variable in classification, and the respective node signifies a value that the node could accept. Instances are classified and arranged based on the importance of the variables, starting at the root node [10]. The tree's root node is the variable that best splits the data set. Internal nodes, also known as split nodes, are the decision-making components that choose to visit future nodes depending on numerous algorithms. The split procedure is completed when a user-defined requirement is fulfilled at the leaf [20].

In the decision tree, the root of the tree denotes condition 1. If the condition is satisfied, then move to another condition in the next level. However, if the condition fails, one will be considered only. In the next level for condition 2, both decision two and decision three exist, depending on condition 2, whether it satisfies or fails, as shown in Fig. 3.

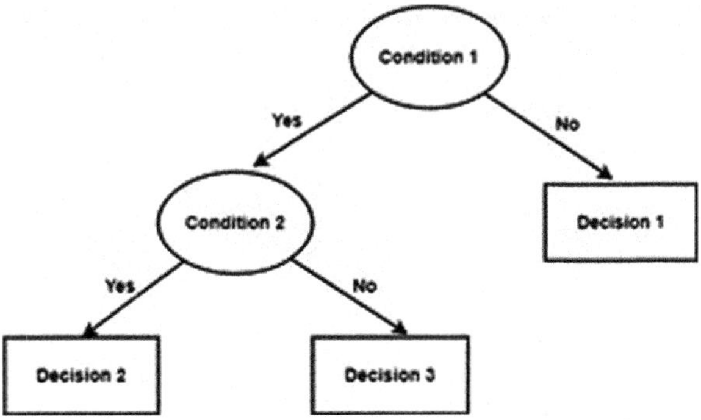

Fig. 3 A decision tree representation

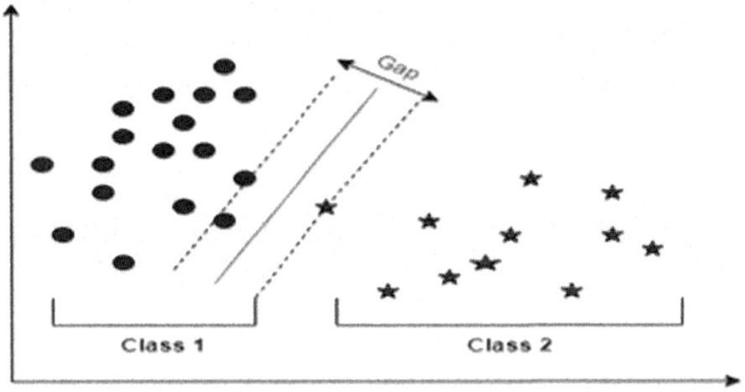

Fig. 4 A support vector machine representation

Random Forest: Random Forest is an ensemble model built on many regression trees, like those seen in a forest where each trained on a portion of the dataset's examples. The final predictions are derived by averaging each tree's forecasts, enhancing prediction accuracy when dealing with unknown data.

Support Vector Machine: In the support vector machine (SVM), classification is done based on the hyper-plane. n-dimensional space is considered to show particular data points, that divide the two classes [21]. New features are utilized to determine the class of a new instance. Figure 4 depicts that the problems arise in separating two groups with a line. The dots belong to class 1, and the stars represent in class 2. An ample margin space predicts the boundaries between the hyper-plane using the support vectors.

K-Nearest Neighbor: K-Nearest Neighbor (KNN) approach is based on a simple non-parametric algorithm. It makes no assumptions about the original data distribution. This method is based on the distance principle. Objects having similar features are more likely to be found in a nearby location. An uncategorized occurrence is assigned a label by examining the class of its nearest neighbor [22]. The label assigned to an object for $k = 3$, using the class of its nearest neighbor, is shown in Fig. 5.

Artificial Neural Network: An artificial neural network (ANN) is constructed by considering the functionality and arrangement of biological neural networks. The weights associated with neurons are fixed, and the network categorizes a different

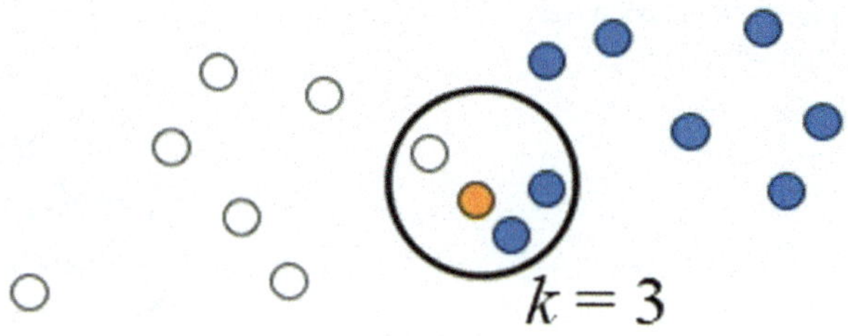

Fig. 5 A k-nearest neighbor representation

data group. In ANN, the blocks are separated into three groups: input that receives information to be processed; output that contains the processing results; and hidden are in the middle to identify input-output relation. The network model is trained initially using a balanced training dataset [23]. After that, the test dataset is utilized to objectively assess the last model appropriate to the training dataset.

3 Cardiovascular Disease Information System

Various heart disease datasets are discussed in this chapter, obtained from the UCI (the University of California, Irvine, CA) Center for the machine learning and intelligent systems repository. An overall view of the information system (datasets) is available in the repository is shown in Fig. 6. The Hungarian, the VA long beach, and Statlog datasets have fewer missing characteristics and more records than the Cleveland dataset. There exist 14 attributes and two classes in Cleveland, Hungarian, the VA long beach, and Statlog dataset containing 303, 294, 200, and 270 instances, respectively [24].

The following are the characteristics considered while analyzing and predicting heart disease. The different characteristics are age, sex, chest pain, resting bps, fasting blood sugar, resting ECG, thalac, cholesterol, old peak, and exang.

The age refers to patient's age whereas sex is either male or female. The gender is described as 1 if the patient is male and 0 if the patient is female. Different types of chest pain present in these features are typical (0), typical angina (1), non-angina pain (2), and asymptomatic (3). In resting bps, the blood pressure level at resting mode is calculated by mm/HG. In fasting blood sugar, the blood sugar levels on fasting $> 120mg/dl$ are symbolized as 1 and 0 otherwise. In resting ECG, the outcome of an

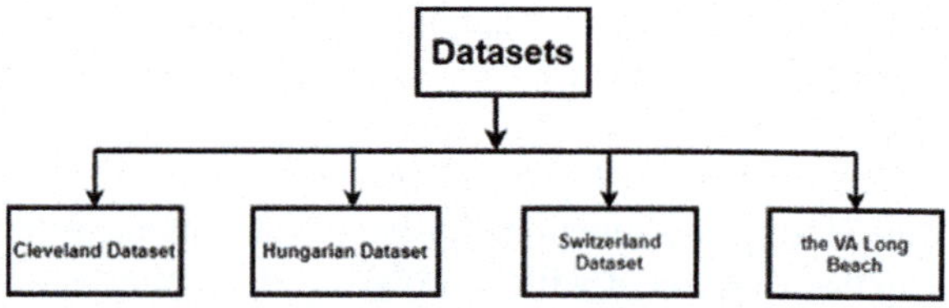

Fig. 6 An overview of cardiovascular disease information system

ECG of the patients is available with different values in this feature such as normal (0), abnormality in ST-T wave (1), and left ventricular hypertrophy (2). The feature thalac describes the patients having high heart rates. Cholesterol refers to serum cholesterol and is calculated for different patients. It is measured in mg/dl. In the old peak, compared to a resting state, exercise-induced ST-depression is observed. Exang refers to exercise angina. The notion 0 denotes exercise angina induced by exercise and 1 for Yes, respectively. The prediction of the target variable is based on two factors. '1' indicates that the patient is at risk for heart disease, whereas '0' indicates that the patient is healthy.

4 Exploratory Data Analysis

The UCI heart disease dataset consists of no null values, which are saved from converting the null values into some data or dropping the data from the dataset. A positive correlation between target and chest pain, thalac, and slope is observed in Fig. 7. A negative correlation between target and sex, observed, exang, ca, thal, old peak is observed. For positive correlation, chest pain is more probable to have heart disease. The correlation graph, scatter plots for data distribution, and different attributes of the dataset are shown in Fig. 7.

Computational Environment: The classification of cardiovascular disease datasets is implemented on a system that supports Jupiter notebook with Intel(R) Core (T.M.) i5-6700K CPU @ 4.00 GHz CPU, NVIDIA GeForce RTX GPU and 32 GB RAM. Anaconda, the pre-package Python distribution that contains all the necessary libraries and software, is installed. For file processing, dataset conversion, or convert-

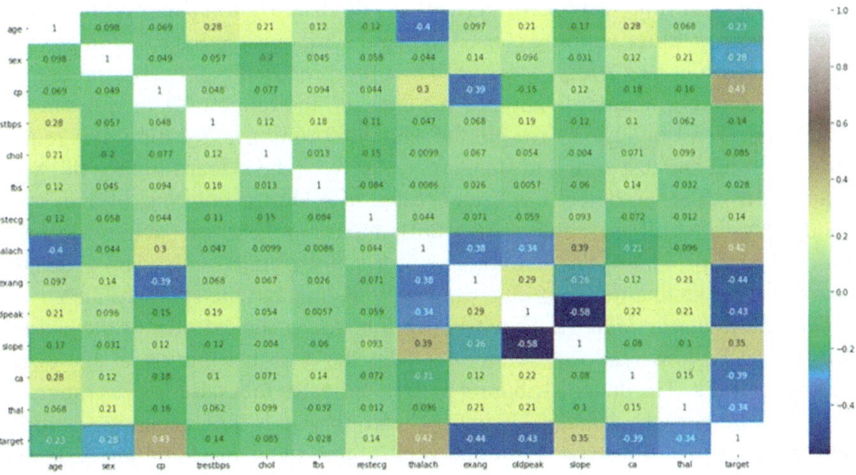

Fig. 7 Correlation graph

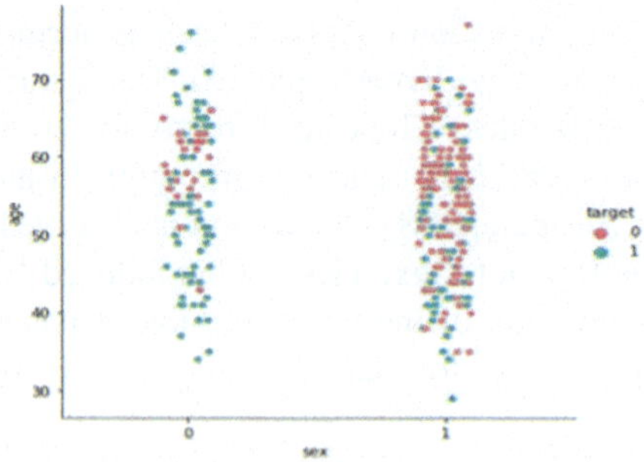

(*a*) *Age verces Sex*

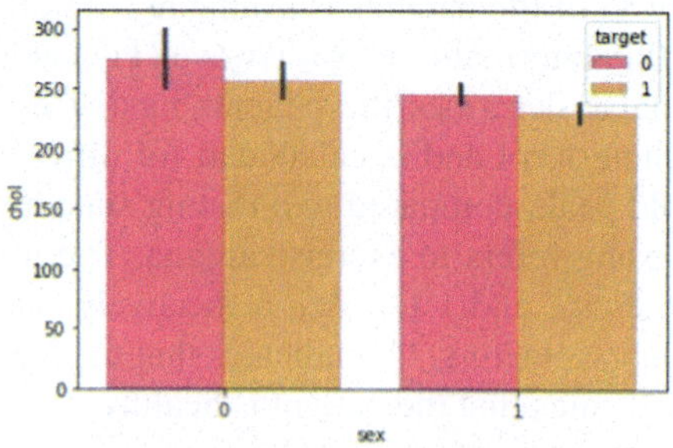

(*b*) *Cholesterol verces Sex*

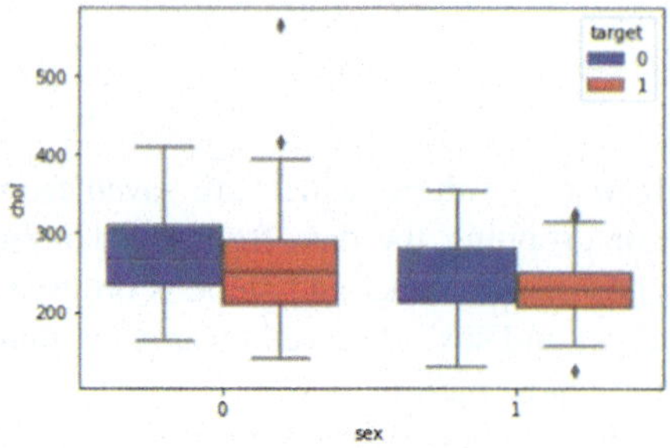

(*c*) *Cholesterol verces Sex in box plot*

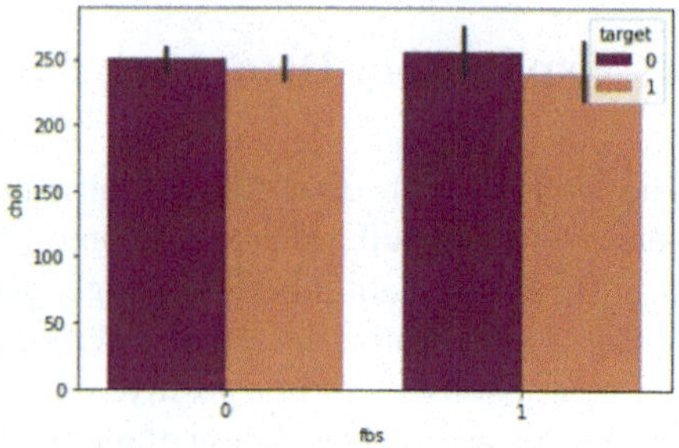

(*d*) *Fasting blood sugar verces Cholesterol*

Fig. 8 Data distribution pattern

ing into a data frame, linear algebra operations, additional libraries like pandas and NumPy are imported. Embedded plotting is done through Numpy, a library used for numerical extensions. Matplotlib, in general, contains bars, pies, lines, and scatter plots. At the same time, seaborn offers a variety of visualization patterns and provides exciting themes. Filter warning is used to avoid some unnecessary warnings.

Data Cleaning and Pre-processing: Data pre-processing converts the data into standard form. For standardization of the data set, the range of data value range is scaled from 0 to 1 or –1 to 1. To split the data into training and testing, it is required to import train and test split and standard scalar from the sklearn library of Python. In Standard scalar, an object has been created to scale the columns of age, resting blood pressure (trestbps), cholesterol (chol), maximum heart rate achieved (thalac), and old peaks present in the dataset. The fit transform function is used to scale the column values so that the algorithm can easily understand the standard data.

Data Distribution Pattern: The visualization among the features of the heart disease dataset is shown below. Figure 8a depicts the data distribution of age in comparison with sex. The data distribution pattern indicates that female ages range from 40 to 70. The portion of males having no heart disease is larger than the portion of females having heart disease. From the value count, the dataset contains 207 males and 96 females. Figure 8b depicts the data distribution of cholesterol in comparison with sex. The bar plot consists of cholesterol level on the Y-axis and gender on

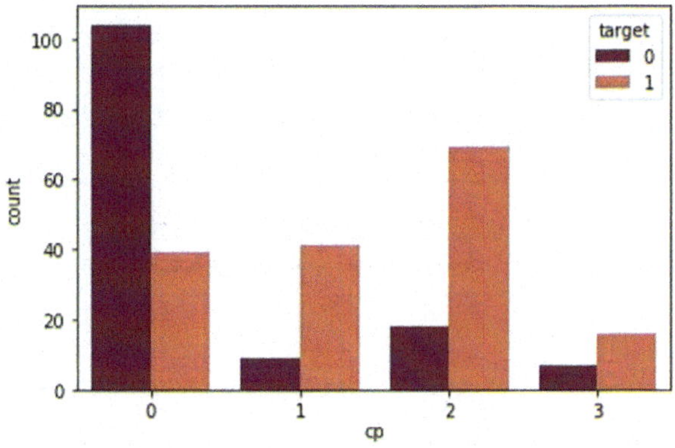

Fig. 9 Shows analysis of chest pain

the X-axis. It shows that the rise in cholesterol levels increases the chance of heart disease.

Figure 8c depicts the data distribution of cholesterol versus sex in the box plot. This figure is drawn by taking sex in X-axis and cholesterol on Y-axis. Concerning cholesterol, the blue box indicates that patients do not have any heart disease, and the red box indicates that patients have heart disease. Here 0 represents females and 1 represents males. Similarly, Fig. 8d depicts the data distribution of fasting blood sugar versus cholesterol. The cholesterol level is plotted concerning fasting blood sugar level, where 0 symbolizes the patients not having heart disease, and 1 symbolizes the patient is having heart disease, respectively.

The chest pain data is depicted in Fig. 9. It shows that level 2 chest pain, i.e., typical angina-type chest pain patients, are more likely to have a heart attack in the near future. The data set contains 87 people who have level 2 typical angina chest pain.

5 Performance Measures

The dataset is split into training and testing set by considering 70 percentage of data for training and the rest for testing. The input features of training data are saved in P and the target value is saved in Z. P train, P test, Z train, and Z test are the objects created to store the input and the target features for training and testing. After the preparation of final data P train = 2756, P test = 1183, Z train = 212 and Z test = 91.

The model evaluation is done based on the validation metric such as cross validation score, confusion matrix, ROC-AUC curve, plotting precision recall curve, sensitivity and specificity, classification error, log loss, and Mathew correlation coefficient. Cross validation is a statistical approach for estimating the skill of machine learning models as shown in Fig. 10.

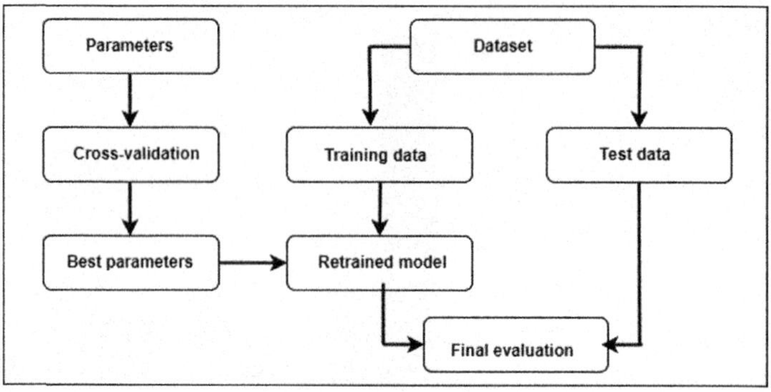

Fig. 10 Demonstrate the cross validation of the dataset

Table 1 Accuracy obtained from cross validation score

Classifiers	Accuracy
KNN	0.85
SVM	0.852
NB	0.84
D-TREE	0.86
RANDOM-FOREST	0.92
ANN	0.86

Manual evaluation of different models using hyper-parameters, increases the chance of overfitting the test data until the estimator performs optimally. The data samples are split into training, validation, and test sets to solve this overfitting problem. The model is validated using the data of training the model. The experiments are performed with the test data after successful validation. The accuracy obtained from cross validation score is shown in Table 1.

Confusion Matrix: Confusion matrix is used to address challenges in binary classification. A confusion matrix is a typical approach to illustrate how well a classifier works on the test data for which the absolute values are recognized in terms of classification. This matrix is based on dualistic probable predicted classes 'YES' or 'NO'. In order to predict the presence of heart diseases, 'YES' would mean patients have the disease, and 'NO' they do not have any heart disease. The predicted analysis of the confusion matrix is TP, TN, FP, and FN, where TP refers to true positive and TN refers to true negative. Similarly, FP refers to false positive and FN refers to false negative. True positive and true negative are the cases in which the model is correctly classified, but false positive and false negative are the circumstances to increase the misclassification rate.

The precision P is defined in Eq. 1 whereas recall R is defined in Eq. 2. Out of all the instances in the dataset, the percentage of accurately predicted instances is known as accuracy. The accuracy A is defined in Eq. 3. Likewise, the $F1$ score is the average reciprocals of precision and recall [25]. It is defined in Eq. 4

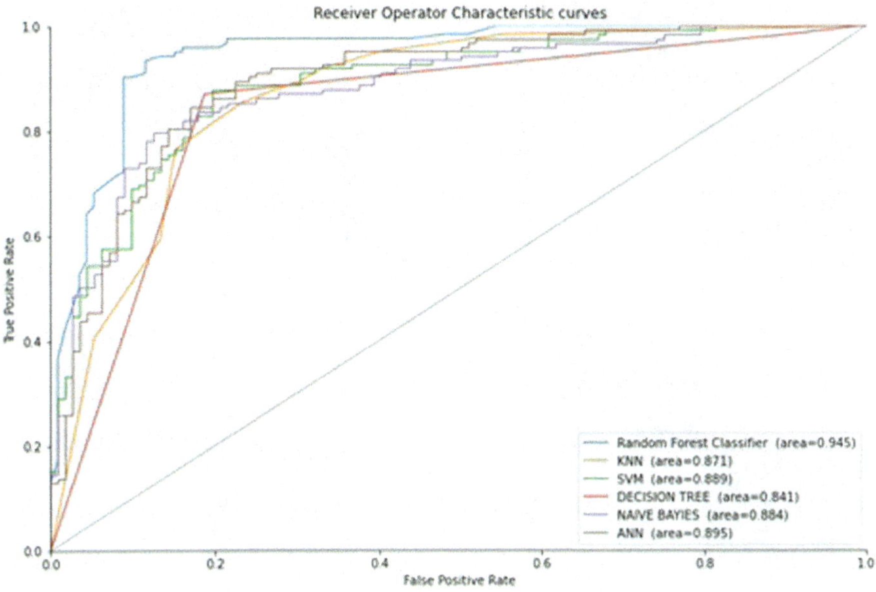

Fig. 11 Receiver operator characteristic curve

$$Precision(P) = \frac{TP}{TP + FP} \tag{1}$$

$$Recall(R) = \frac{TP}{TP + FN} \tag{2}$$

$$Accuracy(A) = \frac{TP + TN}{TP + FP + TN + FN} \tag{3}$$

$$F1score = 2\left(\frac{P \times R}{P + R}\right) \tag{4}$$

AUC and ROC Curve: For multiclass classification, the classifier estimates the likelihood of a data sample fitting into several classes. Hence, an additional procedure is required to compute AUC for multiclass problems. The decision tree may produce one such probability to judge classification performance. Each class produces a hyper-surface with a maximum volume of 1. The Area Under the Curve (AUC), measures the likelihood that a model would randomly rate a positive occurrence higher than a negative occurrence. AUC-ROC (Area Under the Curve Receiver Operating Characteristic) plots is used to visually equate the performance of the model as shown in Fig. 11.

Log Loss and Matthew's Correlation Coefficient: Log loss is a prominent way to assess inaccuracy in machine learning applications. As reducing and finding mistakes

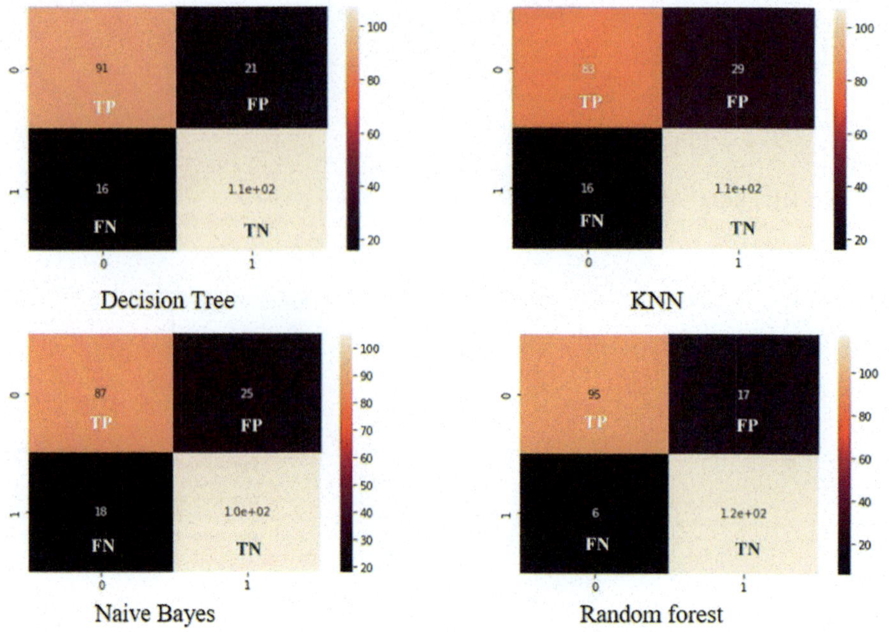

Fig. 12 Confusion matrix of various models

and learning initiative failures enhances accuracy, a machine learning process cannot work successfully without them. This total includes the model's performance. When alternative measures are better suited for specialized research, log loss is a valuable and fundamental way to compare two models. The Matthew's Correlation Coefficient (MCC) is used to assess the validity of binary classifications. Using Eq. 5 below, the MCC may be calculated directly from the confusion matrix.

$$MCC = \frac{TP \times TN - FP \times FN}{\sqrt{(TP + FP)(TP + FN)(TN + FP)(TN + FN)}} \tag{5}$$

The confusion matrix of various machine learning models is depicted in Fig. 12. It can be observed that random forest has the highest true positive value. Table 2 refers to the performance measure of various machine learning models with reference to cardiovascular disease. It indicates that random forest has the highest accuracy, precision, and minimal log loss.

Table 2 Comparison of performance of different models

Model	Accuracy	Precision	Recall	F1 score	MCC	Log loss	ROC
KNN	0.808	0.786	0.869	0.826	0.618	6.61	0.80
SVM	0.825	0.801	0.886	0.841	0.652	6.02	0.822
NB	0.817	0.807	0.853	0.830	0.633	6.319	0.815
Decision tree	0.829	0.826	0.853	0.840	0.658	5.879	0.828
Random forest	0.890	0.860	0.950	0.90	0.790	3.670	0.890
ANN	0.825	0.80	0.886	0.841	0.790	6.026	0.822

6 Conclusion

This study discusses and analyzes various classification techniques with cardiovascular disease datasets. The attribute selection technique is crucial for accurately classifying the heart disease data sets. The characteristics of feature selection are based on the training data. Various performance evaluation metrics are discussed to measure the performance of the machine learning algorithms by considering the training and testing datasets. The accuracy lies between 0.84 and 0.92. It has been observed that random forest has better accuracy as compared to other machine learning algorithms. Further research in this area can enhance feature selection techniques. Advanced data pre-processing and data mining techniques can be associated with machine learning algorithms to improve classification performance.

References

1. Alizadehsani, R., Abdar, M., Roshanzamir, M., Khosravi, A., Kebria, P.M., Khozeimeh, F., Nahavandi, S., Sarrafzadegan, N., Acharya, U.R.: Machine learning-based coronary artery disease diagnosis: a comprehensive review. Comput. Biol. Med. **111**, 103346 (2019)
2. Alić, B., Gurbeta, L., Badnjević, A.: Machine learning techniques for classification of diabetes and cardiovascular diseases. In: Proceedings of the 6th IEEE Mediterranean Conference on Embedded Computing, pp. 1–4 (2017)
3. Dimopoulos, A.C., Nikolaidou, M., Caballero, F.F., Engchuan, W., Sanchez-Niubo, A., Arndt, H., Ayuso-Mateos, J.L., Haro, J.M., Chatterji, S., Georgousopoulou, E.N., Pitsavos, C., Panagiotakos, D.B.: Machine learning methodologies versus cardiovascular risk scores, in predicting disease risk. BMC Med. Res. Methodol. **18**, 1–11 (2018)
4. Rubini, P.E., Subasini, C.A., Katharine, A.V., Kumaresan, V., Kumar, S.G., Nithya, T.M.: A cardiovascular disease prediction using machine learning algorithms. Ann. Romanian Soc. Cell Biol. **25**(2), 904–912 (2021)
5. Wallert, J., Tomasoni, M., Madison, G., Held, C.: Predicting two-year survival versus non-survival after first myocardial infarction using machine learning and Swedish national register data. BMC Med. Inform. Decis. Mak. **17**(1), 1–11 (2017)
6. Yahaya, L., Oye, N.D., Garba, E.J.: A comprehensive review on heart disease prediction using data mining and machine learning techniques. Am. J. Artif. Intell. **4**(1), 20–29 (2020)

7. Ekız, S., Erdoğmuş, P.: Comparative study of heart disease classification. In: Proceedings of the Electric Electronics, Computer Science, Biomedical Engineerings' Meeting, pp. 1–4 (2017)
8. Kanikar, P., Shah, D.R.: Prediction of cardiovascular diseases using support vector machine and Bayesien classification. Int. J. Comput. Appl. **156**(2), 9–13 (2016)
9. Hunter, P.J., Smaill, B.H.: The analysis of cardiac function: a continuum approach. Prog. Biophys. Mol. Biol. **52**(2), 101–164 (1988)
10. Voorhees, A.P., Han, H.C.: Biomechanics of cardiac function. Comprehens. Physiol. **5**(4), 1623 (2015)
11. Ghumbre, S.U., Ghatol, A.A.: Heart disease diagnosis using machine learning algorithm. In: Proceedings of the International Conference on Information Systems Design and Intelligent Applications, pp. 217-225 (2012)
12. Katzenellenbogen, J.M., Ralph, A.P., Wyber, R., Carapetis, J.R.: Rheumatic heart disease: infectious disease origin, chronic care approach. BMC Health Serv. Res. **17**, 1–16 (2017)
13. Solomon, M.D., Tabada, G., Allen, A., Sung, S.H., Go, A.S.: Large-scale identification of aortic stenosis and its severity using natural language processing on electronic health records. Cardiovascul. Digital Health J. **2**(3), 156–163 (2021)
14. Sowmiya, C., Sumitra, P.: Analytical study of heart disease diagnosis using classification techniques. In: Proceedings of the IEEE International Conference on Intelligent Techniques in Control, Optimization and Signal Processing, pp. 1–5 (2017)
15. Ovreiu, M., Simon, D.: Biogeography-based optimization of neuro-fuzzy system parameters for diagnosis of cardiac disease. In: Proceedings of the 12th Annual Conference on Genetic and Evolutionary Computation, pp. 1235–1242 (2010)
16. Acharya, U.R., Meiburger, K.M., Koh, J.E.W., Vicnesh, J., Ciaccio, E.J., Lih, O.S., Tan, S.K., Aman, R.R.A.R., Molinari, F., Ng, K.H.: Automated plaque classification using computed tomography angiography and Gabor transformations. Artif. Intell. Med. **100**, 101724 (2019)
17. Akella, A., Akella, S.: Machine learning algorithms for predicting coronary artery disease: efforts toward an open source solution. Future Sci. OA **7**(6), FSO698 (2021)
18. Meena, G., Chauhan, P.S., Choudhary, R.R.: Empirical study on classification of heart disease dataset-its prediction and mining. In: Proceedings of the IEEE International Conference on Current Trends in Computer, Electrical, Electronics and Communication, pp. 1041–1043 (2017)
19. Jinjri, W.M., Keikhosrokiani, P., Abdullah, N.L.: Machine learning algorithms for the classification of cardiovascular disease-A comparative study. In: Proceedings of the IEEE International Conference on Information Technology, pp. 132–138 (2021)
20. Bhavsar, K.A., Abugabah, A., Singla, J., AlZubi, A.A., Bashir, A.K.: A comprehensive review on medical diagnosis using machine learning. Comput. Mater. Contin. **67**(2), 1997–2014 (2021)
21. Nguyen, T.N., Nguyen, T.H., Vo, D.D., Nguyen, T.D.: Multi-class support vector machine algorithm for heart disease classification. In: Proceedings of the 5th IEEE International Conference on Green Technology and Sustainable Development, pp. 137–140 (2020)
22. Udovychenko, Y., Popov, A., Chaikovsky, I.: Ischemic heart disease recognition by k-NN classification of current density distribution maps. In: Proceedings of the 35th IEEE International Conference on Electronics and Nanotechnology, pp. 402–405 (2015)
23. Khemphila, A., Boonjing, V.: Heart disease classification using neural network and feature selection. In: Proceedings of the 21st IEEE International Conference on Systems Engineering, pp. 406–409 (2011)
24. Meshref, H.: Cardiovascular disease diagnosis: a machine learning interpretation approach. Int. J. Adv. Comput. Sci. Appl. **10**(12), 258–269 (2019)
25. Princy, R.J.P., Parthasarathy, S., Jose, P.S.H., Lakshminarayanan, A.R., Jeganathan, S.: Prediction of cardiac disease using supervised machine learning algorithms. In: Proceedings of the 4th IEEE International Conference on Intelligent Computing and Control Systems, pp. 570–575 (2020)

Automatic Edge Detection Model of MR Images Based on Deep Learning Approach

J. Mehena and S. Mishra

Abstract In medical imaging automatic computer-based diagnosis is taken into account because the most difficult issues within the area of medical image process. Pattern recognition, machine learning, and deep learning approaches primarily influence the problem of automatically finding a decision to detect the fine edges of medical images. Here, deep learning technique is proposed to seek out fine edges automatically of the medical MR images corrupted by noises. Detection based on deep learning approach in medical image processing is employed to detect certain structures or organs and delimit regions of interest from the medical images. The performance of the method proposed in this chapter is evaluated with metrics like peak signal-to-noise ratio (PSNR), structural similarity index measure (SSIM), and edge keeping index (EKI). Experimental results show the approach proposed in this research work that exhibits far better performance than the prevailing techniques as well as generalized soft-computing approaches.

1 Introduction

Edge detection of medical images plays a very crucial role for visual perception of the human body organs. This approach is one of the necessary steps prior to image segmentation and determines the results of the ultimate processed medical images. Conventionally, edge is identified in accordance with the algorithms like Prewitt, Sobel, and Laplacian of Gaussian (LOG) operator; however, in practice, these algorithms be a part of the high-pass filtering process that are not acceptable for

J. Mehena (✉)
Department of Electronics & Telecommunication Engineering, DRIEMS, Cuttack, Odisha, India
e-mail: jibananandamehena@gmail.com

S. Mishra
Department of Electronics and Communication Engineering, Adama Science and Technology University, Adama, Ethiopia
e-mail: satyasis.mishra@astu.edu.et

D. P. Acharjya and K. Ma (eds.), *Computational Intelligence in Healthcare Informatics*, Studies in Computational Intelligence 1132,
https://doi.org/10.1007/978-981-99-8853-2_14

noisy edge detection [1, 2]. In reality, digital images consist of boundaries, shadows of objects, and noise. Hence, that may be tough to differentiate the edges precisely from the noise.

Soft computing is a rising discipline rooted in a group of technologies that aim to use the tolerance for impreciseness and uncertainty in achieving solutions to complicated problem. The principal elements of soft computing are fuzzy logic and neural computing [3, 4]. The technique based on fuzzy logic and neural network will influence the uncertainty and abnormality in digital image processing in a more effective manner as related to the generalized techniques. Once this technique is used for the detection of edges, it provides more robust outcomes related to the generalized techniques like Prewitt, Robert, Sobel, and canny edge detection technique [5, 6]. The character of edges is not constant failing which some of the edges are not detected by the process of detection. The soft-computing approach compromises by permitting the values with none such constraints [7]. This method has rules based on if-then statements and has easily implemented structure within the framework. Simply by adding or changing some of fuzzy rule results are often modified and few approaches that require prior knowledge for which training is essential. The multi-valued logical system that has more than two truth values is the fuzzy logic [8, 9]. Dissimilar thickness edges could be detected by using this fuzzy technique. This technique is very simple to understand and is versatile and is permissive of inexact knowledge data and works successfully in noisy environment and considerably finds a lot of edges with respect to the generalized approaches [10, 11]. However, some of the edges still are not identified properly.

Deep learning encompasses a tremendous impact on varied fields in science within recent years. It has led to significant enhancements in digital image technology [12, 13]. This technology is additionally extremely relevant for medical image analysis [14]. Deep learning could be a directional branch within the emerging area of pattern recognition and machine learning. This approach is using artificial neural networks to perform depiction learning based on knowledge data. This approach will provide the ability to learn and train data and images in the computer. Deep learning has been widely utilized at present in the processing of data, information retrieval, linguistic communication process, customized initiation, and alternative fields, particularly in the analysis of medical images. Therefore, so as to deal with the issues within the generalized methods, this research work developed a medical edge detection model using deep learning approach. This approach has high computational complexity; it extremely improves the detection accuracy on medical images.

This research work organization is represented as follows. Related methods are represented in Sect. 2. Section 3 describes the proposed research design workflow whereas Sect. 4 illustrates the experimental results and its comparison with deep learning approaches for the medical image edge detection. Lastly, Sect. 5 represented conclusions and references.

2 Materials and Methods

Noise is unavoidable in medical images. As a consequence, noise reduction and enhancement are crucial before edge detection of medical MR images. During this research work, generalized approaches in conjunction with soft-computing approaches are compared with deep learning approaches. Deep learning approaches are applied after the pre-processing stage of images that optimizes the problems and provides much better results.

2.1 *Fuzzy Logic Approach*

A significant change in the gray level of the pixel values of the medical images is constituted as edge. In the images, the edges are high-frequency elements. Hence, this is to mention that high-pass filtering will find edges in the images. However, frequency domain will lead the responsibility with further complexity. Thus normally spatial domain based on first- or second-order derivative is used in the edge detection processes [15, 16]. Normally technique like gradient-based approach on first-order derivative is in demand. Here, the fuzzy system is designed by taking into consideration the eight closest neighbors of the pixel value [17]. The vector gradient of the two-dimensional image $f(x, y)$ that incorporates the partial derivatives of $f(x, y)$:

$$\nabla f(x, y) = \begin{bmatrix} G_x \\ G_y \end{bmatrix} = \begin{bmatrix} \frac{\partial f(x,y)}{\partial x} \\ \frac{\partial f(x,y)}{\partial y} \end{bmatrix} \tag{1}$$

where G_x is the gradient along x-axis and G_y is the gradient along y-axis. ∇f is the gradient along x- and y-axes represented as

$$\nabla f(x, y) = mag(\nabla f(x, y)) = \sqrt{{G_x}^2 + {G_y}^2} \tag{2}$$

The magnitude of the edge is designated by edge strength. For easily implementation, the common observation is to approximate the gradient as

$$\nabla f(x, y) = |G_x| + |G_y| \tag{3}$$

where $\nabla f(x, y)$ may be an edge, point, or line at direction x or y. The calculation of the first-order derivative is represented as

$$G_x = f(x) - f(x-1) = f(x+1) - f(x), G_y = f(y) - f(y-1) = f(y+1) - f(y) \tag{4}$$

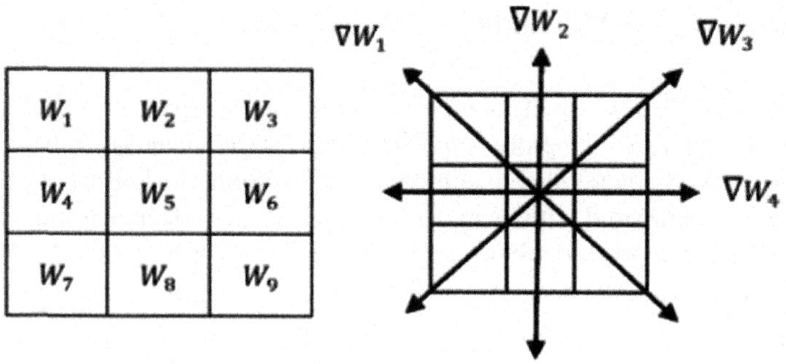

Fig. 1 Direction of edges for (3 × 3) windowing mask

In a (3 × 3) windowing mask, the center pixel with W_5 is at origin and its neighbor pixels with W coefficients, the calculation of the first-order derivative at every directions is represented in Fig. 1.

Possible edge value and first-order derivative values in different direction of Fig. 1 are calculated as follows:

$$\begin{aligned}&\nabla W_1 = \mid W_5 - W_1 \mid + \mid W_9 - W_5 \mid, \nabla W_2 = \mid W_5 - W_2 \mid + \mid W_8 - W_5 \mid \\ &\nabla W_3 = \mid W_5 - W_3 \mid + \mid W_7 - W_5 \mid, \nabla W_4 = \mid W_5 - W_4 \mid + \mid W_6 - W_5 \mid \\ &\nabla W = \nabla W_1 + \nabla W_2 + \nabla W_3 + \nabla W_4\end{aligned} \quad (5)$$

In this research work, ∇W_1, ∇W_2, ∇W_3, and ∇W_4 are calculated and placed in the fuzzy-based system which employs the membership function of triangular-type and applied Sugeno fuzzy inference system. The membership functions of triangular-type input are reducible as

$$\mu_{\Delta}(z) = \begin{cases} \frac{z-q}{r-q} \text{ if } q \le z < r \\ \frac{s-z}{s-r} \text{ if } r \le z < s \end{cases} \quad (6)$$

where q, r and s are constant parameters which characterize membership functions' shape. Five inputs, low (LO), very low (VL), medium (MD), high(HI), and very high (VH), are employed to the fuzzy system. Sugeno constant K values are specified for output constant values that make the edges' sharpness and thickness. R_p represents the output of the pth rule. The output can be found by taking into consideration the weighted average of the individual rule as

$$Y = \frac{\sum_{p=1}^{4} (\nabla W_p) R_p (\mu_{VL} + \mu_{LO} + \mu_{MD} + \mu_{HI} + \mu_{VH})}{\sum_{p=1}^{4} \mu_{VL}(\nabla W_p) + \mu_{LO}(\nabla W_p) + \mu_{MD}(\nabla W_p) + \mu_{HI}(\nabla W_p) + \mu_{VH}(\nabla W_p)}. \quad (7)$$

2.2 Neuro-Fuzzy Approach

Neuro-fuzzy (NF) subdetectors and post-processor are used in this approach. The NF edge detector structure is shown in Fig. 2. All subdetectors within the structure operate upon the (3 × 3) windowing mask. Every subdetector estimates a distinct neighborhood relation among its neighbors and the center pixel of the window [18, 19]. Figure 2 represents the coefficients and different edge directions for (3 × 3) windowing mask. The performance of the edge detection is better, if more number of NF subdetectors is utilized and however computational cost is more.

Here, the triangular membership function, singleton fuzzifier, and Sugeno inference system are used. Each subdetector has 3-inputs and 1-output of Sugeno-type first-order fuzzy inference system. Though the NF subdetector has 3-inputs and every input has 3 membership functions, the rule base contains a complete set of 27 rules. The subdetector outputs are given to the post-processor that produces the final output of the NF edge detector. In reality, the average value computation can be performed by taking the outputs of the subdetector and compares this value with a threshold. Half of the available dynamic range for the pixel values is constituted as threshold value. The post-processor output value is also the edge detector output and designate the pixel in the center is an edge or not.

The subdetector training model is shown in Fig. 3. Each and every subdetectors are trained one after another and the edge detector parameters are optimized under training. The parameters are adjusted iteratively so that convergence occurs to the perfect edge detector. The subdetectors parameters during training are tuned by backpropagation learning algorithm, so that learning error can be minimized [20, 21].

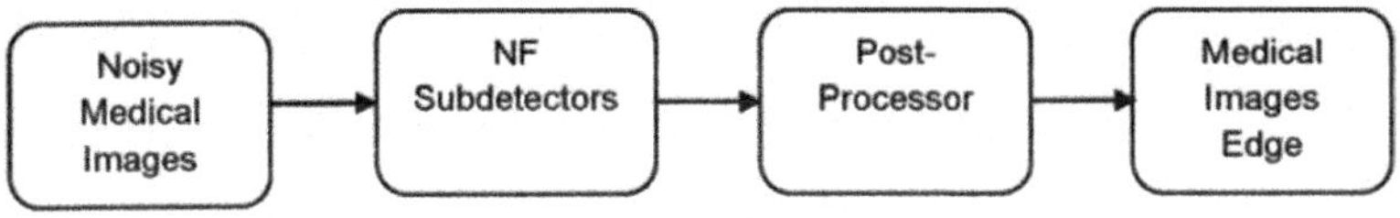

Fig. 2 Neuro-fuzzy edge detector

Fig. 3 Subdetector training model

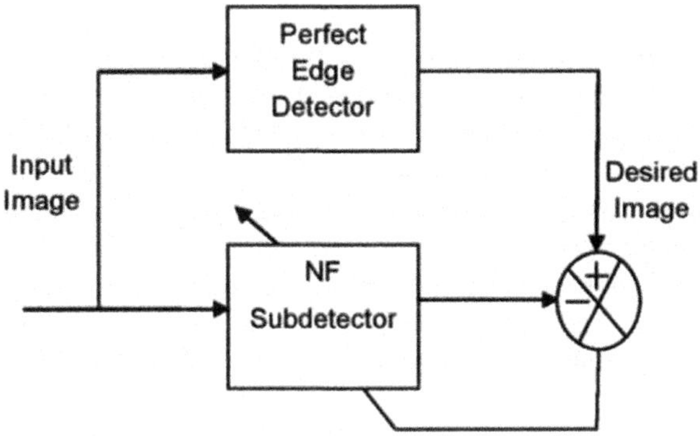

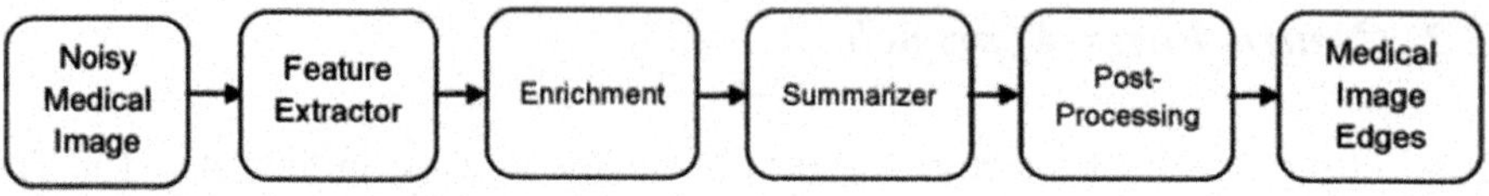

Fig. 4 Deep learning-based edge detection model

3 Proposed Research Design Workflow

The model proposed in this research incorporates the extracted feature, enrichment along with summarizer. Figure 4 shows the deep learning model of edge detection. The noisy medical images given to the feature extractor that consist of (3 × 3) size convolutional neural network (CNN) layer. The gradient operator is initialized with eight gradient directional kernels. The zero mean with unity variance CNN layers is initialized in this work. The feature map outputs indicate edges of the medical image after the training process. During this research work, dilated convolution is used to shape filtering in multiple scale environments as compared to the generalized edge detector. This convolution has fewer parameters with larger receptive capability. This module is known as enrichment. This module extracts edges along with object information. The output of each and every dilated filter is joined simultaneously at the end.

The outcome of the enrichment modules provides various features that are given to the summarizer module. This module summarizes the features to provide edges in the images. Sigmoid activation functions with eight (1 × 1) convolutional layers are employed in this process. In the post-processing stage, non-maximum suppression (NMS) and multiple scale testing are used. The aim is to find prediction maps of the medical image edge. Different resolutions of the input image can be obtained and resized to provide into the network. The final medical image edges can be obtained by averaging the outputs in the original image.

4 Experimental Results and Analysis

In this research work, the experimental performance analysis can be achieved using MATLAB software. Deep learning along with soft-computing techniques is applied on medical MR images taken from large medical databases [19]. The medical images with 5% noise density are considered here. The gray-level values of the images under test are of 8 bits. Figure 5 shows the noisy medical images. The performance of the method proposed in this research work is evaluated with metrics like peak signal-to-noise ratio (PSNR), structural similarity index measure (SSIM), and edge keeping index (EKI) [22]. The experimental results obtained with different edge detection approaches are shown in Figs. 6, 7, 8, 9, and 10.

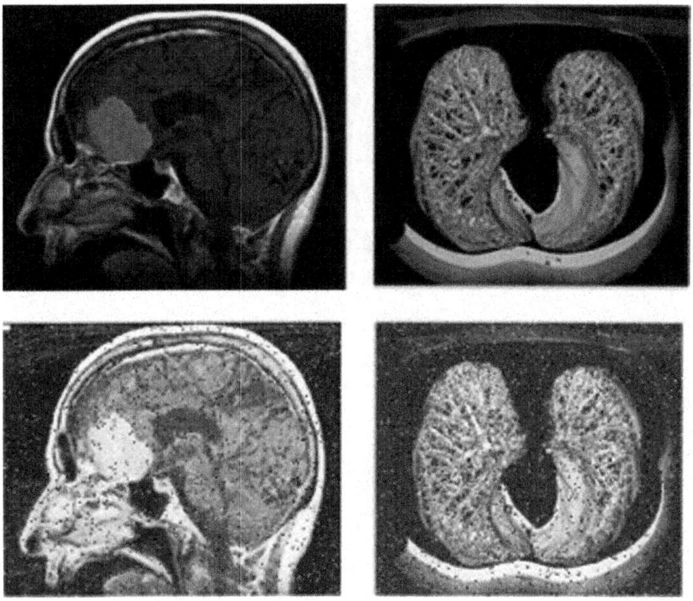

Fig. 5 MR images (Image 1-Left, Image 2-Right) original image on the top and noisy image on the bottom

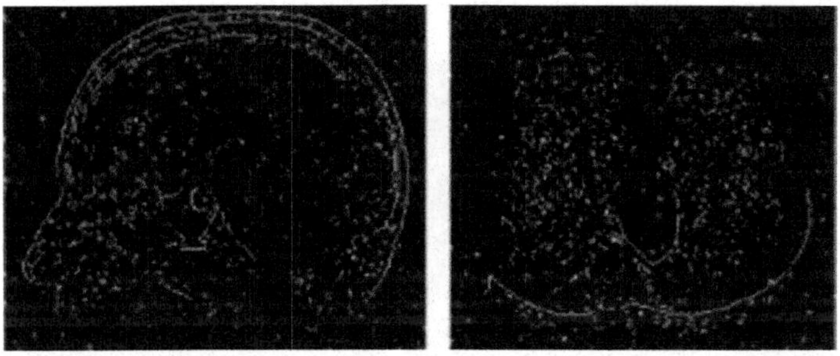

Fig. 6 Sobel operator processed MRI (Image 1-Left, Image 2-Right)

Fig. 7 Canny operator processed MRI (Image 1-Left, Image 2-Right)

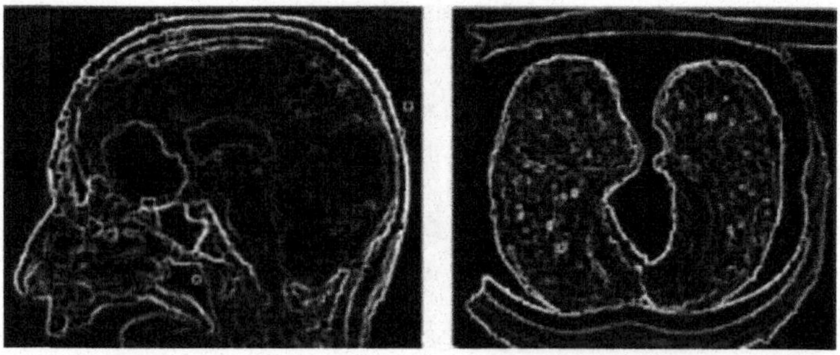

Fig. 8 Fuzzy logic approach processed MRI (Image 1-Left, Image 2-Right)

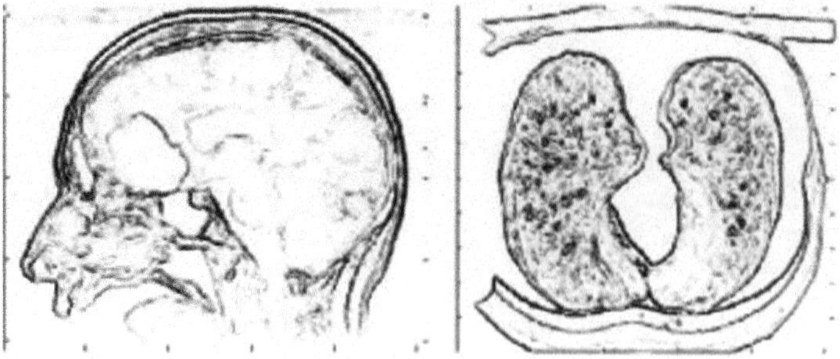

Fig. 9 Neuro-fuzzy approach processed MRI (Image 1-Left, Image 2-Right)

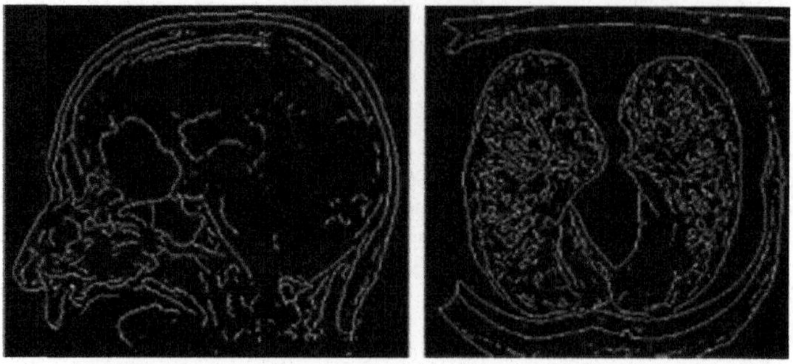

Fig. 10 Proposed deep learning approach processed MRI (Image 1-Left, Image 2-Right)

PSNR is the ratio between the maximum possible power of a signal and the power of distorting noise that affects the quality of its representation. SSIM index is a method to measure the similarity between two images. The SSIM index can be viewed as a quality measure of one of the images being compared provided the other image is regarded as of perfect quality. The structural similarity index correlates with human visual system and is used as a perceptual image quality evaluation metric. The SSIM is defined as function of luminance, contrast, and structural components. EKI is used to evaluate and establish how well the edges are maintained during the

Table 1 Quality assessment parameters of the medical MRI

MR image	Assessment parameter	Sobel	Canny	Fuzzy logic	Neuro-fuzzy	Proposed
1	$SNR_{Peak}(dB)$	46.41	56.38	57.12	58.28	60.32
	SSIM	0.704	0.791	0.801	0.814	0.891
	EKI	0.516	0.625	0.629	0.632	0.712
2	$SNR_{Peak}(dB)$	50.65	60.40	64.12	64.80	68.32
	SSIM	0.763	0.814	0.816	0.876	0.902
	EKI	0.575	0.748	0.757	0.792	0.825

detection process. The thicknesses of various cardiac boundaries are better extracted using the edges of corresponding muscles.

The observation made by the experiment that the performance is very poor in case of the Sobel operator. Its experimental output images are critically affected by noise; several of the noises are wrongly identified as edges. Much better performance can be obtained by using canny edge detection technique that detects correctly numerous edges of the MR images. Still, the issue of noise is not identified clearly. The edge detector based on fuzzy logic and neuro-fuzzy discards almost all noises; still the edges of the images are not identified clearly. Furthermore, the proposed approach based on deep learning detects successfully most of the fine edges from the MR images.

On the other hand, the comparative study and quantitative analysis of medical MR images along with 5% noise density is represented in Table 1. This shows the values of various quality assessment parameters of two medical MR images. The performance evaluation of these methods on medical MR images for noise suppression and edge preservation has been carried out and is presented for comparison in Table 1. The graphical representation of various quality assessment parameters with 5% noise density is shown in Fig. 11. It is observed that the proposed deep learning approach yields higher PSNR SSIM and EKI with different amount of noise density. This is evidence of the maximum noise suppression with significant edges and fine preservation.

5 Conclusions

This research work proposed deep learning edge detection model to detect automatically fine edges from noisy medical MR images. The model proposed is incorporating the extracted feature, enrichment along with summarizer. Here, dilated convolution is used to shape filtering in multiple scale environments as compared to the generalized edge detector. This convolution has fewer parameters with larger receptive capability. The enrichment module provides various features that are given to the summarizer

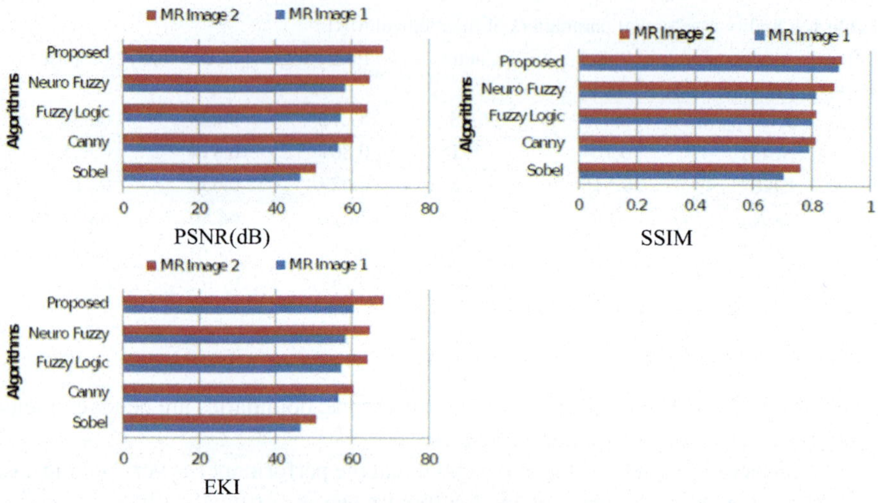

Fig. 11 Assessment of various quality parameters

module. This module summarizes the features to provide edges in the images. In the post-processing stage, NMS and multiple scale testing are used. Experimental results and analysis show the approach proposed in this research work exhibits far better performance than the prevailing techniques as well as generalized soft-computing methods. The experimental results clearly show that the method proposed in this research work yields higher PSNR SSIM and EKI with noisy environment. This is evidence of the maximum noise suppression with significant edges and fine preservation.

References

1. Husein, H.A., Zainab, R.M.: Edge detection of medical images using Markov basis. Appl. Math. Sci. **11**(37), 1825–1833 (2017)
2. Canny, J.: A computational approach to edge detection. IEEE Trans. Pattern Anal. Mach. Intell. **8**(6), 679–687 (1986)
3. Liang, L.R., Looney, C.G.: Competitive fuzzy edge detection. Appl. Soft Comput. **3**, 123–137 (2003)
4. Rajab, M.I., Woolfson, M.S., Morgan, S.P.: Application of region-based segmentation and neural network edge detection to skin lesions. Comput. Med. Imaging Graph. **28**, 61–68 (2004)
5. Mehena, J.: Medical image edge detection based on mathematical morphology. Int. J. Comput. Commun. Technol. **4**(1), 07–11 (2013)
6. Raman, M., Himanshu, A.: Study and comparison of various image edge detection techniques. Int. J. Image Process. **3**(1), 1–12 (2010)
7. Mehena, J., Adhikary, M.C.: Medical image edge detection based on neuro-fuzzy approach. Int. J. Comput. Inf. Eng. **10**(1), 229–232 (2016)

8. Choi, Y.S., Krishnapuram, R.: A robust approach to image enhancement based on fuzzy logic. IEEE Trans. Image Process. **6**(6), 808–825 (1997)
9. Russo, F.: Edge detection in noisy images using fuzzy reasoning. IEEE Trans. Instrum. Measur. **47**, 802–808 (1998)
10. Richard, A.P.: A new algorithm for image noise reduction using mathematical morphology. IEEE Trans. Image Process. **4**(3), 554–568 (1995)
11. Jing, X., Nong, Y., Shang, Y.: Image filtering based on mathematical morphology and visual perception principle. Chin. J. Electron. **13**(4), 612–616 (2004)
12. Firoz, R., Ali, M.S., Khan, M.N.U., Hossain, M.K., Islam, M.K., Shahinuzzaman, M.: Medical image enhancement using morphological transformation. J. Data Anal. Inf. Process. **4**(1), 1–12 (2016)
13. Hassanpour, H., Asadi, S.: Image quality enhancement using pixel wise gamma correction. Int. J. Eng.-Trans. B: Appl. **24**(4), 301–311 (2011)
14. Ker, J., Wang, L., Rao. J., Lim, T.: Deep learning applications in medical image analysis. IEEE Access **6**, 9375–9389 (2018)
15. Abdallah, A.A., Ayman, A.A.: Edge detection in digital images using fuzzy logic technique. World Acad. Sci., Eng. Technol. **5**(4), 264–269 (2009)
16. Singh, S.K., Pal, K., Nigam, M.J.: Novel fuzzy edge detection of seismic images based on bi-level maximum. Int. J. Signal Imaging Syst. Eng. **3**(3), 169–178 (2010)
17. Pushpajit, A.K., Nileshsingh, V.T.: A fuzzy set approach for edge detection. Int. J. Image Process. **6**(6), 403–412 (2012)
18. Yuksel, M.E.: Edge detection in noisy images by neuro-fuzzy processing. AEU Int. J. Electron. Commun. **61**(2), 82–89 (2007)
19. Bohern, B.F., Hanley, E.J.: Extracting knowledge from large medical databases. Autom. Approach Comput. Biomed. Res. **28**(3), 191–210 (1995)
20. Vasavada, J., Tiwari, S.: An edge detection method for grayscale images based on BP feedforward neural network. Int. J. Comput. Appl. **67**(2), 23–28 (2013)
21. Hagan, M.T., Menhaj, M.B.: Training feed forward networks with the Marquardt algorithm. IEEE Trans. Neural Netw. **5**(6), 989–993 (1994)
22. Wang, Z., Bovik, A.C., Sheikh, H.R., Simoncelli, E.P.: Image quality assessment: From error measurement to structural similarity. IEEE Trans. Image Process. **13**(4), 01–14 (2004)

Lung Disease Classification Based on Lung Sounds—A Review

Vishnu Vardhan Battu, C. S. Khiran Kumar, and M. Kalaiselvi Geetha

Abstract Sounds of hearts and lungs are one of the important human physiological signals. The lungs produce noises, when you breathe in and out. These noises can be identified with a stethoscope. The noises of your breath might be normal or abnormal. Based on the noise, a disease can also be classified. This chapter explains the different ways to identify lung diseases such as chronic obstructive pulmonary disease (COPD) and pneumonia by classifying the lung sound. Initially, basic symptoms are considered to recognize pneumonia. Further, how computational intelligence in health care addressing the treatment is discussed. There are many machine learning algorithms to recognize lung diseases based on lung sounds. This chapter throws light on various classifiers for the detection of lung diseases.

1 Introduction

Sounds of hearts and lungs are one of the important human physiological signals. The lungs produce noises, when you breathe in and out. These noises can be identified with a stethoscope or by just breathing. The noises of your breath might be normal or abnormal. The abnormal sounds usually indicate some kinds of diseases of related organ. Abnormal breath sounds can state a lung problem, such as obstruction, inflammation, infection, fluid in the lungs, and asthma. The sound of a typical breath is comparable to that of air known scientifically as vesicular and adventitious noises. Adventitious lung sounds are rhonchi, wheezes, crackles, squeak, pleural rub, and stridor [1]. Rhonchi is a low-pitched breath sound whereas crackles is a high-pitched

V. V. Battu (✉) · M. Kalaiselvi Geetha
Annamalai University, Chidamvaram, Tamil Nadu, India
e-mail: bvishnuvardhan84@gmail.com

M. Kalaiselvi Geetha
e-mail: geesiv@gmail.com

C. S. Khiran Kumar
University of Maryland Baltimore County, Catonsville, Maryland, United States
e-mail: sivkhiran3@gmail.com

D. P. Acharjya and K. Ma (eds.), *Computational Intelligence in Healthcare Informatics*, Studies in Computational Intelligence 1132,
https://doi.org/10.1007/978-981-99-8853-2_15

breath sound. Similarly, wheezing is a high-pitched whistling sound caused by constriction of the bronchial tubes and stridor is an example of aberrant breath sounds. These sounds are harsh, vibratory sounds caused by narrowing of the upper airway when air tries to travel via bronchial tubes that are filled with fluid or mucus. Rhonchi occurs when the tiny airways in the lungs are filled with fluid and if there is any air movement in the sacs. When you breathe crackles occur. When a person develops pneumonia or heart failure, the air sacs are filled with fluid. When the bronchial tubes become inflamed and narrowed, wheezing ensues. When the upper airway narrows, stridor ensues. An irregularity in the respiratory tract causes a pulmonary crackling sound. The pulmonary crackling sound is a short-duration, discontinuous sound that can occur during the inspiratory, expiratory, or both phases of breathing.

Although airflow restriction is a necessary condition for the generation of wheezes, airflow restriction can occur in the absence of wheezes. In 1967, Forgacs postulated that wheezes are caused by oscillations of the bronchial walls induced by airflow, and that the pitch of the wheeze is determined by the bronchial walls' mechanical qualities. The pitch of the wheeze is determined by the bulk and flexibility of the airway walls, as well as the flow velocity, but not by the airway's length or size. Wheezes can also be characterized as polyphonic or monophonic. Monophonic wheezing is characterized by the occurrence of a single musical note that begins and ends at different times. It can be caused by a local pathology such as bronchial blockage caused by a tumor, broncho stenosis.

When there is a rigid obstruction, wheeze is heard throughout the respiratory cycle. When there is a flexible obstruction, wheezing may be inspiratory or expiratory. The severity may fluctuate with changes in posture, as is the case in patients with tumor-induced partial bronchial blockage. Fixed monophonic wheeze has a stable frequency and a long duration, whereas random monophonic wheeze has a variable frequency and duration during the respiratory cycle. Asthma patients may exhibit random monophonic wheezes [2, 3].

Likewise squawks, occasionally referred to as squeaks, are brief inspiratory wheezes lasting less than 200 ms. According to acoustic research, the fundamental frequency ranges between 200 and 300 Hz. Squawks are frequently heard in patients with pulmonary fibrosis caused by a variety of causes, most notably hypersensitivity pneumonitis. There have been additional causes of pneumonia and bronchiolitis obliterans identified. Squawks are most common during late inspiration, and they are frequently preceded by crackles during late inspiration. If there is no evidence of restrictive lung disease, those with squawks should be suspected of pneumonia. Crackles are non-musical, discontinuous, explosive lung sounds that occur during inspiration and occasionally during expiration. The duration, loudness, pitch, timing in the respiratory cycle, and relationship to coughing and shifting body posture are used to classify crackles as fine or coarse. Continuous lung sounds last 250 ms or longer, and discontinuous sounds last 25 ms or less. Crackles are short-lived sounds that last less than 20 ms. The time expanded waveform analysis is proposed to objectively differentiate the crackles.

Upper respiratory tract obstruction causes stridor. It is a loud, high-pitched melodic sound. The following factors distinguish it from wheezing. It is louder above

Table 1 Classification of abnormal lung sounds

Type	Characteristics	Acoustics related	Diseases
Fine crackles	Discontinuous high-pitched inspiratory	Rapidly dampened wave deflection frequency: about 650 Hz shorter duration	Lung fibrosis, pneumonia, congestive heart failure
Coarse crackles	Discontinuous low-pitched inspiratory	Rapidly dampened wave deflection frequency about 350 Hz longer duration	Same as fine crackles but usually more advanced disease
Wheezes	Continuous high-pitched expiratory than the inspiratory	Sinusoid frequency greater than 100–5000 Hz, duration more than 80 ms	Asthma, COPD, Tumor, Foreign body
Rhonchus	Continuous low-pitched expiratory than the inspiratory	Sinusoid frequency about 150 Hz, duration more than 80 ms	Bronchitis, Pneumonia
Stridor	Continuous high-pitched inspiratory	Sinusoid frequency more than 500 Hz	Epiglottitis, after extubation, Foreign body
Pleural friction rub	Continuous low-pitched inspiratory and expiratory	Rhythmic succession of short sound frequency less than 350 Hz, duration more than 15 ms	Pleurisy pericarditis, pleural tumor

the neck than it is against the chest wall. Second, stridor is primarily an inspiratory phenomenon. The turbulent flow passing through a constricted portion of the upper respiratory tract causes stridor. On the other hand, pleural rub is a non-musical, short explosive sound that is grating, rubbing, creaky, or leathery in nature and can be heard during both stages of breathing. In most cases, the expiratory component is identical to the inspiratory component. It arises as a result of irritated pleural surfaces rubbing against one another when breathing. It is crucial to distinguish it from crackles in the clinic. Table 1 shows the classification of abnormal lung sounds and disorders associated with them. The most common noises are crackles, wheezes, and rhonchi. Hearing those sounds substantially aids in the diagnosis of pulmonary disorders.

Viruses, bacteria, and even fungus can cause infections in the lungs. One of the most common types of lung infections is pneumonia. Pneumonia is an infection of one or both lungs caused by bacteria, viruses, or fungus. A person becomes infected by breathing the germs or virus after a nearby infected individual sneezes or coughs. In some circumstances, pneumonia is contagious. People over 65 years of age and infants under the age of two who have a weakened immune system are prone to get illness. Pneumonia appears to be the most common indication of a serious illness in case of lung or heart disease, neurological conditions that make swallowing difficult, smoking or drinking of alcohol, and exposed to toxic fumes, chemicals, or secondhand smoke. In pregnant women, pneumonia appears to be the most common indication of a serious illness.

Fever was present in 99% of cases, fatigue was present in 70% of cases, dry cough was present in 59% of cases, anorexia was present in 40% of cases, myalgias was present in 35% of cases, dyspnea was present in 31% of cases, and sputum was present in 31% of cases in 138 patients with COVID-19 pneumonia in Wuhan. Tachypnea is the most common symptom of dyspnea, which affects 31% of persons. At first, breath sounds may appear distinct and rapid. As the problem worsens, you may hear minor wheezing (expiratory wheezing continuous, musical sound). If you have mild pneumonia, you may hear fine crackles (Rales) and bronchial breath noises. It sounds like someone is burning wood in a fireplace. In some persons, COVID-19 can induce acute respiratory distress syndrome.

For an accurate study, lung sounds should be captured at various locations in both lungs. Second, the patient's effort to breathe has a significant impact on the quality of lung sound. Lung sounds were recorded for 10 s at six different locations on the chest, three on each side; on the back between the spine and the medial border of the scapula at T4–T5; on the midpoint between the spine and the mid axillary line at T9–T10; and on the front where the medioclavicular line crosses the second rib. The prime objective of this chapter is to identify lung diseases by classifying sounds using deep learning.

The rest of the chapter is organized as follows: Sect. 2 provides in-depth knowledge to recognize symptoms naturally. Further, clinical ways to recognize symptoms of pneumonia are discussed in Section 3 followed by details of datasets in Sect. 4. Various computational intelligence techniques used for sound classification are discussed in Sect. 5. The conclusion is drawn in Sect. 6.

2 Natural Ways to Recognize Symptoms

Lung disease signs and symptoms vary depending on the ailment. They can differ from one individual to the next and evolve through time. Signs and symptoms of chronic illnesses often appear gradually and worsen with time. Acute diseases develop quickly and can range from mild to severe. While each lung illness has its own characteristics, many lung disorders share certain common signs and symptoms. For example, coughing is a common symptom. Breathing difficulty (dyspnea), wheezing and gasping for air, coughing up mucus, blood, or sputum are common symptoms. Pain in the chest and exhalation difficulties found in case of obstructive lung conditions such as COPD. A bluish tinge on the skin (due to lack of oxygen), clubbing of the fingertips, and aberrant fingernail development are signs of long-term oxygen deprivation [4].

Work-Related Lung Diseases: Lung disorders that are aggravated by particular work conditions are known as work-related lung illnesses. Long-term exposure to some irritants inhaled into the lungs causes them to develop. Even after the exposure has ended, certain lung disorders may have long-term consequences. A single exposure to a harmful agent can damage the lungs in some situations. This problem can be exacerbated by smoking. Lung difficulties are caused by particles in the air that come from a variety of sources. Factories, smokestacks, exhaust, fires, mining, building, and agriculture are among these sources. The smaller the particles, the more damage to the lungs they might do. Smaller particles are more easily inhaled and transported deep into the lungs.

Coughing, shortness of breath that worsens with physical activities like abnormal breathing patterns, chest discomfort, and chest tightness are the symptoms of work-related lung disorders which often mimic those of other illnesses or ailments.

Long-Term Lung Diseases: Feeling progressively out of breath is the most common symptom. Breathing problems in certain people may worsen considerably more quickly, spanning weeks or months. This is especially true in the case of interstitial lung disorders. Breathing gets noticeably poorer for patients in the terminal stages of a lung ailment. Their lung function does not exactly return to normal after each exacerbation, and breathing becomes more difficult. As long-term lung disease progresses, the lungs become less efficient. Any activity, even simple movements like changing positions, chatting, eating, might cause shortness of breath. Lying flat makes it difficult to breathe, one might try sleeping in a more upright position. Low amounts of oxygen in the blood may be caused by a reduction in lung function. This might lead to fluid retention in your legs and stomach, which is unpleasant. Exacerbations restrict the amount of oxygen in your bloodstream even more, exacerbating your symptoms.

Other signs and symptoms include a persistent cough, a loss of appetite, chest pain, and sleep disturbances. The following are the most typical physical symptoms: becoming more out of breath, having frequent exacerbations, finding it difficult to maintain a healthy body weight due to loss of appetite, and becoming more anxious and sad.

Breathlessness: Inhalers, pills, and occasionally nebulizers can help with breathlessness. When you are out of breath, a hand-held fan can come in handy. The sensation of air on your face can make breathing easier. However, if your breathlessness is severe and your blood oxygen level is low, long-term oxygen therapy may help you breathe better and improve your quality of life.

When your lungs can no longer keep adequate oxygen in your blood, one need long-term home oxygen. This oxygen is usually provided via an oxygen concentrator, a machine that concentrates oxygen from the air. This machine must be used for at least 15 h every day. The amount of oxygen required is carefully calculated and monitored, and it may be necessary to raise the amount over time.

Anxiety and Depression: If your breathing becomes more difficult, anxiety and despair are common side effects that might make you feel much worse. Isolation and loneliness are also prevalent issues. Appetite loss is a typical problem that occurs as a natural part of the disease process.

Signs and Symptoms of Pneumonia: Lung infections, such as pneumonia, are typically small, but they can be hazardous, especially for patients with weakened immune systems or chronic conditions such as COPD. COPD is made up of two diseases: chronic bronchitis and emphysema. Chronic bronchitis generates a mucus-filled cough, which is the main difference between the two conditions. The most common symptom of emphysema is shortness of breath. Emphysema development can be influenced by genetics. Coughing up phlegm (mucus), fever, sweating, or chills, shortness of breath during normal activities or even while resting, chest pain that gets worse when you breathe or cough, tiredness or exhaustion, loss of appetite, nausea, and headaches are all symptoms of pneumonia.

3 Clinical Process to Recognize Pneumonia

The simplest and most common way for diagnosing clinical diseases from chest auscultation is to use a stethoscope. Advanced procedures are heavily used by doctors [5]. These are blood test, computerized tomography (CT) scan, spirometry and diffusion capacity, oximetry test, echocardiogram test, and biopsy.

Blood Test: Proteins, antibodies, and other markers of auto-immune illnesses and inflammatory reactions to environmental exposure can be detected using certain blood tests.

Computerized Tomography (CT) Scan: This imaging test is critical for diagnosing interstitial lung disease and is frequently the initial step. CT scanners combine X-ray pictures obtained from a range of angles to create cross-sectional scans of inside structures. A high-resolution CT scan can be quite beneficial in determining the extent of lung damage caused by interstitial lung disease. It may exhibit fibrosis characteristics, which can assist in narrowing the diagnosis and guiding treatment decisions.

Spirometry and Diffusion Capacity: In this test, you must exhale rapidly and forcefully via a tube connected to a machine that measures the amount of air your lungs can hold and the rate at which you can exhale it. Additionally, it evaluates the ease with which oxygen may be transported from the lungs to the bloodstream.

Oximetry Test: In this simple test, a little device is placed into one of your fingers to determine the oxygen saturation of your blood. It can be performed at rest or while exercising to monitor the course and severity of lung disease.

Echocardiogram Test: An echocardiogram is a type of ultrasonography that employs sound waves to see the heart. It can create still photographs of your heart's architecture as well as films that demonstrate how your heart works. This test can determine how much pressure is present in your right side of the heart.

Analyses of Lung Tissue: Often, pulmonary fibrosis can be definitively diagnosed in the laboratory only after testing a small amount of lung tissue (biopsy).

The stethoscope is an acoustic device that transfers sounds from the chest piece to the listener's ears through an air-filled hollow tube. Higher frequency sounds will be transmitted by the diaphragm, while lower frequency sounds will be transmitted by the bell. As a result, the sound transmission of the acoustic stethoscope will be proportional to the frequency of the heart sounds. Low-frequency noises, such as those below 50 Hz, may not be heard due to variations in sensitivity of the human ear. Because of the acoustic stethoscope's limitations, an electronic gadget that is significantly more advanced than the original stethoscope has emerged. When comparing an electronic stethoscope to a regular stethoscope, there are some significant differences.

4 Data Availability

The dataset was publicly available from the ICBHI Challenge database. The goal is to provide a more equitable distribution of respiratory disease cases across disease groups [6, 7]. The core dataset includes 70 individuals with a variety of respiratory ailments, including pneumonia, asthma, heart failure, bronchitis (BRON disorders), and COPD. A total of 35 healthy controls were also surveyed. The age of the participants in this study was not a variable of interest in order to ensure a fair investigation. The participants ranged in age from youngsters through adults to the elderly. After thoroughly comprehending the parameters of the study and the technique involved, all participants signed a written consent form. The study protocol was created in accordance with the declaration of Helsinki and was approved by the King Abdullah University Hospital's Institutional Review Board (IRB).

The diagnostic and recording procedures were carried out by two thoracic surgeons. Depending on the diagnostic requirements of each patient, lung sounds (LS) were recorded from one of the following conventional anterior or posterior chest positions: upper left or right, or middle left or right. Patients were recorded in a supine posture, with the stethoscope snugly positioned over the region of interest, to reduce artifacts generated by people. A single-channel stethoscope-based acquisition approach was used to capture the audio signals. An embedded ambient and frictional reduction technology is included in the system. Using a 16-bit quantizer, all signals were gathered at a sample rate of 4 kHz and bandlimited to a frequency range of 20–2 kHz. Additionally, during the capture stage, the resonance frequency between 50 and 500 Hz was lowered to prevent tampering from heartbeat sounds. There were 308 5-s lung sound recordings in the whole dataset. This period is long enough to encompass a minimum of one respiratory cycle and has been used in previous studies based on individuals' typical resting respiration rates. Generally, adopting tiny data windows makes it easier to gather medical data while simultaneously increasing the model's computing efficiency. Furthermore, training the model on lung sound signals rather than respiratory cycles makes data curation and labeling much simpler.

As part of the scientific challenges offered at the International Conference on Biomedical and Health Informatics (ICBHI), the ICBHI challenge database is released openly accessible for research purposes [8]. The database as a whole had 126 individuals and 920 lung sound recordings from individuals of varying ages and respiratory illnesses. One of the following stethoscope systems was used to capture the audio signals. These are Littmann 3200, Littmann C—II SE, AKG Harman C417 L, and Welch Allyn meditron. The trachea, as well as the anterior, posterior, and lateral left and right chest regions, were chosen as recording sites. Only recordings related to one of the relevant respiratory disorders were selected for this inquiry to strengthen the main dataset. The respiratory recordings were separated into 5s non-overlapped frames consecutively, without removing breathing cycles, to retain compatibility with the original dataset. In addition, all sounds are re-sampled at a 4000 Hz sampling rate. There were 110 patients in all, with 1176 lung sound recordings in this collection. In the training-validation data, all recordings were treated as an independent sample.

A key difficulty in this sector is the lack of publicly accessible large databases that can be utilized to construct algorithms and compare results. Despite the fact that most works used in-house data collectors, the data sources for the 77 papers included by pramono comprehensive examination included 13 publicly available databases [9]. The databases from online repositories and audio CD companion books are given below. The most often utilized databases are the RALE lung sounds.

- RALE repository of lung sounds.
- ICBHI 2017 challenge respiratory sound database.
- Littmann repository.
- Auscultation skills: breath and heart sounds.
- Fundamentals of lung and heart sounds.
- Heart and lung sounds reference library.
- RD repository by East Tennessee State University.
- Secrets heart & lung sounds workshops.
- Sound cloud repository.
- Understanding heart sounds and murmurs.
- Understanding Lung Sounds.

Effect of Positioning on Recorded Lung Sound: An increase in lung sound intensity during auscultation is considered indicative of lung expansion since lung sound intensity is generally connected to lung capacity [10, 11]. There would be no differences in the data recorded in corresponding regions between (1) the left and right lungs in the sitting position; (2) the dependent and non-dependent lungs in the side-lying position, (3) the sitting position and the dependent position, or (4) the sitting position and the non-dependent position.

Sitting in a chair or on the side of the bed is the best position for chest auscultation. During the examination, however, the patient's clinical condition and comfort must be addressed, and some patients may only tolerate lying at a 45 deg angle. The ways to listening to breath sounds on the anterior and posterior chest auscultation are shown in Fig. 1.

5 Computational Intelligence in Lung Sound Classification

Listening to pulmonary sounds in the chest using a stethoscope is a common, cost-effective, and non-invasive method of evaluating respiratory disorders [12]. However, due to its inherent subjectivity and poor frequency sensitivity, the stethoscope

Fig. 1 Chest auscultation

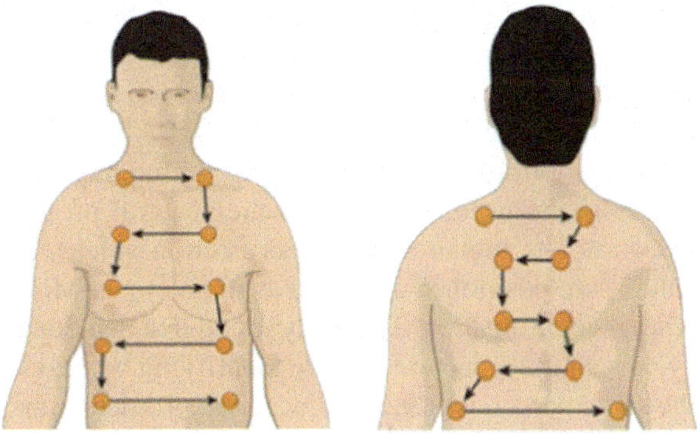

(*a*)*Anterior* (*b*)*Posterior*

is regarded as an unsatisfactory diagnostic tool for respiratory disorders. The stethoscope attenuates frequencies beyond 120 Hz [13]. Sound waves with frequencies equivocating from 20 to 20,000 Hz can be perceived by the human ear. LS normally below 100 Hz, with a noticeable decrease in sound strength between 100 and 200 Hz, but sensitive microphones will be able to detect them. By extracting features first and then applying multiple categorization models, Machine Learning techniques help to address the restrictions.

5.1 Feature Extraction Methods and the Classification

Some of the commonly used techniques to extract features [6, 7, 14–16] are Hilbert–Huang transform, spectrogram, Cepstrum, and Mel-Cepstrum,

Hilbert–Huang Transform Technique: The Hilbert–Huang Transform (HHT) is a technique for decomposing any signal into Intrinsic Mode Functions (IMFs) and extracting instantaneous frequency data. It is capable of handling both non-stationary and non-linear data. Rather than being a theoretical tool, the HHT algorithm is more like an empirical approach that can be applied to a data collection. The Hilbert–Huang Spectrum's (HHS) performance on fine and coarse crackles both simulated and genuine. The feature extraction procedure for distinguishing Ronchi and crackles from regular sounds is made up of three signal-processing components: fmin/fmax, the frequency ratio calculated using the Welch method from the Power Spectral Density (PSD). The HHT estimated the IF's exchange time as well as the average IF. Singular spectrum analysis was used to determine the eigenvalues. To distinguish between the rhonchus, crackles, and normal lung sound, an SVM classifier was utilized.

Spectrogram and 2D Representations: As a foundation for automated wheezing recognition, Riella presented "spectrogram image processing" of LS cycles. The captured LS signal was used to create the spectrogram. To adjust the contrast and separate out the higher frequency elements, the resultant spectrogram image is passed through a bi-dimensional convolution filter and a limiter. The acquired spectrogram is used to calculate the spectral average, which is then fed into a Multi-layer Perceptron (MLP) Artificial Neural Network (ANN). To boost the contrast and distinguish the highest amplitude parts in the spectrogram image, a 2D convolution filter and a half-threshold are used. In addition, the spectral projection is calculated and saved as an array. The higher spectrum projection values, as well as their relevant spectral values, are positioned and used as inputs to an MLP ANN, which allows automatic wheeze identification. The study found that detecting wheezing for an isolated breath cycle is 84.82% accurate, while identifying wheezes from groups of breath cycles of the same person was 92.86% accurate. To detect the presence of wheeze in filtered narrowband lung sound signals, sample entropy histograms are used.

Cepstrum and Mel-Cepstrum: Further, a cepstral approach is developed to study the respiratory system's acoustic transformation behavior. Initially, the LS signal is divided into segments, each of which is represented by a smaller number of cepstral coefficients. Using the Vector Quantization (VQ) method, such segments are classified as wheezes or normal LS. To classify wheeze and normal LS, cepstral analysis using Gaussian Mixture Models (GMM) is proposed. Mel-Frequency Cepstral Coefficients (MFCC) are used to differentiate overlapped regions in the LS signal. The Mel-Frequency Cepstrum (MFC) is a representation of a sound's short-term power spectrum based on the linear cosine transform of a log power spectrum on a non-linear MFC scale. The coefficients that make up an MFC are known as MFCCs. They are made up of a cepstral representation of the audio clip. A spectrogram is a graphic representation of the frequency spectrum of a sound or other signal as it changes with time or another variable. They are popular in fields including music, sonar, radar, speech processing, and seismology. Because MFCC features are frequently utilized in audio detection systems, it is possible to determine a baseline value for accuracy, precision, recall, sensitivity, and specificity using the MFCC features. Audio detection also uses spectrogram pictures. However, they are never put to the test with CNNs in terms of respiratory audio. To distinguish between wheeze and non-wheeze, Auto-regressive (AR), Wavelet Transform (WT), and MFCC are employed for feature extraction. Besides, LS is divided into two categories: normal and wheezing. Linear predictive coding (LPC) is a technique for extracting features from signals in the time domain. It is a linear combination of prior samples weighted by LPC coefficients. The coefficients and prediction error of a 6th order LPC are used to construct a feature vector. The features are extracted from 51.2ms segments of each event with 12.8ms overlap, WT, Fourier transform, and MFCC, as well as classification algorithms based on VQ, ANN, and GMM, which are evaluated using Receiver Operating Curves (ROC). In LS recorded from asthma and COPD patients, another method is described to recognize wheeze and non-wheeze epochs. Ratio irregularity, mean-crossing irregularity, and Renyi entropy are among the features derived from LS.

Analysis Based on Spectral: Self-organizing Maps (SOM) based on FFT spectra were used to collect lung sound from emphysema, asthma, fibrosis alveolitis, and healthy patients. The spectrum analysis is used to investigate the effects of bronchodilators in asthma patients. Using multi-scale Principal Component Analysis (PCA), the major variability of Fourier power spectra was found, and Fourier spectra were used for categorization using the 1-nearest neighbor technique. The optimization of Fourier spectra was carried out using a power raising transformation. The chest wall and tracheal LS spectral properties were investigated [17]. In comparison to crackles previously reported, crackles recorded according to the guidelines of Computerized Respiratory Sound Analysis (CORSA) exhibit shorter 2CD scores and higher frequencies. Lung sounds were first decomposed in the Time-Frequency domain, and then AR averaging, sample entropy histograms distortion, and recursively measured instantaneous kurtosis were used to extract features from a particular frequency bin with defined signal properties. Using the Support Vector Machine (SVM) classifier, the study found mean classification accuracies of 97.7% for inspiratory and 98.8% for expiratory segments, respectively.

Analysis of Time Expanded Waveforms: Lung sounds visualized by Time Expanded Waveform Analysis (TEWA) studied lung crackling features in individuals with asbestosis (AS), Asbestos-Related Pleural Disease (ARPD), and Left Ventricular Failure (LVF). Interstitial Pulmonary Fibrosis (IPF), COPD, Congestive Heart Failure (CHF), and pneumonia were all studied using lung sound mapping and TEWA. Each group looked at 20 different subjects. It also looked at 15 people who did not have any signs of lung disease. Differences in time, character, and location were observed, allowing these groups to be distinguished. The bootstrap method was used to construct and test several logistic regression models. Using regression models, 79% of the subjects were correctly classified. The area under the ROC curve for IPF and CHF ranged from 0.96 to 0.80 for COPD.

Wavelet-Based Approaches: Further, SVM is used to combine wavelet-based feature extraction and classification. The Discrete Wavelet Transform (DWT) and a classifier based on a Radial Basis Function (RBF) neural network were used to classify adventitious LS. The LS signals were decomposed into frequency sub-bands using WT, and statistical characteristics were derived from those sub-bands to reveal the wavelet coefficient distribution. An ANN-based system that was trained using the resilient back-propagation algorithm was used to classify LS as wheeze, squawk, stridor, crackle, rhonchus, or normal. For wheeze analysis, the Continuous Wavelet Transform (CWT) was used. To describe the analytical domain, CWT was combined with third-order spectra. Bi-coherence and instantaneous wavelet bi-spectrum, along with bi-phase and bi-amplitude curves, were used to discern the non-linear interactions of wheeze harmonics and their changes over time.

For TF analysis, many types of windows and wavelets were utilized, including Blackman, Gaussian, Hanning, Bartlett, Hamming, Rectangular, and Triangular, and for TS analysis, Mexican Hat, Paul, and Morlet. The frequency components with no information were removed using the Dual-Tree Complex Wavelet Transform (DTCWT) to improve crackling detection. MLP, k-Nearest Neighbor (k-NN), and SVM classifiers were given the ensemble and individual features as input. These features were used to compare and assess classification results across non-pre-processed and pre-processed data. Moreover, an integrated method is demonstrated for crackling recognition. The three serial modules in this system are crackles separation from LS using a wavelet packet filter, crackle detection using Fractal Dimension (FD), and crackling classification using GMM [18]. Two thresholds were used in the frequency and temporal domains to identify crackles from the LS. To locate each crackling using its FD, the output of the crackle peak detector and WPSTNST was sent through a denoising filter. The features estimated from LS were the Gaussian Bandwidth (GBW), Peak Frequency (PF), and Maximal Deflection Width (MDW). GMM was used to distinguish between coarse and fine crackles. Further, Wavelet Networks (WN) are used to quantify and parameterize crackles with the goal of representing the waveform with a smaller number of parameters. The complicated Morlet wavelets were used to represent the waveforms at the nodes of double- and single-node networks, with the double-node generating a lesser modeling error. WN was employed to define and classify fine and coarse crackles. The first node of the wavelet function was trained to fit the crackles, whereas the second node represented the presence of error in the first node. Scaling, frequency, time-shifting, and two weight factors were employed as crackles classification characteristics in the WN node parameters of cosine and sine components.

Hidden Markov Models for Maximum Likelihood Estimation: A maximum likelihood technique using hidden Markov models to classify normal and abnormal LS is discussed in the literature. The spectral and power properties of the LS signal were recovered, and a stochastic technique was used to detect anomalous LS.

Auto-regressive Models: This model compares various parameterization approaches for LS collected from the posterior thoracic position for the classification of healthy versus adventitious lung diseases. The eigenvalues of the covariance matrix, Multivariate Auto-regressive Models (MAR), and Univariate Auto-regressive (UAR) were utilized to create feature vectors. A Supervised Neural Network (SNN) was fed these feature vectors as input. AR models were used to extract feature characteristics from overlapping parts of lung sound. AR was used to analyze the lung sound of ill and healthy patients. Using the vectors, reference libraries were created. The performance of two classifiers, a quadratic and a k-NN classifier, was correlated and validated for multiple model orders. To classify individuals as healthy, interstitial lung disease, or bronchiectasis, the resulting model attributes were fed into an SVM classifier. Using ANN and k-NN classifiers, several classification experiments were conducted on each respiratory phase to classify pathological and normal LS. The AR model parameters and frequency spectrum were used to investigate the LS of healthy and ill people in order to create an LS-based diagnosis.

Audio Spectral Envelope Flux: Audio Spectral Envelope (ASE) was used as a feature vector to identify wheezes from regular respiratory sounds. The MPEG-7 standard contains a description of ASE. Isolated sound occurrences were divided into frames of 32 ms width and 8 ms shift. The features of each segment were extracted individually and then averaged to represent a single occurrence.

5.2 Miscellaneous Methods

A model based on a biomimetic multi-resolution investigation of spectro-temporal modulation features in pediatric auscultations in the LS is suggested. For parametric representation and automatic categorization of the LS, the signal coherence method is used. To determine the morphological embedding complexity, the lacunarity, skewness, kurtosis, and sample entropy were calculated. SVM and Extreme Learning Machine (ELM) were used to calculate such features from 20 aberrant and 10 normal subjects. Using these feature vectors, this method achieved a classification accuracy of 92.86%, a sensitivity of 86.90%, and a specificity of 86.30%. An iterative kurtosis-based detector was created using kurtosis. An iterative kurtosis-based detector was used to find the main kurtosis peaks. Those peaks were found within a sliding window along the LS signal, indicating that the raw LS signal had non-Gaussianity.

L. J. Hadjileontiadis [19] was the first to introduce automatic discrimination of discontinuous lung sounds, such as Squawks (SQ), Coarse Crackles (CC), and Fine Crackles (FC). This study proposed a texture-based classification method that used the lacunarity of lung sounds to distinguish the distribution of SQ, CC, and FC throughout the respiratory cycle. Hadjileontiadis used FD to analyze Bowel Sounds (BS) and Lung Sounds (LS) from healthy and bowel and pulmonary pathology patients, respectively. Despite fluctuations in time duration and/or amplitude, the study was able to determine the time location as well as the duration of LS and BS. Children's LS frequently differ from baseline LS in terms of amplitude and pattern behavior after bronchoconstriction. To test these ideas, pre- and post-methacholine challenge LS signals from eight children aged 9 to 15 years were analyzed using fractal-based and time-domain analysis (MCh). Sakaiat demonstrated a sparse representation-based approach for detecting unexpected LS. Because noise cannot be shown sparsely by any bases, the sparse representation can be used to distinguish clear LS and anomalous LS from noisy LS. These distinct sound components could be used to identify the degree of abnormality. Crackles, which were observed in the accidental LS, were identified as the pulsating waveforms.

To identify the existence of tuberculosis reported by Beckerat, the most essential aspects in the frequency and time domain were determined using the Statistical Overlap Factor (SOF). These characteristics were used to train an ANN to classify auscultation recordings as normal or TB-origin. Habukawa used two sensors to record LS from the trachea and the right anterior chest position, then calculated the acoustic transfer behavior between the two places, revealing a link between attenuation and frequencies during LS propagation. The features retrieved from those transfer characteristics were the tracheal sound index (TRI) and chest wall sound index.

Another metric, breath sound intensity (BSI), was developed to see if BSI might detect asthma better than TRI or CWI. There was a significant difference in TRI and BSI values between asthmatic and non-asthmatic children ($p = 0.007$, p 0.001), as well as a significant difference in CWI and TRI values between well-controlled and poorly controlled groups ($p = 0.001$). With a specificity of 84.2% and a sensitivity of 83.6%, BSI was able to distinguish between the two groups. The authors expressed their belief that BSI may be used to assess asthma control. The results of an automated Discrimination Analysis (DA) method for distinguishing coarse crackles (CCs), fine crackles (FCs), and squawks (SQs) are also studied. The oscillatory patterns of SQs, CCs, and FCs were found to be dissimilar using the proposed technique.

To distinguish the parameterized segments from each phase, MLP classifiers were utilized. To get a final view on each subject, the decision vectors from distinct phases were combined using a non-linear decision combination function. The Fisher Discriminant technique (FD) was utilized to distinguish between the wheeze and non-wheeze windows in LS recorded from COPD and asthma patients. A new method for assessing pneumonia patients' respiration sounds is also demonstrated. The feature extraction was done using a short time Fourier transform, and the dimensionality reduction was done with PCA. The classification was done with a Probabilistic Neural Network (PNN), which was found to be 77.6% accurate. Further, employing autocorrelation is proposed to extract respiratory sound features, which was done to apply the fast Fourier transform, and classification for diagnosis was done with 98.6% accuracy using ANN and Adaptive Neuro-Fuzzy Inference System (ANFIS).

The ANN and AdaBoost methods, as well as the Random Forest method, were utilized to classify asthma, with a classification accuracy of 90%. ANFIS correctly identified LSs as wheeze, crackles, stridor, and normal, providing a quantitative basis for aberrant respiratory sound detection. In patients with Interstitial Pulmonary Fibrosis (IPF), an automated method is presented for crackling analysis. This approach was presented in order to differentiate the crackling in patients with IPF from those with CHF and pneumonia. Crackles were recognized and tallied automatically by the software. SVMs and ANNs were used to classify diseases. IPF crackles were detected with an accuracy of 86%, while CHF crackles were discovered with an accuracy of 82%.

Spectrogram and 2D wavelet transforms, principle component analysis, sample entropy (Sampen) histograms, adaptive crisp active contour models (ACACM), Hilbert–Huang transform (HHT), cepstral coefficient, TVAR model, Time Expanded Waveform Analysis (TEWA), Vector Auto-regressive (VAR) process, Wavelet Packet Decomposition (WPD), Fractional Hilbert Transform, Dual-Tree Feature Extraction Wavelet-based time-frequency analysis can be used to extract features more successfully, as this method can readily be applied to non-stationary data.

These feature extraction algorithms have been combined with well-known machine learning and deep learning techniques, including naive Bayes classifiers, k-nearest neighbors, SVN, ANN, CNN, and RNN. In general, exactness findings for wheeze ranged from 97% to 70.2%; 97.5% to 86% for crackle; and 99% for normal sound types. Respiratory tract infections, pneumonia, bronchiectasis, bronchiolitis, asthma, and COPD were among the disorders investigated. For binary, ternary, and multi-class classification, these investigations demonstrated accuracy of up to 93.3%, 99%, and 98%, respectively [20].

It is indeed worth noting that hybrid deep learning algorithms produced the most strong results for multi-class categorization. While it shows promise without requiring the use of complex classification engineering methods, training a viable deep network architecture may be time-consuming and computationally expensive. Furthermore, the training procedure is iterative and involves a large number of model parameters and datasets. The shortage of abundant good, variety, and labeled training data is one of the most critical challenges in clinical settings. Previous research has employed a variety of data augmentation approaches to solve the issue of data availability, including variational autoencoders, adaptive synthetic sampling, and synthetic minority oversampling. Minority oversampling, on the other hand, might lead to data loss throughout the testing process, skewing the results toward high false accuracies.

Indeed, the majority of data augmentation strategies focus on improving classification performance rather than on the fundamental need of using fine-grained datasets that correctly reflect the population under investigation. A strong Deep Learning framework was built in a separate study. The trials tested the model's ability to classify noises from the ICBHI 2017 Challenge. The length of the respiratory cycle, the temporal resolution, and the network design are all elements that affect the ultimate prediction accuracy. CNN-MoE is an unique CNN that employs a variety of trained models. The model had an correctness of 0.80 and 0.86 for the 4-class and 2-class subtasks, correspondingly, when it came to classifying respiratory anomalies.

By creating their own stethoscope and recording their own patients, a study team in Turkey employed SVM and CNN to classify lung sounds into multiple categories. There were 17,930 sounds in all, with 1630 themes. To obtain a base value for accuracy, feature abstraction was done using MFCC augmented with STFT. Using open-source tools and Pylab, a spectrogram (800 $\times$ 600) RGBA, then (28 $\times$ 28) was created. According to Aykanat, spectrogram image classification using CNN performs as well as SVM.

The model's input consists of MFCCs extracted in the frequency range of 50 Hz to 2,000 Hz, as well as their first derivatives. According to the ICBHI score, the approach produces performance results of up to 39.37%. Class ensemble received the highest official score of 39.56. With 6 states and full covariance matrix type, the second-best result was 0.4232 sensitivity, 0.5669 specificity, and a 39.37 official score. The classifier's sensitivity is dropping, indicating that it was unable to resolve unexpected sound types.

Following feature extraction and segmentation, different classification methods such as MFCC, SVM, MLP, Extreme Learning Machine Neural Network (ELMNN), GMM, Maximum Likelihood (ML) approach using Hidden Markov Models (HMM), k-NN classifiers, GAL, ANN, Constructive Probabilistic Neural Network (CPNN), and Supervised Neural Network (SNN) is carried out. For the inspiratory and expiratory segments, the SVM had the maximum accuracy of 97.7% and 98.84% as shown in Figs. 2 and 3, respectively [20].

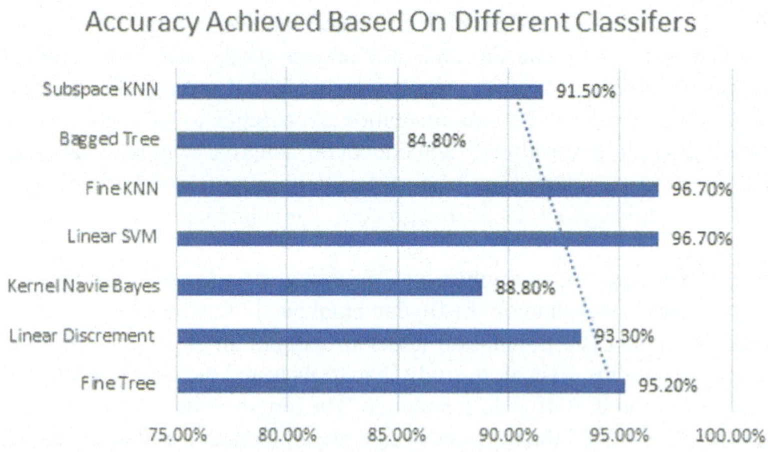

Fig. 2 Accuracy of different classifiers

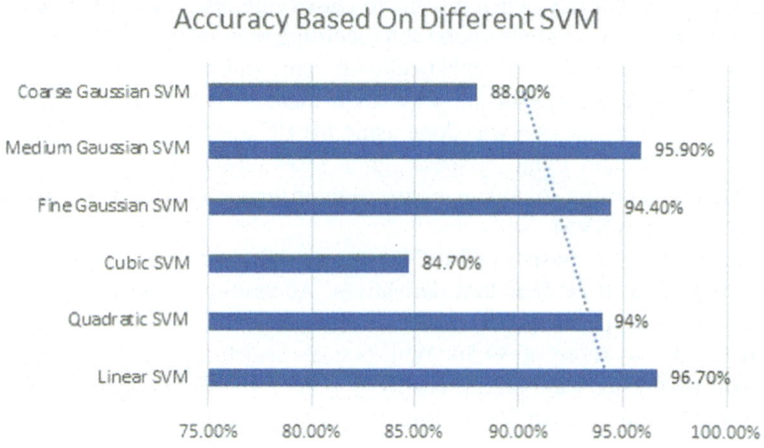

Fig. 3 Accuracy of SVM types

6 Conclusion

For diagnosing the type of lung ailment, a lot of work has been carried out to extract features and classify lung sounds. Many studies have shown that lung sound can be used to diagnose lung disease. However, some chronic diseases have symptoms that are quite similar, and as a result, they are frequently misdiagnosed. There is still work to be done in terms of diagnosing and discriminating against chronic diseases. It will also assist clinicians in making accurate diagnosis decisions in chronic diseases. It has also been discovered that the best disease classification method for making diagnostic decisions is a support vector machine. Existing methods have attempted to cover the detection and discrimination of various lung sounds, but more work remains to be done in order to identify and distinguish lung disorders.

References

1. Moorthy, D.P., Harikrishna, M., Mathew, J., Sathish, N.: Sound classification for respiratory diseases using machine learning technique. Int. Res. J. Engin. Technology **8**(4), 3779–3782 (2021)
2. Kim, Y., Hyon, Y., Jung, S.S., Lee, S., Yoo, G., Chung, C., Ha, T.: Respiratory sound classification for crackles, wheezes, and rhonchi in the clinical field using deep learning. Sci. Rep. **11**(1), 17186 (2021)
3. Jácome, C., Aviles-Solis, J.C., Uhre, Å.M., Pasterkamp, H., Melbye, H.: Adventitious and normal lung sounds in the general population: comparison of standardized and spontaneous breathing. Respir. Care **63**(11), 1379–1387 (2018)
4. Jácome, C., Aviles-Solis, J.C., Uhre, Å.M., Pasterkamp, H., Melbye, H.: Adventitious and normal lung sounds in the general population: comparison of standardized and spontaneous breathing. Respir. Care **63**(11), 1379–1387 (2018)
5. Sengupta, N., Sahidullah, M., Saha, G.: Lung sound classification using cepstral-based statistical features. Comput. Biol. Med. **75**, 118–129 (2016)
6. Nguyen, T., & Pernkopf, F.: Lung sound classification using snapshot ensemble of convolutional neural networks. In: Proceedings of the 42nd Annual International Conference of the IEEE Engineering in Medicine and Biology Society, pp. 760–763 (2020)
7. Shi, L., Du, K., Zhang, C., Ma, H., Yan, W.: Lung sound recognition algorithm based on vggish-bigru. IEEE Access **7**, 139438–139449 (2019)
8. Villanueva, C., Vincent, J., Slowinski, A., Hosseini, M.P.: Respiratory sound classification using long short term memory. arXiv preprint arXiv:2008.02900
9. Pramono, R.X.A., Bowyer, S., Rodriguez-Villegas, E.: Automatic adventitious respiratory sound analysis: a systematic review. PLoS ONE **12**(5), e0177926 (2017)
10. Jones, A., Jones, R.D., Kwong, K., Burns, Y.: Effect of positioning on recorded lung sound intensities in subjects without pulmonary dysfunction. Phys. Ther. **79**(7), 682–690 (1999)
11. Sarkar, M., Madabhavi, I., Niranjan, N., Dogra, M.: Auscultation of the respiratory system. Ann. Thoracic Med. **10**(3), 158–168 (2015)
12. Ponte, D.F., Moraes, R., Hizume, D.C., Alencar, A.M.: Characterization of crackles from patients with fibrosis, heart failure and pneumonia. Med. Engin. Phys. **35**(4), 448–456 (2013)
13. Serbes, G., Sakar, C.O., Kahya, Y.P., Aydin, N.: Pulmonary crackle detection using time-frequency and time-scale analysis. Digital Signal Process. **23**(3), 1012–1021 (2013)
14. Asatani, N., Kamiya, T., Mabu, S., Kido, S.: Classification of respiratory sounds using improved convolutional recurrent neural network. Comput. Electr. Engin. **94**, 107367 (2021)
15. Rocha, B.M., Pessoa, D., Marques, A., Carvalho, P., Paiva, R.P.: Automatic classification of adventitious respiratory sounds: a (un) solved problem? Sensors **21**(1), 57 (2020)
16. Reyes, B. A., Charleston-Villalobos, S., Gonzalez-Camarena, R., Aljama-Corrales, T.: Analysis of discontinuous adventitious lung sounds by Hilbert-Huang spectrum. In: Proceedings of the 30th Annual International Conference of the IEEE Engineering in Medicine and Biology Society, pp. 3620–3623 (2008)
17. Jin, F., Sattar, F., Goh, D.Y.: New approaches for spectro-temporal feature extraction with applications to respiratory sound classification. Neurocomputing **123**, 362–371 (2014)
18. Lu, X., Bahoura, M.: An integrated automated system for crackles extraction and classification. Biomed. Signal Process. Control **3**(3), 244–254 (2008)

19. Hsu, F.S., Huang, S.R., Huang, C.W., Huang, C.J., Cheng, Y.R., Chen, C.C., Chen, Y.T., Lai, F.: Benchmarking of eight recurrent neural network variants for breath phase and adventitious sound detection on a self-developed open-access lung sound database-HF Lung V1. PLoS ONE **16**(7), e0254134 (2021)
20. Kim, Y., Hyon, Y., Jung, S.S., Lee, S., Yoo, G., Chung, C., Ha, T.: Respiratory sound classification for crackles, wheezes, and rhonchi in the clinical field using deep learning. Sci. Rep. **11**(1), 17186 (2021)

Analysis of Forecasting Models of Pandemic Outbreak for the Districts of Tamil Nadu

P. Iswarya, H. Sharan Prasad, and Prabhujit Mohapatra

Abstract The research is conducted based on the primary data available on the data portal which is gathered from different sources of the Government and the Private. There have been several efforts for analyzing and predicting future COVID-19 cases based on primary data. The present study is based on an inferential methodology which is one of the most widely used data science techniques in the study of events like COVID-19 time-series analysis. Analyzing and predicting the COVID-19 cases in upcoming months utilizing SIR, ARIMA models, and forecasting. The implementation of the proposed approach is demonstrated on real-time data of districts in Tamil Nadu. The current work serves to be of great importance in the prediction of the COVID-19 crisis in day-to-day life.

1 Introduction

A novel coronavirus (COVID-19) is a new respiratory virus, firstly discovered in humans in November 2019 that has spread worldwide, having been more than 124 million infected people so far. In India, it has been more than 11 million, and in Tamil Nadu, 8 lakhs 68 thousand positive cases have been registered. It has constrained us to receive another way of life which is not the same as the one we are utilized to. Due to this pandemic, our economic survival is a big question mark. And also, while growth rate of corona cases 0.5% at the national level, the development rate in Tamil Nadu is 0.4%. World Wide, the most elevated recuperation rate is 90.85%. In Tamil Nadu, 93.36% have recovered based on earlier data. The rate of recovery went sharply from mid-July, almost in parallel to new cases getting reported. For instance,

P. Iswarya · H. Sharan Prasad · P. Mohapatra (✉)
School of Advanced Sciences, Vellore Institute of Technology University, Vellore, India
e-mail: prabhujit.mohapatra@vit.ac.in

P. Iswarya
e-mail: iswarya.p2019@vitstudent.ac.in

H. Sharan Prasad
e-mail: sharanprasad.h2019@vitstudent.ac.in

D. P. Acharjya and K. Ma (eds.), *Computational Intelligence in Healthcare Informatics*, Studies in Computational Intelligence 1132,
https://doi.org/10.1007/978-981-99-8853-2_16

recoveries reported daily were higher than new cases reported on 40 of the past 60 days. Despite cases coming down, Tamil Nadu was among the six states reporting more than thousand cases per day. So, in this study, the prediction report can be seen for the various districts in Tamil Nadu and can be aware those districts priorly.

The rest of the chapter is prepared as follows: Sect. 2 studies the research on literature survey. Section 3 shows light on the methodology used for the prediction followed by result analysis in Sect. 4. Finally, Sect. 5 concludes this research work.

2 Literature Survey

Behl and Mishra [1] used SIR model on COVID-19 life cycle: predictive modeling of states in India. It also uses forecasting technique of ARIMA to predict cases prone to COVID-19 in every state. It utilized SIR model for predictive accuracy on Python Jupyter notebook and Matplotlib 3.2.1 platforms. Further it has been analyzed the SEIR model to outbreak, forecast the number of cases in China and outside China [2]. The model forecasted the cases for 240 days from January 2020. Likewise, 4 distinct models are utilized to forecast the excepted number of affirmed cases over the upcoming 4 weeks in Saudi Arabia [3]. They evaluated the presentation of the models the information was part into two subsets and was utilized for training the models. The data was partitioned into training and testing sets in 70 and 30 percent, respectively. Similarly, Khan and Gupta [4] researched on ARIMA- and NAR-based predicting model used to forecast the COVID-19 cases in India from April 2020. It was predicted using ARIMA model and NAR neural networks. They predicted it was 1500 cases per day. The prediction was made with existing conditions. Again, a research paper uses time-dependent SIR model for COVID-19 with undetectable infected persons using SIR model to track the spread and recovery rate at time t [5]. It used the information of China jurisdiction and predicted the values. The prediction uses SIR model to dissect the effect of the unnoticeable contaminations on the outspread of the infection.

Bohner et al. [6] used dynamic epidemic model based on Bailey classical differential system and derive its exact solution. They end the conversation by investigating the solidness of the output to the dynamic model on account of steady coefficient. Further, a research work presents a model on dynamic tracking with model-based forecasting for the spread of COVID-19 pandemic by utilizing SIR model by the arrangement of three coupled non-linear ordinary differential equations describing the evaluation [7]. It took the data from Italy, India, South Korea, and Iran. Additionally, research on Corona Virus (COVID-19): ARIMA-based time-series analysis to forecast near future utilized the ARIMA model to forecast COVID-19 incidences which is discussed [8]. The parameters are estimated for the ARIMA (2, 2, 2) model, and thus the p-value is found.

3 Research Methodology

As mentioned before, the primary data have been used throughout the research analysis. Data is collected from Tamil Nadu government website. The main objective of this research work is to study the possibility of high positive cases in the districts of Tamil Nadu. It will help researcher to think on various research models over data analysis and predictive models.

3.1 SIR Model

First, in the SIR model, one must see $s(t)$ should vary with time or $r(t)$ vary with time or $i(t)$ vary with time. Time has to be explained $s(t) + i(t) + r(t) = 1$. For the above equation, individual formulas for susceptible $s(t)$, infected $i(t)$, and recovered $r(t)$ in SIR are defined. The susceptible expression is defined in Eq. 1 whereas recovered expression is defined as in Eq. 2. The infected expression is defined in Eq. 3.

$$\frac{ds}{dt} = -bs(t).I(t) \tag{1}$$

$$\frac{dr}{dt} = k.i(t) \tag{2}$$

$$\frac{di}{dt} = b.s(t).i(t) - k.i(t) \tag{3}$$

These equations lead to $S'(t) = -rSI$, $I'(t) = rSI - aI$, and $r'(t) = aI$. The r and a in the above equations, the differential condition is known as the disease rate and recuperation pace of COVID-19 of Tamil Nadu. The proposed investigation of the normal season of COVID-19 flare-up of Tamil Nadu is around 2 months. The mathematical estimation of r and a is valuable in the underlying level for settling the three differential conditions of coronavirus episode in Tamil Nadu.

$$\frac{ds}{dt} = -rSI \tag{4}$$

$$\frac{dI}{dt} = rSI - aI \tag{5}$$

$$\frac{dR}{dt} = aI \tag{6}$$

Equations 4–6 are known as the Kermack–Mckendrick equations. This model SIR is exceptionally helpful for the data examination of coronavirus in India and in Tamil Nadu. Again, adding Eqs. 4–6 it is obtained as

$$\frac{ds}{dt} + \frac{dI}{dt} + \frac{dR}{dt} = -rSI + rSI - aI + aI = 0 \tag{7}$$

On solving, $S' + I' + R' = N$, where N is known as the consistent of joining which is to gauge the absolute size of the number of inhabitants in COVID-19 at introductory level. All the qualities at the underlying state begin with 0 that is $S(0), I(0), R(0)$. There are two cases that are involved in SIR model. The two cases are as follows:

Case1: If S_0 is less in ratio $\frac{a}{r}$, then the infection I will be decreased or on the other hand essentially goes to zero after some time.

Case2: If S_0 is greater than the ratio $\frac{a}{r}$, then the infection I will be epidemic.

3.2 *ARIMA Model*

It uses the data in the past estimations of the time arrangement which can alone be utilized to foresee future qualities. Here, in ARIMA (p, d, q), it is to be noted that p is autoregressive, d is difference value, and q is moving average.

3.3 *Forecasting*

In an exponential smoothing model, at every period, the model will take in a piece from the latest request perception and recall a part of the last estimation. Therefore, the last estimate the model did was added a part of past request perception and a piece of the past prediction. That suggests that this previous estimate fuses everything the model adapted so far subject to ask for request history. The mathematical representation of exponential smoothing is

$$ft = \alpha dt - 1 + (1 - \alpha) ft - 1; \quad 0 < \alpha \leq 1 \tag{8}$$

In future prediction, when crossed the recorded period, one need to populate a prediction for future. Therefore, $Ft > t* = ft*$, i.e., f(t *). The last prediction that could make dependent on request history.

4 Results and Discussions

By implementing the models, discussed in the previous section for the dataset the output is given below. SIR model's output from MATLAB is depicted in Fig. 1.

According to Fig.1, the recovery rate is higher and the susceptible graph and infected are slightly increasing in the span of 8 weeks. The SIR model is very use-

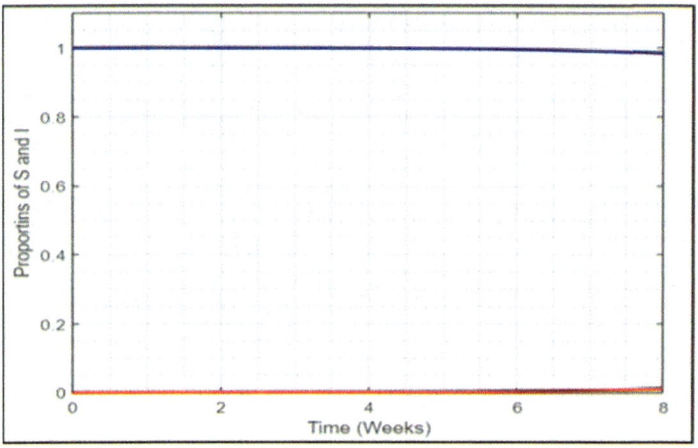

Fig. 1 Output of SIR model

Table 1 ARIMA model of Chennai

Iteration	SSE	p−value	d−value
0	9535283	0.100	–54.542
1	8595767	0.250	–45.472
2	8112659	0.400	–36.545
3	8042207	0.475	–32.440
4	8041374	0.483	–32.294
5	8041363	0.484	–32.292
6	8041363	0.484	–32.292

ful for the pandemic situations like this. By the result that the hospitals and the government can be cautious and can take necessary actions.

Using coefficient variation ranking and predicted value average, among the districts of Tamil Nadu, Chennai, Karur, Erode, Chengalpattu, and Tiruppur Districts have registered the greatest number of positive cases till February 2021. According to ARIMA model, at Chennai, the estimates at each iteration are presented inbreak Table 1, where relative change in each estimate is less than 0.001. The final estimates of parameters are presented in Table 2. It is to be noted that in differencing, one standard distinction is considered, whereas in number of perceptions: original arrangement is 229 and subsequent to differencing is 228.

According to residual sums of squares (SS), the SS is 7993736 and mean sums of squares (MS) is 35370.5 where degree of freedom is 226. Adjusted box-pierce (Ljung–Box) Chi-square statistic is presented in Table 3. The results relating to ARIMA model for the Chennai City are presented in Fig. 2. From the results and Fig. 2, it is seen that ARIMA model fitting and moving average fitting is with difference 1. Chennai has the highest residual sum of squares with 7993736.

Table 2 Final estimates of parameters of Chennai

Type	Coefficient	SE coefficient	T-value	P-value
AR 1	0.4838	0.0581	8.32	0.000
Constant	–32.3	12.5	–2.59	0.010

Table 3 Adjusted box-pierce chi-square statistic for Chennai

Parameters	Case 1	Case 2	Case 3	Case 4
Lag	12	24	36	48
Chi-square	38.65	57.40	75.19	108.77
DF	10	22	34	46
P-value	0.000	0.000	0.000	0.000

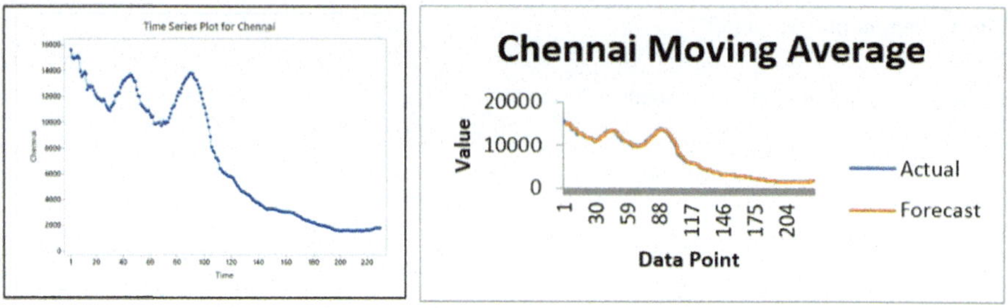

Time series plot Moving average

Fig. 2 ARIMA model results of the Chennai city

Table 4 ARIMA model of Tiruppur

Iteration	SSE	p–value	d–value
0	1615605	0.100	–0.439
1	1484642	–0.050	–0.563
2	1422062	–0.200	–0.703
3	1416185	–0.259	–0.770
4	1416169	–0.262	–0.778
5	1416169	–0.262	–0.779

Like Chennai, according to ARIMA model, at Tiruppur the estimates at each iteration are presented in Table 4, where relative change in each estimate is less than 0.001. The final estimates of parameters are presented in Table 5. It is to be noted that in differencing, one standard distinction is considered whereas in number of perceptions: original arrangement is 229 and subsequent to differencing is 228.

According to residual sums of squares (SS), the SS is 1416101 and MS is 6265.93 where degree of freedom is 226. Adjusted box-pierce (Ljung–Box) chi-square statis-

Table 5 Final estimates of parameters of Tiruppur

Type	Coefficient	SE coefficient	T-value	P-value
AR 1	–0.2623	0.0642	–4.09	0.000
Constant	–0.78	5.24	–0.15	0.882

Table 6 Adjusted box-pierce chi-square statistic for Tiruppur

Parameter	Case 1	Case 2	Case 3	Case 4
Lag	12	24	36	48
Chi-square	12.31	26.04	37.12	48.73
DF	10	22	34	46
P-value	0.265	0.250	0.327	0.364

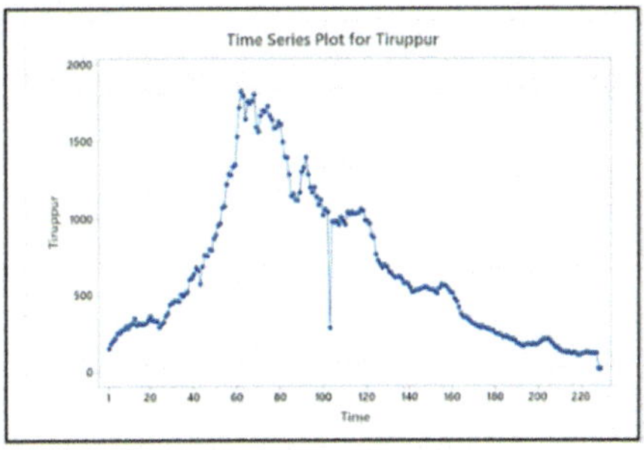

Time series plot

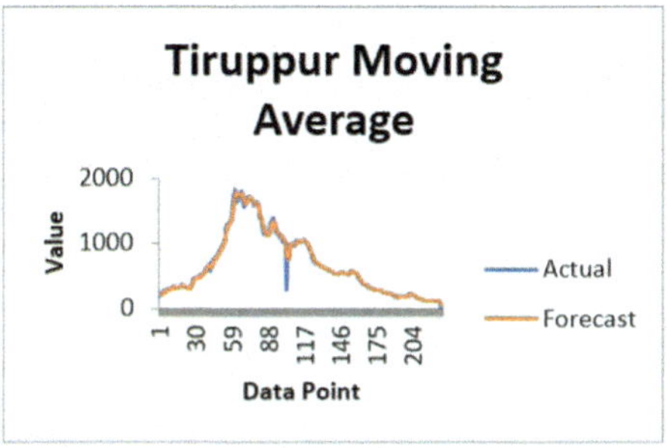

Moving average

Fig. 3 ARIMA model results of the Tiruppur City

tic is presented in Table 6. The results relating to ARIMA model for the Tiruppur City are presented in Fig. 3. From the results and Fig. 3, it is seen that ARIMA model fitting and moving average fitting is with difference 1. Tiruppur has the less residual sum of squares with 1416101.

Further for forecasting using SPSS, the model statistics is computed. It is presented in Figs. 4 and 5. From Figs. 4 and 5, it is obtained that model fitting difference is 14, and Tiruppur difference has the greatest increase with significance of 0.741 in upcoming weeks and where Chengalpattu has the least or no increase with 0.001 among these five districts. The parameters are estimated for the ARIMA (3,1,1) (0,0,0) model and thus the p-value is calculated.

Model Statistics

Model	Number of Predictors	Model Fit statistics		Ljung-Box Q(18)			Number of Outliers
		Stationary R-squared	Normalized BIC	Statistics	DF	Sig.	
Chennai-Model_1	0	.300	10.515	28.946	14	.011	0
Karur-Model_2	0	.114	6.901	19.375	14	.151	0
Erode-Model_3	0	.142	7.099	27.058	14	.019	0
Chengalpattu-Model_4	0	.095	9.038	36.109	14	.001	0
Tiruppur-Model_5	0	.100	8.841	10.288	14	.741	0

Fig. 4 Model statistics

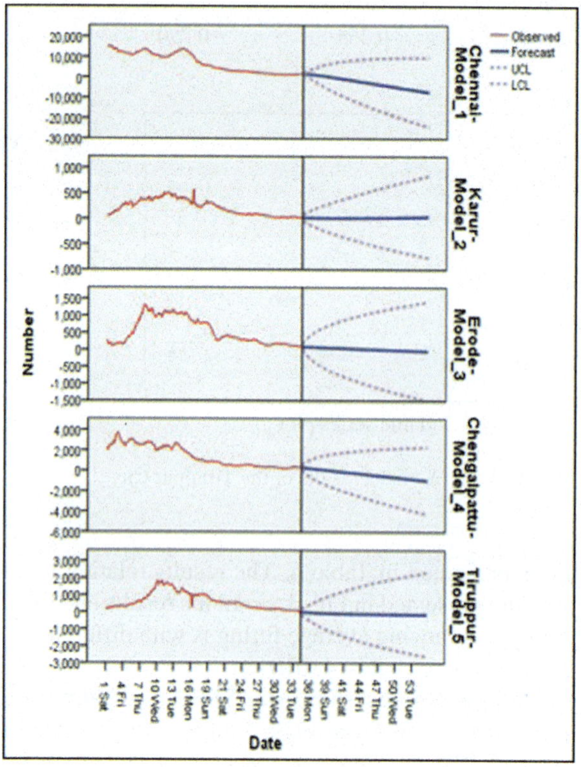

Fig. 5 Moving average of Tiruppur

5 Conclusion

The models implemented in this chapter are SIR model which is a mathematical model and ARIMA model which is a time-series analysis. The data which is used in this study passed all the test cases and validated because all the results in three software are the same. The software used in this chapter are SPSS, Minitab, MATLAB, and MS-Excel. Based on the result Tiruppur District had a drastic increase in the positive cases as compared to all other districts. Where Chengalpattu District has much lower cases among the four other districts. Implementation of ARIMA forecasting in Minitab and Tableau is the future extension of this research work.

References

1. Behl, R., Mishra, M.: COVID-19 lifecycle: predictive modelling of states in India. Glob. Bus. Rev. **21**(4), 883–891 (2020)
2. Gade, D.S., Aithal, P.S.: Smart cities development during and post COVID-19 pandemic-a predictive analysis. Int. J. Manage., Technol., Soc. Sci. **6**(1), 189–202 (2021)
3. Alzahrani, S.I., Aljamaan, I.A., Al-Fakih, E.A.: Forecasting the spread of the COVID-19 pandemic in Saudi Arabia using ARIMA prediction model under current public health interventions. J. Infect. Public Health **13**(7), 914–919 (2020)
4. Khan, F.M., Gupta, R.: ARIMA and NAR based prediction model for time series analysis of COVID-19 cases in India. J. Safety Sci. Resilience **1**(1), 12–18 (2020)
5. Alzahrani, S.I., Aljamaan, I.A., Al-Fakih, E.A.: Forecasting the spread of the COVID-19 pandemic in Saudi Arabia using ARIMA prediction model under current public health interventions. J. Infect. Public Health **13**(7), 914–919 (2020)
6. Bohner, M., Streipert, S., Torres, D.F.: Exact solution to a dynamic SIR model. Nonlinear Anal. Hybrid Syst **32**, 228–238 (2019)
7. Cooper, I., Mondal, A., Antonopoulos, C.G.: Dynamic tracking with model-based forecasting for the spread of the COVID-19 pandemic. Chaos, Solitons Fractals **139**, 110298 (2020)
8. Tandon, H., Ranjan, P., Chakraborty, T., Suhag, V.: Coronavirus (COVID-19): ARIMA-based time-series analysis to forecast near future and the effect of school reopening in India. J. Health Manag. **24**(3), 373–388 (2022)
9. Roy, S., Bhunia, G.S., Shit, P.K.: Spatial prediction of COVID-19 epidemic using ARIMA techniques in India. Model. Earth Syst. Environ. **7**, 1385–1391 (2021)

Suppression of Artifacts from EEG Recordings Using Computational Intelligence

Bommala Silpa, Malaya Kumar Hota, and Norrima Mokthar

Abstract Brain–computer interface system will be useful for physically disabled people to analyze and diagnose different health problems. The signal processing module is a major part of the brain–computer interface system. It is divided into four sub-modules notably, pre-processing, feature extraction, feature selection, and classification. EEG captures small amounts of brain activity and is a well-known signal acquisition method, due to its good temporal resolution, cheap cost, and no significant safety concerns. The objective of the EEG-based brain–computer interface system is to extract and translate brain activity into command signals which helps physically disabled people. The EEG recordings are contaminated by a variety of noises generated from different sources. Among these, the eye blinks have the greatest influence on EEG signals because of their high amplitude. This chapter provides a detailed review of the basic principles of various denoising methods, which also succinctly presents a few of the pioneer's efforts. Further, the comparative analysis is carried out using EMD, AVMD, SWT, and VME-DWT methods for filtering eye blink artifacts. The VME-DWT method is found to perform better than the SWT, AVMD, and EMD methods in terms of signal information retention, which perfectly encapsulates the relevance of our quantitative study. Computational intelligence develops a new approach for identifying and analyzing discriminating characteristics in signals. An EEG-based brain–computer interface system should use computational intelligence to reduce the noises from EEG data proficiently.

Keywords EEG · Brain-computer interface · Computational intelligence · EMD · AVMD · SWT · VME-DWT

B. Silpa · M. K. Hota (✉)
School of Electronics Engineering, Vellore Institute of Technology, Vellore, Tamil Nadu, India
e-mail: malayakumar.h@vit.ac.in

B. Silpa
e-mail: bommala.silpa2020@vitstudent.ac.in

N. Mokthar
Faculty of Engineering, University of Malaya, Kuala Lumpur, Malaysia
e-mail: norimmamokthar@um.edu.my

D. P. Acharjya and K. Ma (eds.), *Computational Intelligence in Healthcare Informatics*, Studies in Computational Intelligence 1132,
https://doi.org/10.1007/978-981-99-8853-2_17

1 Introduction

Brain–Computer Interface (BCI) research was initially started in the 1970s, addressing a different communication route that is independent of the brain's usual ancillary nerve and muscle output channels [1]. An early notion of the BCI system advocated monitoring and interpreting the brain signals to operate a prosthetic arm and perform a particular movement [2]. The term "BCI" is then described as a systematic communication route between the human brain and an external device. Human BCIs have received a lot of interest during the last decade, striving to translate human cognitive patterns using brain activity. It communicates with the computer via recorded brain activity to control external objects or environments in a way that is consistent with self-indulge.

BCIs are classified into two categories. The first one is active and reactive BCI. For operating a device, the active BCI develops a pattern from brain activity that is physically and knowingly manipulated by the clients, independent of external circumstances [3]. The reactive BCI extracts outputs from brain processes in response to external stimulus, which is then manipulated indirectly by the user to operate an application. The second one of BCI is passive BCI, which investigates the appearance, consciousness, and intelligence despite requiring intentional control to supplement a Human–Computer Interaction (HCI) with implicit data [4].

Several investigations have discovered that brain signals may be used to do any activity requiring muscular movement's control [5]. Small quantities of electromagnetic waves generated by neurons in the brain produce EEG signals [6]. Because of its non-invasiveness, flexibility, adequate temporal resolution, and expenditure in comparison to other brain signal recording methods, EEG is one of the most often utilized signal collection techniques in present BCI systems [7]. EEG signals are produced by a stochastic biological mechanism that is very complicated, non-linear, and non-stationary, which includes disruptions from different sources.

Researchers frequently use a variety of EEG signal collection devices [8]. In recent times, portable wireless dry or wet electrode EEG recording equipment that is low-cost and simple to use, have been employed by researchers in several investigations. The goal of BCI systems is to take out certain signatures of brain impulses and turn them into directions that may be used to control external devices [9]. The signal processing section of the BCI system is divided into four modules, namely, pre-processing, feature extraction, feature selection, and classification. The feature selection module [10] is frequently skipped by researchers since it's only useful when the quantity of the features extracted by the feature extraction stage are quite large. Slower execution times occur from large feature sets, rendering many BCI systems, particularly online BCI systems, utterly worthless. As a result, the feature selection phase is used to save processing time by reducing the dimensionality of the data.

BCI systems based on EEG have a wide range of applications such as brain controlled wheelchair, robot controlling, emotion classification, rehabilitation of locked-in patients, cognitive task classification, neuro-prosthesis, P300 spellers, and gaming. Therefore, every level of the BCI system requires the application of computational intelligence, beginning with noise removal, progressing to feature extraction and selection, and finally signal classification.

The book chapter is organized as follows: Sect. 2 discusses computational intelligence followed by the background of the EEG signal in Sect. 3. Section 4 briefly discusses the artifacts removal techniques and Sect. 5 describes performance evaluation and discussion. Finally, the conclusion is presented in Sect. 6.

2 Computational Intelligence

Computational Intelligence (CI) is the design, theory, implementation, and development of physiologically and linguistically inspired methods [11]. It is a new, emerging computer discipline, and defining it is difficult. Besides, it employs novel models, many of which have a high machine learning quotient. Computational intelligence's primary applications are engineering, computer science, bio-medicine, and data analysis. CI is a system that works with low-level quantitative data; has a pattern-recognition component but does not use knowledge in the sense of artificial intelligence; and has fault tolerance, simulation adaptability, and error rates comparable to human performance.

The primary pillars of computational intelligence are Evolutionary Computing (EC), Swarm Intelligence (SI), Artificial Neural Networks (ANN), and Fuzzy Systems (FS). The beginnings of each computational intelligence paradigm are from biological systems. Artificial neural networks simulate biological neural systems, evolutionary computing simulates genetic evolution, swarm intelligence simulates the social norms of organisms that live in swarms, and the fuzzy system is based on research into how organisms interact with their surroundings.

Evolutionary algorithms are based on principles from biological evolution, in which organisms adapt in changing environmental conditions over many generations. Swarm intelligence evolved from the study of colonies of social creatures. The study of the social behavior of organisms in swarms spurred the development of highly effective optimization and clustering algorithms. Both EC and SI can handle certain kinds of optimization issues and are part of the metaheuristics family, which provides methods to solve a wide range of optimization problems. Metaheuristics are a broad framework for optimizing different types of problems that explore search space regions with efficient learning operators and exploit accumulative search controlled by given criteria.

Metaheuristics are a comprehensive optimization framework that uses efficient learning operators to explore search space regions and perform accumulative searches controlled by specified parameters to solve a range of optimization issues. Metaheuristics are used to solve problems for which no efficient algorithm exists. Metaheuristics success and processing time are vitally reliant on a thorough mapping of the problem to the metaheuristics stages and the effective execution of each step. These algorithms work iteratively developing a set of candidate solutions. One significant benefit is that, it can often be stopped after any iteration step since they have at least some solution candidates available at all times. The details of evolutionary computing and swarm intelligence are discussed in the following section.

2.1 Evolutionary Computing

Evolutionary computing is a natural evolution-inspired computational intelligence approach. This approach begins with the formation of a group of people who responds to a problem. The first population can be generated randomly by an algorithm and a new population will be created based on the fitness value of a previously developed population. This process continues until the termination requirements are satisfied, and many classes of EC algorithms have been developed. The components which have an impact on the evolutionary search process are encoding problem-solving solutions, generating an initial population, fitness function to assess population, selection based on fitness, genetic operators for population modification, criteria for search termination, and values for different parameters.

The Harmony Search Algorithm (HSA) is an evolutionary metaheuristics algorithm motivated by the harmonic process of seeking a good state of harmony [12]. HSA is simple to implement, quickly converges to the best solution, and produces a great sufficient solution in a reasonable time. The procedure of HSA has the following steps:

1. Initialize harmony search parameters.
2. Initialize a Harmony Memory (HM).
3. Create a new harmony from HM. The generation of new harmony is based on three parameters such as Harmony Memory Consideration Rate (HMCR), Pitch Adjustment Rate (PAR), and random consideration (r_1 and r_2). If the random variable $r_1 < HMCR$, the new vector can be formed randomly changing the values in the current HM, and another random variable $r_2 < PAR$, then the pitch adjustment is calculated using Eq. 1 where x_{ij}^{new} is a new harmony, x_{ij} is current harmony, *rand* represents the random number, and FW is the fret width.

$$x_{ij}^{new} = x_{ij} + rand(0, 1) + FW \tag{1}$$

4. If the new harmony is superior to the minimal harmony in HM, update HM with the new harmony.
5. If the halting requirements are not met, go to step 2.

2.2 Swarm Intelligence

A swarm is an organized cluster of interacting creatures. Swarm intelligence evolved from the investigation of colonies of social species. The study of the social conduct of organisms (individuals) in swarms inspired the development of highly efficient optimization algorithms like the choreography of bird flocks inspired the development of the particle swarm optimization technique [13]. Besides, many swarm intelligence techniques are studied in healthcare applications [14].

Particle Swarm Optimization (PSO) is a universal optimization algorithm developed based on the behavior of the flocks of birds. PSO is a technique where a group of particles moves iteratively in search space for finding better results. The procedure of the PSO algorithm is presented below:

1. Initialize PSO parameters.
2. Generate a random population equal to the particle size.
3. Determine the objective function values for all particles. The objective function values reflect the local best (*pBest*) for each particle, where the particle with the best objective function value being the global best (*gBest*).
4. Update the velocity of each particle using Eq. 2, where v_{ij}^{t+1} and v_i^t represent particles next and current velocities with iteration i and position t, a_1 and a_2 are acceleration constants, and u_1 and u_2 are random numbers. The notion pb_i^t and gb^t are local and global best values of the particles.

$$v_{ij}^{t+1} = v_i^t + a_1 u_1^t (pb_i^t - p_i^t) + a_2 u_2^t (gb^t - p_i^t) \tag{2}$$

5. Update the positions of particles using Eq. 3, where p_i^{t+1} and p_i^t are particles next and current positions.

$$p_i^{t+1} = p_i^t + v_i^{t+1} \tag{3}$$

6. The new particles with the best fitness value replace the value of *gBest* in the population.
7. Save the obtained best results for further process.
8. Repeat the steps (beginning with step 4) till the requirement is met.

3 Characteristics of the EEG Signal

The amplitudes of EEG signals are relatively tiny, measuring between 30 and 100 microvolts, but usually less than 50 microvolts [15]. The frequency range of an EEG signal is from 0.1 Hz to 100 Hz, and distinct frequency bands δ, θ, α, β, and γ are based on particular relationships with human behavior. The frequency range of bands differs from one to another as shown in Table 1. The source of an aberration in human behavior can be explained by analyzing human EEG waves.

Table 1 EEG frequency bands with their frequencies

Frequency band	Range of frequency band	Conditions and mental states
Delta (δ)	0.1 Hz to 4 Hz	Deep sleep
Theta (θ)	4 Hz to 8 Hz	Relaxed state and meditation
Alpha (α)	8 Hz to 13 Hz	A relaxed state of conscious, but not drowsy
Beta (β)	13 Hz to 30 Hz	Problem solving, judgment, and decision making
Gamma (γ)	30 Hz to 100 Hz	Higher mental activity, Motor functions

3.1 Types of Artifacts

Artifacts are signals observed by an EEG that do not originate in the brain. They can occur at any time during the recording procedure. Sometimes the magnitude of the artifacts is bigger than the amplitude of the EEG signals obtained from the brain, resulting in distorted EEG signals. Artifacts are classified into two types based on their origin: non-physiological and physiological [16]. Non-physiological artifacts are caused by power lines, inadequate contact of electrodes or sudden discharge of electrodes called "electrode pops", and movements of surroundings around the patient. Physiological artifacts are caused by the body itself and include ocular artifacts, muscle artifacts, and cardiac artifacts [17].

Non-physiological Artifacts: The different types of non-physiological artifacts are power line interference, electrode artifacts, and environmental artifacts.

Power line interference: To eliminate this effect from power lines, appropriate grounding is always supplied to the patient. However, recording devices might still be affected by electromagnetic interference from power lines. Because this interference loads the amplifiers of the electrodes. The ground of the amplifier becomes an active electrode, and the EEG recording captures the frequency components of the power line. This sort of artifact is identified by a rise in the time base.

Electrode artifacts: The electrode artifacts occur by an abrupt impedance change between the electrode and the skin. Waveform can have single or multiple sharp shifts. Until shown differently, a rapid discharge that happens at electrodes is considered an artifact. The impedance change can be less abrupt sometimes. This artifact may look like a delta wave because of its low voltage.

Environmental artifacts: Different types of movements surrounding the patient may be possible, which leads to environmental artifacts. This might be caused because of other people's movements near the patient, the patient's head movement, and electrode movement during excessive jelly application. There is also an artifact from respirators, which has a wide range of appearance and frequency. To identify the artifact, ventilator rate monitoring in a separate track might be employed. EEG amplifiers can be overloaded by high-frequency radiation from nearby equipment, or unusual waveforms might be created if electrodes are touched or damaged.

Physiological Artifacts: The physiological artifacts are classified as ocular artifacts, muscle artifacts, and cardiac artifacts.

Ocular artifacts: Ocular artifacts can be observed in EEG recordings when the electrodes are positioned close to the eyes. The human eye is a dipole, with a positive charge in the front and a negative charge in the back. Artifacts in EEG are caused by horizontal eyeball movement and vertical eyelid movement. Another type is eye flutter, which is not the same as repeated blinking but has a faster movement of the eyelid with a smaller amplitude. Such types of signals can be recorded by Electrooculogram (EOG). EOG signals have much higher amplitude than EEG signals [18], and their frequency is similar to that of EEG signals.

Muscle artifacts: Muscle artifacts can be created by any muscular contraction and stretching near signal recording locations, as well as the individual talking, sniffing, and swallowing [19]. The amplitude and waveform of artifacts will be affected by the degree of muscular contraction and stretch, characterized by short duration and a high-frequency range of 50 Hz to 100 Hz. An electromyogram (EMG) can be used to measure muscle artifacts.

Cardiac artifacts: Cardiac artifacts can occur when electrodes are positioned close to a blood artery [20] and record the movement of the heart's expansion and contraction. Such abnormalities, known as pulse artifacts, have a frequency of roughly 1.2 Hz and exist within an EEG signal. As a result, it is difficult to remove. ECG, or electrocardiogram, is another type of cardiac activity that measures the electrical signal produced by the heart. ECG, unlike pulse artifacts, may be monitored consistently and recorded apart from brain activity.

4 Artifact Removal Techniques

The EEG-based BCI has many applications, but the presence of the artifact in the EEG signal spectrum components leads to improper interfacing. Basic filtering algorithms are inadequate for removing these artifacts from EEG data. Various techniques have been presented in recent decades for the efficient reduction of artifacts from EEG data. This section discusses different approaches presented by the authors, for suppression of both physiological and non-physiological artifacts.

4.1 Filtering Methods

Different filtering algorithms were applied in the cancellation of EEG artifacts, the two most notable of which are adaptive filtering and Wiener filtering. Nonetheless,

W is a weighting coefficient that will be adjusted to reduce the error rate between the anticipated and primary EEG signals. The following section provides a quick overview of two widely used filtering techniques.

Adaptive Filtering: The basic process of an adaptive filtering technique is to calculate the degree of artifact interference in the EEG signal by continuously modifying the weights under the optimization process, and subtracting it from EEG signals [21]. The primary EEG input signal is represented as a combination of noise-free EEG data and a noise signal using the Eq. 4, where EEG_{pri} and EEG_{pure} are the primary and desirable signals, and N is the noise signal that must be removed in accordance with the artifact.

$$EEG_{pri} = EEG_{pure} + N \tag{4}$$

As one of the filter's inputs, a reference channel is provided. To get the pure signal, the Least Mean Squares (LMS) method is employed to assist the filter in upgrading its weight parameter. The Recursive Least Squares Method (RLMS) is an alternate approach that converges quicker than the LMS algorithm, but its calculation cost is high [22]. Disadvantages include the necessity for additional sensors to give reference inputs.

Wiener Filtering: The Wiener filtering approach is an optimum filtering technique, similar to adaptive filtering. The Wiener filter technique calculates the estimated EEG data to minimize the error rate between the clean EEG signal and the filtered EEG signal [23]. The linear filter is established by computing the power spectral densities of the filtered and polluted signals, as there is no fundamental information on statistics. Although Wiener filtering eliminates the need for an additional reference, the calibration required increases the complexity of its use.

4.2 Regression Methods

Regression techniques are conventional ways of eliminating artifacts. It is used on the presumption that each channel reflects the entirety of the clean EEG signal plus some artifacts [24]. The regression methodology first determines the amplitude relationship between the reference signals and the noisy EEG signals and then subtracts the estimated artifacts from the noisy EEG signals. As a result, the regression technique requires external reference channels such as EOG, ECG to exclude various artifacts. The general form of regression to eliminate artifacts from the EEG data is expressed in Eq. 5, where β is the regression coefficient calculated between EEG and EOG. EEG_{corr} and EEG_{raw} are corrected EEG signals and raw contaminated EEG signals, and EOG_{est} is the contaminating ocular activity, respectively.

$$EEG_{corr} = EEG_{raw} - \beta(EOG_{est}) \tag{5}$$

Hillyard and Gallambos [25] used regression algorithms based on the time domain to reduce ocular artifacts from EEG signals. Regression methods are further pioneered in the frequency domain and successfully coupled to EEG signal analysis [26].

4.3 *Wavelet Transform*

The Wavelet Transform (WT) is one of the popular methods in signal analysis. Wavelet transform decomposes an input signal into its collection of small functions, which are referred to as wavelets. These wavelets are created through dilations and shifting of the mother wavelet as defined in Eq. 6, where t represents a signal in the time domain, u and s are shifting and dilation parameters.

$$\Psi_{u,s}(t) = \frac{1}{\sqrt{s}}\Psi(\frac{t-u}{s}) \tag{6}$$

The wavelet transform can be done by computing the inner products $W_t(u, s) = \langle t, \psi_{u,s} \rangle$. If s is a continuous variable then $W_t(u, s)$ is known as the Continuous Wavelet Transform (CWT). If $s = a^j$ then $W_t(u, s) = W_t(u, j)$ is known as a Discrete Wavelet Transform (DWT). The CWT technique with adaptive thresholding was suggested to suppress cardiac artifacts from contaminated EEG signals [27]. Further, a combination of CWT and Independent Component Analysis (ICA) is used to identify and eliminate artifacts in cardiac-related potentials by leveraging both temporal and spatial features [28].

The DWT is an extremely effective wavelet transform for generating discrete wavelet representations of signals [29]. The main disadvantage of DWT is time variance, this is especially essential in analytical signal processing applications like EEG. Stationary Wavelet Transform (SWT) eliminates DWT's time-variance drawback, although it has repeated information and is rather slow [30]. The filter's structure at each step distinguishes DWT from SWT. Each level of decomposition produces approximation and detailed sequences that are identical to the original sequence length. The procedure up samples the coefficients of a filter by the factor 2^{j-1} after collecting the coefficients at the jth level [31]. Further, skewness-based SWT is used to remove eye blink artifacts from EEG signals [32]. Similarly, DWT and SWT methods are used for the suppression of ocular artifacts and comparing the performance [33].

4.4 *Blind Source Separation*

The Blind Source Separation (BSS) approach employs several unsupervised algorithms that do not need prior knowledge of reference channels. Let X represent the measured signals, and S represent the source signals which include both original and artifact signals. An unknown matrix A linearly combines these source signals to obtain the observed signals as defined in Eq. 7.

$$X = AS \tag{7}$$

The BSS algorithm is a reversal of the original, i.e., $U = WX$, where U represents the estimated sources, and the reverse of X is W. Then the components that depict artifacts can be ignored, and the other components rebuild EEG data to fulfill the goal of denoising.

Principal Component Analysis: Principal Component Analysis (PCA) is the most basic and extensively utilized BSS approach. It uses a methodology based on the eigenvalues of the covariance matrix. Using orthogonal transformation, this approach first transforms correlated values into uncorrelated variables. These variables are known as principal components and can be implemented using Single Value Decomposition (SVD).

PCA was initially used to eliminate ocular artifacts from the contaminated EEG recordings, and variance is used to extract the principal components describing blinks and eye movements. Then cleaned EEG signal is produced by using an inverse process after discarding the relevant components. It was observed that, PCA outperforms linear regression approaches in terms of processing efficiency [34]. However, it is often difficult to meet the criterion required that the components of artifacts be unrelated to EEG signals. Furthermore, when the probability of artifacts and EEG data is identical, PCA methods are ineffective to disentangle the artifacts.

Independent Component Analysis: Independent Component Analysis (ICA) is an extension of PCA, confined to transform orthogonal directions. It is more efficient and adaptable for separating artifact sources from EEG signals assuming that the source signals are linearly separable from one another and the observation signal must have a dimension larger than or equal to that of the original signal.

ICA can deconstruct an observable signal into independent components, and the clean EEG signal can be rebuilt by rejecting the independent components that contain artifact sources. Further, extended ICA is used to eliminate artifacts from EEG, and the results were effective when compared to the regression technique [35]. A method described for automatically extracting and removing eye movement artifacts using the ICA technique to prevent inaccuracies generated by manually selecting components [36]. Besides, five variations of ICA were developed to separate artifactual information from the signals [37].

Canonical Correlation Analysis: Canonical Correlation Analysis (CCA) is a popular BSS approach. In contrast, the ICA approach employs higher level analytics, whereas CCA employs statistics of second order resulting in a reduced processing time. The difference between CCA and ICA is in terms of the requirements for separating sources. CCA technique distinguishes components from unrelated sources and the ICA technique distinguishes components from statistically independent sources. The CCA was initially used to suppress muscle artifacts from the EEG signals [38].

The autocorrelation differences between EEG data and muscle artifacts were investigated using CCA to eliminate muscle artifacts.

4.5 Mode Decomposition Methods

An extensive range of mode decomposition methods were developed for denoising EEG signals in the last two decades. They are empirical mode decomposition, ensemble empirical mode decomposition, and variational mode decomposition.

Empirical Mode Decomposition: The Empirical Mode Decomposition (EMD) is an iterative decomposition technique that converts non-linear and non-stationary signals into a set of Intrinsic Mode Functions (IMF). Every IMF has to fulfill two requirements such as the number of extrema and zero crossings must be the same or differ at most by one; the IMF created by the extrema must have a mean value of zero. The procedure of EMD includes the following steps:

1. Calculate the local maxima and minima from the signal $x(t)$ in time domain t.
2. Connect extrema by a cubic spline to get envelopes.
3. Find the mean value $m(t)$ and subtract it from original as defined in Eq. 8

$$h_1(t) = x(t) - m(t) \tag{8}$$

4. Check the IMF conditions, if the difference signal $(h_1(t))$ does not fulfill the conditions. Repeat steps from 1 to 3.
5. If the desired stopping criterion is met, consider that $c_1(t)$ is the first IMF and the residue can be obtained by using Eq. 9.

$$r_1(t) = x(t) - c_1(t) \tag{9}$$

6. Now the residue signal $r_1(t)$ is used as a signal for further decomposition and the above process repeats to find the number of modes.

Some techniques using EMD are developed to eliminate ocular artifacts from EEG recordings. The EMD method can adaptively decompose a signal into modes, and then artifactual modes can be identified using different approaches. Also, EMD was applied to suppress blinks from EEG without using a reference EOG [39]. Although the EMD approach was beneficial for non-linear and non-stationary signals but suffering from a mode mixing effect, which occurs due to the signal's intermittency.

Ensemble Empirical Mode Decomposition: The Ensemble Empirical Mode Decomposition (EEMD) algorithm was suggested to solve the problem of mode mixing. Gaussian white noise with finite amplitude is included in the input signal before performing the EMD process and the IMFs average is collected across many trials to eliminate the mode mixing problem. Because there is no association between the noises generated in consecutive trials, the noise introduced in each trial can be canceled once the ensemble average is obtained [40].

Moreover, a combination of EEMD and PCA algorithms is used for suppressing ocular artifacts from the EEG signal. The EEMD algorithm in conjunction with CCA was used for the suppression of muscle artifact sources from EEG signals [41, 42].

Variational Mode Decomposition: Variational Mode Decomposition (VMD) decomposes a non-stationary signal X into many sub-signals x_k, also known as Band-Limited Intrinsic Mode Functions (BLIMF) because they are compact and centered on a frequency ω_k [43]. In VMD, parameter selection enables application-specific decomposition. The VMD algorithm for determining the BLIMFs is furnished below:

1. First, a Hilbert transform is used to derive the analytical signal for each mode.
2. Second, an exponential term is added to each mode's analytical signal to change its estimated center frequency, with each mode's spectrum modulated to the associated base frequency band.
3. Finally, the bandwidth is determined using the demodulated signals' Gaussian smoothness, i.e., the squared L^2-norm of the gradient.

When compared to EMD and EEMD, the modes generated by VMD exhibit reduced instantaneous frequency fluctuations. In terms of noise robustness, tone detection and separation, and signal reconstruction, VMD outperforms EMD and EEMD.

Variational Mode Extraction: The computational complexity is more in the VMD algorithm because it extracts the feasible IMFs from the signal. Due to this, an extension to VMD which is Variational Mode Extraction (VME) has been developed [44]. VME algorithm extracts a desired mode based on center frequency from the input signal, which means that the signal is divided into desired mode and the residual signal. Based on VME, the VME-DWT [45] algorithm is developed for the detection and elimination of eye blink from EEG data. First, the eye blink interval is selected from the desired mode obtained by VME. Then, the selected interval is filtered using the DWT method.

5 Performance Evaluation and Discussion

Performance evaluation is primarily required to estimate an algorithm's capability in completing some particular tasks. The performance of the artifact elimination method indicates how well the algorithm can remove artifacts from EEG signals. To be considered robust, an artifact reduction technique must perform well in both simulation and real-time healthcare conditions. The EEG signal has a broad range of medical applications in the diagnosis of different health problems and it is extremely important in neuronal communication like brain–computer interface systems. Generally, the recorded EEG signals are typically marred by different artifacts, which may originate from both physiological and non-physiological sources, because high-amplitude eye blinks have the most effect on EEG data. As a result, removing eye blinks is a step that must be taken before analyzing EEG signals.

In this book chapter, the simulation results of the EMD [39], VMD [46], SWT [32], and VME-DWT [45] filtering techniques have been compared to evaluate the denoising approaches since these decomposition methods do not require the extra artifact reference channel. To analyze and assess how well each method is working, 10 real EEG signals contaminated with eye blinks from the Cyclic Alternating Pattern (CAP) sleep database [47] and 15 synthetic EEG signals from the EEG fatigue database are chosen. The synthetic EEG signals contaminated with eye blinks are generated as described in [48]. The performance of these filtering techniques is analyzed based on the parameter: change in power spectral density (ΔPSD). The (ΔPSD) is calculated between contaminated and filtered EEG signals of corresponding frequency bands using Eq. 10, where k represents a particular frequency band of EEG signal. The lower value of ΔPSD indicates the effective suppression of eye blinks from EEG signals.

$$\Delta PSD_k = \Delta PSD_k^{contaminatedEEG} - \Delta PSD_k^{filteredEEG} \tag{10}$$

For visually analyzing the performance of filtering methods, the 2s durations of EEG signals contaminated with eye blinks are taken in this study. The simulations are carried out in the environment of MATLAB 2021a. In the EMD method, the contaminated EEG signal is divided into IMFs, and then the noisy IMFs are identified and eliminated to produce a filtered EEG signal. In the VMD method, the initial parameters α and K are set to 10000 and 12. First, the contaminated EEG signal is divided into 12 IMFs based on K values, and then the noisy IMFs were identified. Finally, the regression analysis is used to eliminate the eye blink component from the contaminated EEG signal. In the SWT method, the mother wavelet "db4" is used due to its resemblance with the EEG signal, and the decomposition level is selected based on the skewness value of approximate coefficients. Similarly, in the VME-DWT method, the desired mode is extracted using VME to detect the eye blink interval. Then, the skewness-based decomposition level of DWT is used to filter the eye blink intervals.

The graphical depiction of the VME-DWT, SWT, VMD, and EMD filtering techniques is shown in Fig. 1, Fig. 2, and Fig. 3, respectively. As per visual inspection, the VME-DWT method retains most of the signal information, whereas in the case of SWT, VMD, and EMD methods some of the signal information is lost.

Tables 2 and 3 illustrate the ΔPSD values on average of these filtering techniques, the lower ΔPSD values imply that the cleaned EEG signals were acquired with less distortion. The frequency band between δ and θ, where eye blinking is focused, exhibits the greatest changes. Comparatively, the α and β frequency ranges are less impacted. The VME-DWT method performs better than SWT, VMD, and EMD methods by preserving δ, θ, α, and β bands.

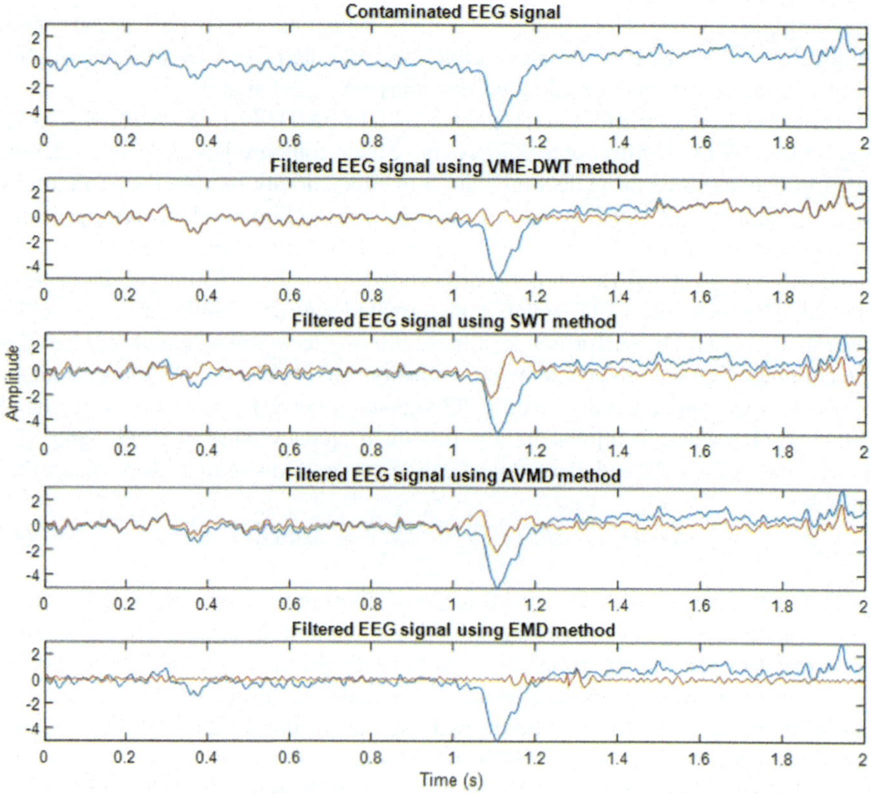

Fig. 1 Contaminated and filtered EEG signals of real data

The advantage of the VME-DWT method is that it concentrates on identifying and filtering only the eye blink intervals, while the SWT, VMD, and EMD methods are processing the entire signal. Still, there is a scope to enhance the artifact correction technique's performance with metaheuristics algorithms. Further, different types of metaheuristics algorithms are used to improve the quality of the DWT method for denoising EEG signals [49, 50]. This is done by optimizing the DWT parameters like mother wavelet selection, thresholding type, selection rules, and rescaling methods. The use of metaheuristics algorithms can be extended to other EEG signal denoising algorithms to enhance performance.

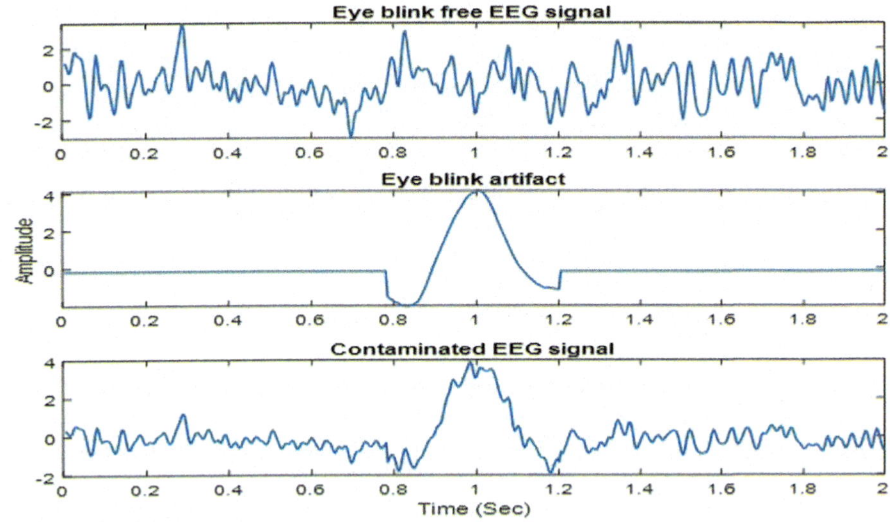

Fig. 2 Synthetic EEG signal

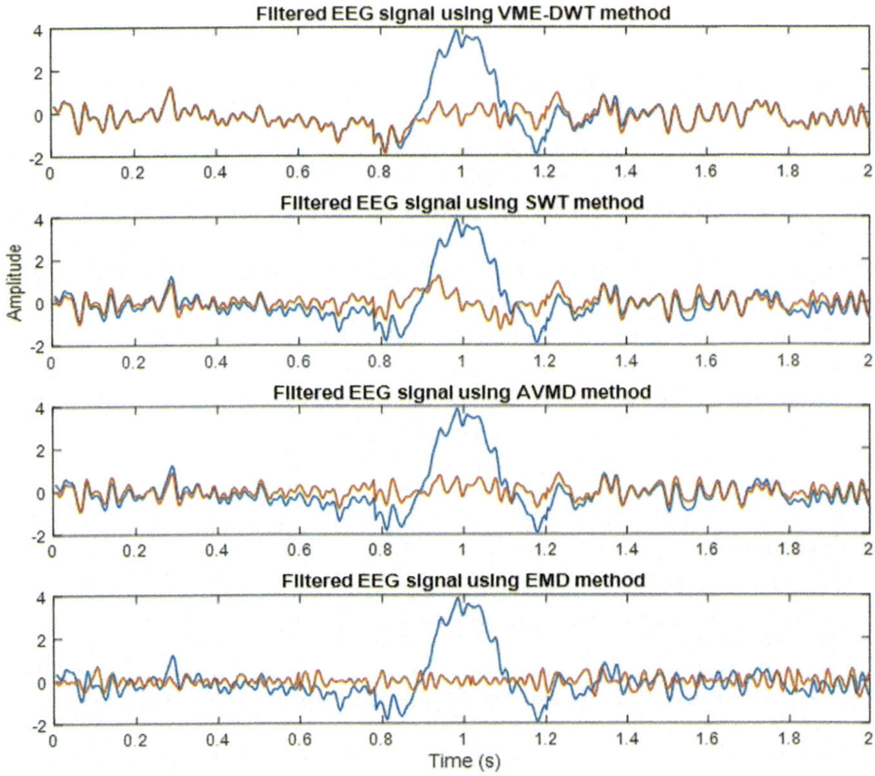

Fig. 3 Filtered EEG signals

Table 2 *ΔPSD* values of real EEG data

Parameter	EMD	VMD	SWT	VME-DWT
ΔPSD_δ	10.996 ± 1.233	7.85 ± 0.874	7.758 ± 0.863	5.410 ± 0.612
ΔPSD_θ	5.193 ± 0.548	3.883 ± 0.413	5.193 ± 0.548	3.240 ± 0.341
ΔPSD_α	1.888 ± 0.201	1.28 ± 0.137	1.156 ± 0.124	1.266 ± 0.136
ΔPSD_β	1.035 ± 0.118	0.181 ± 0.023	0.346 ± 0.040	0.827 ± 0.095

Table 3 *ΔPSD* values of synthetic EEG data

Parameter	EMD	VMD	SWT	VME-DWT
ΔPSD_δ	8.849 ± 0.915	7.993 ± 0.917	7.945 ± 0.911	7.053 ± 0.811
ΔPSD_θ	6.042 ± 0.644	6.010 ± 0.680	5.660 ± 0.640	5.477 ± 0.620
ΔPSD_α	0.991 ± 0.261	0.499 ± 0.056	0.460 ± 0.052	0.317 ± 0.036
ΔPSD_β	0.424 ± 0.145	0.150 ± 0.017	0.139 ± 0.015	0.057 ± 0.007

6 Conclusion

The EEG signal is a record of brain activity that is used to diagnose various health conditions, and it plays a major role in brain–computer interfacing. The EEG-based BCI system's goal is to extract and interpret brain activity into command messages that assist physically impaired persons, but the EEG signals are always tainted by artifacts from many sources. Many researchers have worked in recent years on creating strategies for dealing with the removal of artifacts from EEG signals. In this chapter, we provided a detailed assessment of the numerous available approaches for identifying and removing artifacts, as well as a comparison of EMD, VMD, SWT, and VME-DWT methods for removing eye blink artifacts. The graphical representation and ΔPSD results validate that the VME-DWT method performs better and retains the signal information than EMD, VMD, and SWT methods. In future work, computational intelligence can be used to improve the performance of noise removal techniques for better identification and suppression of artifacts from the EEG signal.

References

1. Vidal, J.J.: Toward direct brain-computer communication. Annu. Rev. Biophys. Bioeng. **2**(1), 157–180 (1973)
2. Guger, C., Harkam, W., Hertnaes, C., Pfurtscheller, G.: Prosthetic control by an EEG-based brain-computer interface (BCI). In: Proceedings of 5th European Conference for the Advancement of Assistive Technology, pp. 3–6 (1999)
3. Fetz, E.E.: Real-time control of a robotic arm by neuronal ensembles. Nat. Neurosci. **2**(7), 583–584 (1999)

4. Zander, T.O., Kothe, C.: Towards passive brain-computer interfaces: applying brain-computer interface technology to human-machine systems in general. J. Neural Eng. **8**(2), 025005 (2011)
5. Wolpaw, J.R., Birbaumer, N., McFarland, D.J., Pfurtscheller, G., Vaughan, T.M.: Brain-computer interfaces for communication and control. Clin. Neurophysiol. **113**(6), 767–791 (2002)
6. Liu, N.H., Chiang, C.Y., Chu, H.C.: Recognizing the degree of human attention using EEG signals from mobile sensors. Sensors **13**(8), 10273–10286 (2013)
7. Mutasim, A.K., Tipu, R.S., Bashar, M.R., Islam, M.K., Amin, M.A.: Computational intelligence for pattern recognition in EEG signals. In: Pedrycz, W., Chen, SM. (eds) Computational Intelligence for Pattern Recognition. Studies in Computational Intelligence, vol. 777, pp. 291–320 (2018)
8. Mao, X., Li, M., Li, W., Niu, L., Xian, B., Zeng, M., Chen, G.: Progress in EEG-based brain robot interaction systems. Comput. Intell. Neurosci. **2017**, 1742862 (2017)
9. Schalk, G., McFarland, D.J., Hinterberger, T., Birbaumer, N., Wolpaw, J.R.: BCI2000: a general-purpose brain-computer interface (BCI) system. IEEE Trans. Biomed. Eng. **51**(6), 1034–1043 (2004)
10. Kevric, J., Subasi, A.: Comparison of signal decomposition methods in classification of EEG signals for motor-imagery BCI system. Biomed. Signal Process. Control **31**, 398–406 (2017)
11. Baroroh, D.K., Chu, C.H., Wang, L.: Systematic literature review on augmented reality in smart manufacturing: collaboration between human and computational intelligence. J. Manuf. Syst. **61**, 696–711 (2021)
12. Geem, Z.W., Kim, J.H., Loganathan, G.V.: A new heuristic optimization algorithm: harmony search. SIMULATION **76**(2), 60–68 (2001)
13. Poli, R., Kennedy, J., Blackwell, T.: Particle swarm optimization: an overview. Swarm Intell. **1**, 33–57 (2007)
14. Kumari, N., Acharjya, D.P.: Data classification using rough set and bioinspired computing in healthcare applications-an extensive review. Multimedia Tools Appl. **82**(9), 13479–13505 (2023)
15. Minguillon, J., Lopez-Gordo, M.A., Pelayo, F.: Trends in EEG-BCI for daily-life: requirements for artifact removal. Biomed. Signal Process. Control **31**, 407–418 (2017)
16. Tatum, W.O., Dworetzky, B.A., Schomer, D.L.: Artifact and recording concepts in EEG. J. Clin. Neurophysiol. **28**(3), 252–263 (2011)
17. Lee, S., Buchsbaum, M.S.: Topographic mapping of EEG artifacts. Clin. EEG **18**(2), 61–67 (1987)
18. Schlögl, A., Keinrath, C., Zimmermann, D., Scherer, R., Leeb, R., Pfurtscheller, G.: A fully automated correction method of EOG artifacts in EEG recordings. Clin. Neurophysiol. **118**(1), 98–104 (2007)
19. Urigüen, J.A., Garcia-Zapirain, B.: EEG artifact removal-state-of-the-art and guidelines. J. Neural Eng. **12**(3), 031001 (2015)
20. Goncharova, I.I., McFarland, D.J., Vaughan, T.M., Wolpaw, J.R.: EMG contamination of EEG: spectral and topographical characteristics. Clin. Neurophysiol. **114**(9), 1580–1593 (2003)
21. Marque, C., Bisch, C., Dantas, R., Elayoubi, S., Brosse, V., Perot, C.: Adaptive filtering for ECG rejection from surface EMG recordings. J. Electromyogr. Kinesiol. **15**(3), 310–315 (2005)
22. He, P., Wilson, G., Russell, C.: Removal of ocular artifacts from electro-encephalogram by adaptive filtering. Med. Biol. Eng. Compu. **42**, 407–412 (2004)
23. Somers, B., Francart, T., Bertrand, A.: A generic EEG artifact removal algorithm based on the multi-channel Wiener filter. J. Neural Eng. **15**(3), 036007 (2018)
24. Sweeney, K.T., Ward, T.E., McLoone, S.F.: Artifact removal in physiological signals-Practices and possibilities. IEEE Trans. Inf Technol. Biomed. **16**(3), 488–500 (2012)
25. Hillyard, S.A., Galambos, R.: Eye movement artifact in the CNV. Electroencephalogr. Clin. Neurophysiol. **28**(2), 173–182 (1970)
26. Whitton, J.L., Lue, F., Moldofsky, H.: A spectral method for removing eye movement artifacts from the EEG. Electroencephalogr. Clin. Neurophysiol. **44**(6), 735–741 (1978)

27. Jiang, J.A., Chao, C.F., Chiu, M.J., Lee, R.G., Tseng, C.L., Lin, R.: An automatic analysis method for detecting and eliminating ECG artifacts in EEG. Comput. Biol. Med. **37**(11), 1660–1671 (2007)
28. Hamaneh, M.B., Chitravas, N., Kaiboriboon, K., Lhatoo, S.D., Loparo, K.A.: Automated removal of EKG artifact from EEG data using independent component analysis and continuous wavelet transformation. IEEE Trans. Biomed. Eng. **61**(6), 1634–1641 (2013)
29. Mallat, S.G.: A theory for multiresolution signal decomposition: the wavelet representation. IEEE Trans. Pattern Anal. Mach. Intell. **11**(7), 674–693 (1989)
30. Kumar, P.S., Arumuganathan, R., Sivakumar, K., Vimal, C.: An adaptive method to remove ocular artifacts from EEG signals using wavelet transform. J. Appl. Sci. Res. **5**(7), 711–745 (2009)
31. Nason, G.P., Silverman, B.W.: The stationary wavelet transform and some statistical applications. In: Wavelets and Statistics, pp. 281–299, Springer, New York (1995)
32. Shahbakhti, M., Maugeon, M., Beiramvand, M., Marozas, V.: Low complexity automatic stationary wavelet transform for elimination of eye blinks from EEG. Brain Sci. **9**(12), 352 (2019)
33. Khatun, S., Mahajan, R., Morshed, B.I.: Comparative study of wavelet-based unsupervised ocular artifact removal techniques for single-channel EEG data. IEEE J. Transl. Eng. Health Med. **4**, 1–8 (2016)
34. Casarotto, S., Bianchi, A.M., Cerutti, S., Chiarenza, G.A.: Principal component analysis for reduction of ocular artefacts in event-related potentials of normal and dyslexic children. Clin. Neurophysiol. **115**(3), 609–619 (2004)
35. Jung, T.P., Makeig, S., Westerfield, M., Townsend, J., Courchesne, E., Sejnowski, T.J.: Removal of eye activity artifacts from visual event-related potentials in normal and clinical subjects. Clin. Neurophysiol. **111**(10), 1745–1758 (2000)
36. Joyce, C.A., Gorodnitsky, I.F., Kutas, M.: Automatic removal of eye movement and blink artifacts from EEG data using blind component separation. Psychophysiology **41**(2), 313–325 (2004)
37. Frølich, L., Dowding, I.: Removal of muscular artifacts in EEG signals: a comparison of linear decomposition methods. Brain Inf. **5**(1), 13–22 (2018)
38. De Clercq, W., Vergult, A., Vanrumste, B., Van Paesschen, W., Van Huffel, S.: Canonical correlation analysis applied to remove muscle artifacts from the electroencephalogram. IEEE Trans. Biomed. Eng. **53**(12), 2583–2587 (2006)
39. Patel, R., Janawadkar, M.P., Sengottuvel, S., Gireesan, K., Radhakrishnan, T.S.: Suppression of eye-blink associated artifact using single channel EEG data by combining cross-correlation with empirical mode decomposition. IEEE Sens. J. **16**(18), 6947–6954 (2016)
40. Wu, Z., Huang, N.E.: Ensemble empirical mode decomposition: a noise-assisted data analysis method. Adv. Adapt. Data Anal. **1**(01), 1–41 (2009)
41. Patel, R., Gireesan, K., Sengottuvel, S., Janawadkar, M.P., Radhakrishnan, T.S.: Common methodology for cardiac and ocular artifact suppression from EEG recordings by combining ensemble empirical mode decomposition with regression approach. J. Med. Biol. Eng. **37**(2), 201–208 (2017)
42. Chen, X., Xu, X., Liu, A., McKeown, M.J., Wang, Z.J.: The use of multivariate EMD and CCA for denoising muscle artifacts from few-channel EEG recordings. IEEE Trans. Instrum. Meas. **67**(2), 359–370 (2017)
43. Dragomiretskiy, K., Zosso, D.: Variational mode decomposition. IEEE Trans. Signal Process. **62**(3), 531–544 (2013)
44. Nazari, M., Sakhaei, S.M.: Variational mode extraction: a new efficient method to derive respiratory signals from ECG. IEEE J. Biomed. Health Inform. **22**(4), 1059–1067 (2017)
45. Shahbakhti, M., Beiramvand, M., Nazari, M., Broniec-Wójcik, A., Augustyniak, P., Rodrigues, A.S., Wierzchon, M., Marozas, V.: VME-DWT: an efficient algorithm for detection and elimination of eye blink from short segments of single EEG channel. IEEE Trans. Neural Syst. Rehabil. Eng. **29**, 408–417 (2021)
46. Dora, C., Biswal, P.K.: An improved algorithm for efficient ocular artifact suppression from frontal EEG electrodes using VMD. Biocybern. Biomed. Engin. **40**(1), 148–161 (2020)

47. Terzano, M.G., Parrino, L., Smerieri, A., Chervin, R., Chokroverty, S., Guilleminault, C., Hirshkowitze, M., Mahowaldf, M., Moldofskyg, H., Rosah, A., Thomas, R., Walters, A.: Atlas, rules, and recording techniques for the scoring of cyclic alternating pattern (CAP) in human sleep. Sleep Med. **3**(2), 187–199 (2002)
48. Maddirala, A.K., Veluvolu, K.C.: SSA with CWT and k-means for eye-blink artifact removal from single-channel EEG signals. Sensors **22**(3), 931 (2022)
49. Alyasseri, Z.A.A., Khader, A.T., Al-Betar, M.A., Abasi, A.K., Makhadmeh, S.N.: EEG signals denoising using optimal wavelet transform hybridized with efficient metaheuristic methods. IEEE Access **8**, 10584–10605 (2019)
50. Phadikar, S., Sinha, N., Ghosh, R.: Automatic eyeblink artifact removal from EEG signal using wavelet transform with heuristically optimized threshold. IEEE J. Biomed. Health Inform. **25**(2), 475–484 (2020)

Rough Computing in Healthcare Informatics

Madhusmita Mishra and D. P. Acharjya

Abstract Data analytics and health care are combined in health informatics. It focuses on using information technology to efficiently collect, secure, and provide high-quality health care, all of which have a favorable impact on the patient–physician interaction. Classification, clustering, feature selection, diagnosis, prediction, and decision support systems are just a few of the outcomes of information technology's concentration on healthcare data analysis. The primary focus of this chapter is on rough computing methods used in healthcare informatics. In turn, it supports the doctors' decision-making.

1 Introduction

In the modern world information technology is used to serve and connect people. It is possible by margining computing and communication all together. Nowadays, information technology is treated as a fact of life. It creates an information system to manage an organization. The major application areas of information technology are banking, industry, administration, education, daily life, entertainment, health care, and many more. With the rapid growth of the population and the variety of newly born diseases, physicians are facing a lot of challenges in diagnosing a disease correctly. This is because the healthcare dataset contains many irrelevant data that do not have any influence on disease diagnosis. At the same time, a physician simply cannot eliminate those data without knowing their significance in the healthcare dataset. At this juncture, the introduction of knowledge discovery and information technology plays a vital role in healthcare informatics.

M. Mishra
Dr. Sudhir Chandra Sur Institute of Technology and Sports Complex, Kolkata, India
e-mail: madhu.smita7@gmail.com

D. P. Acharjya (✉)
School of Computer Science and Engineering, Vellore Institute of Technology University, Vellore, India
e-mail: dpacharjya@gmail.com

D. P. Acharjya and K. Ma (eds.), *Computational Intelligence in Healthcare Informatics*, Studies in Computational Intelligence 1132,
https://doi.org/10.1007/978-981-99-8853-2_18

The intended concepts of a dataset are typically described by a group of characteristics in machine learning and data mining. It is anticipated that the features will contain as much valuable information as feasible to create accurate models for tackling challenges like classification, feature selection, and prediction. Additionally, the features might be as little as feasible. However, it is challenging to separate features that are significant from those that are not because there is frequently minimal prior knowledge about the dataset. As a result, many features, including many unnecessary and duplicate characteristics, must often be taken into account. Unfortunately, irrelevant and redundant features, which are mostly brought on by the curse of dimensionality, will not only impair the training efficiency but also significantly affect the effectiveness of machine learning thus taught to them. Several algorithms have been proposed to deal with these superfluous features [1]. Rough computing is one of those techniques that reduces superfluous features on a healthcare dataset [2].

Another important concept of healthcare informatics is data classification and clustering. The healthcare data generated from various sources contains uncertainties. For example, in liver disease diagnosis if bilirubin is 2.4 then it is moderate and not risky whereas it is risky when bilirubin is 2.5. But, the values 2.4 and 2.5 are almost equal. Analyzing such uncertainties using the human brain is difficult and thus intelligent algorithms are necessary to deal with such uncertainties. Many intelligent classification algorithms relating to fuzzy logic [3] and rough set [4] are introduced in the literature. However, rough set classification algorithms are relatively new. In many cases, rough set and fuzzy set both are hybridized and classification is carried out [5]. Likewise, rough-fuzzy clustering algorithms are also studied over various real-life problems [6].

Information retrieval and decision support systems are another two prime concepts in healthcare informatics. Computers are now utilized to analyze vast amounts of healthcare data. It opens up the difficult field of data analytics. As a result, it is crucial to adopt effective data structures for data storage and information retrieval. Information retrieval is a method for finding obscure patterns in information systems. Additionally, it aids decision-makers in selecting the appropriate course of action. Simultaneously, intelligent algorithms are introduced to process such an information system further while handling uncertainties. With the invention of such algorithms and methods, the machine may now think somewhat artificially. Additionally, information retrieval is a multidisciplinary field that draws on linguistics, computer science, statistics, and mathematics [7]. Rough computing in information retrieval and decision support systems is relatively new. A computerized system known as a decision support system collects, evaluates, and then synthesizes data to create detailed information reports. But, the evaluation of healthcare data to make precise decisions requires intelligent algorithms. Hybridized rough computing plays a vital role in decision support systems [8].

Apart from feature selection, classification, clustering, and development of decision support systems, there are several other concepts also associated with healthcare informatics. However, the objective of this chapter is to put forth the developments of rough computing in perception with healthcare informatics.

The remainder of the chapter is organized as follows. Following the introduction, Sect. 2 presents information system concepts. In Sect. 3, the fundamentals of rough computing are covered. Rough computing that is hybridized with other computing methodologies is covered in Sect. 4. Additionally, Sect. 5 provides an introduction to healthcare informatics in terms of feature selection, classification, clustering, and decision support systems. Applications in health care are covered in Sect. 6. In Sect. 7, the chapter's conclusion is delivered.

2 Information System

The main goal of information retrieval and data mining is to gain knowledge about classification. However, straightforward classification cannot solve real-world issues. The sequence of items is one of these major issues. An information system aids in interpreting the raw data and portrays a significant function in the preliminary set. Using a rough set, further aids in clustering, classification, prediction, and developing a decision support system. A countable number of elements make up an information system, and these elements are distinguished by a countable number of parameters. These parameters are also known as features or characteristics. It is possible to create a tabular version of this information system. The information table's columns and rows each contains a parameter and an object, respectively. The parameter values are contained in the respective table cell.

An information table is a two-dimensional table that contains a definable set of data about the universe in terms of objects, parameters, and parameter values. A decision table is an information table that contains conditions and decisions [9]. A countable, non-void universal collection of objects $Q = \{q_1, q_2, \ldots, q_n\}$ and a countable, non-void set of parameters $P = \{p_1, p_2, \ldots, p_m\}$ are both parts of an information table, denoted as $I = (Q, P, V, f)$. The possible parameter values are denoted as $V = \cup_{p \in P} V_p$. In addition, $f : (Q \times P) \rightarrow V$ is the information function. Additionally, the information table I is known as a decision table if $P = C \cup D$, where C is the subset of parameters known as conditions and D is the subset of parameters known as decision.

Table 1 provides an example of a standard decision system with $Q = \{q_1, q_2, q_3, q_4, q_5, q_6\}$ representing a non-void observable set of toys; $P = \{p_1, p_2, p_3\}$ representing a countable collection of conditional parameters with design, dimension, and color; and D representing the decision parameter. In Table 1, the toy q_1 is specifically described as having a square shape, a big size, a green color, and a high price. It establishes the specifics of the toy q_1. All of the parameter values in the decision table shown in Table 1 are discrete and categorical, making it a qualitative information system [10].

Table 2 provides an example of a standard decision system with $Q = \{q_1, q_2, q_3, q_4, q_5\}$ representing a non-void observable set of patients; $P = \{p_1, p_2, p_3\}$ representing a countable collection of conditional parameters with albumin, bilirubin, and prothronbin; and D representing the decision parameter. In Table 2, the patient q_1 is

Table 1 A standard qualitative decision system

Toys	Design (p_1)	Dimension (p_2)	Color (p_3)	Retail price in INR (D)
q_1	Square	Big	Green	300
q_2	Square	Small	Green	100
q_3	Round	Small	Pink	100
q_4	Square	Small	Green	300
q_5	Triangular	Small	Green	700
q_6	Square	Medium	Red	100

Table 2 A standard quantitative decision system

Patients	Albumin (p_1)	Bilirubin (p_2)	Prothronbin (p_3)	Diagnosis (D)
q_1	1.3	0.3	8	Fatty liver
q_2	2.3	4.2	14.3	Cirrhosis
q_3	1.7	2.1	11.2	Fibrosis
q_4	4.3	3.4	13.2	Cirrhosis
q_5	5.4	1.3	11.5	Fibrosis

specifically described as having a albumin 1.3, bilirubin 0.3, prothronbin 8, and fatty liver. It establishes the specifics of the patient q_1. All of the parameter values in the decision table shown in Table 2 are continuous, making it a quantitative information system [10].

3 Rough Computing

Pawlak's rough set theory is a technique for data analysis; knowledge discovery; and acquisition from incomplete, ambiguous, and vague information systems [11]. The rough set's main benefit is that it doesn't require any prior understanding of the elements, excluding their properties. Equivalence relations are the central concept in the rough set. The equivalence relation is later generalized to many extents, including fuzzy equivalence relation, fuzzy tolerance relation, binary relation, and many more. Similar to this, the rough set design for a single universe has been expanded to two universes. Numerous rough set variations have been produced based on these modifications. To name a few: fuzzy rough set, rough set on fuzzy approximation space, etc. All these variations of the rough set are included in rough computing [12].

3.1 Rough Set

A classification-related analytical technique called the rough set is used to extract knowledge from an information system. Later, it was expanded to include feature reduction and rule creation based on the information table's core and reducts. In general, using the partition facilitates learning from an information table. The equivalence relation and its attributes are used to create the information system's partition. The indiscernibility relation used to get the partitions is crucial to the examination of the rough collection of data [11].

A classification-related analytical technique called the rough set is used to extract knowledge from an information system. Later, it was expanded to include feature reduction and rule creation based on the information table's core and reducts. In general, using the partition facilitates learning from an information table. The equivalence relation and its features are used to create the information system's partition. The indiscernibility relation is crucial to the examination of the rough collection of data. The analytical foundation for rough set theory is the classification of the members using the equivalence relation. The basic granule or equivalence class produced from a single feature is what constitutes the fundamental knowledge of the universe. The combination of the rudimentary concepts is considered to be either a classical (crisp) set or an imprecise (rough) set. Assume $A \subseteq P$, and $q_i, q_j \in Q$. We say q_i and q_j are indistinguishable by the collection of parameters A in P is necessary and sufficient for the given Eq. 1 to hold.

$$f(q_i, p) = f(q_j, p) \quad \forall p \in A \tag{1}$$

The indiscernibility relation derived from Table 1 is produced below. The basic granules obtained for various features are as follows:

$$Q/\{A = p_1\} = \{\{q_1, q_2, q_4, q_6\}, \{q_3\}, \{q_5\}\}$$
$$Q/\{A = p_2\} = \{\{q_1\}, \{q_2, q_3, q_4, q_5\}, \{q_6\}\}$$
$$Q/\{A = p_3\} = \{\{q_1\}, \{q_2, q_4\}, \{q_3\}, \{q_5\}, \{q_6\}\}$$
$$Q/\{A = \{p_1, p_2, p_3\}\} = \{\{q_1\}, \{q_2, q_4\}, \{q_3\}, \{q_5\}, \{q_6\}\}$$

The equivalency relation divides the universe's collection of elements into indiscernible classes, as shown by the analysis above. The toys belonging to each class cannot be distinguished from one another for the components of each class A. Figure 1 shows a graphic representation of the principles.

A countable collection of items is represented by the universe Q. Consider $R \subseteq Q \times Q$ to be an equivalence relation on Q. The relation R is hence transitive, symmetric, and reflexive. It then classifies the information system, yielding equivalence classes. These indistinguishable classes are not connected. The partitions thus generated are known as fundamental granules. A precise set is a group of fundamental granules. The void set is regarded to be a determinable set. Therefore, all of the determinable sets comprise the Boolean algebra [13]. An approximation space is the phrase used to describe the ordered pair (U, R).

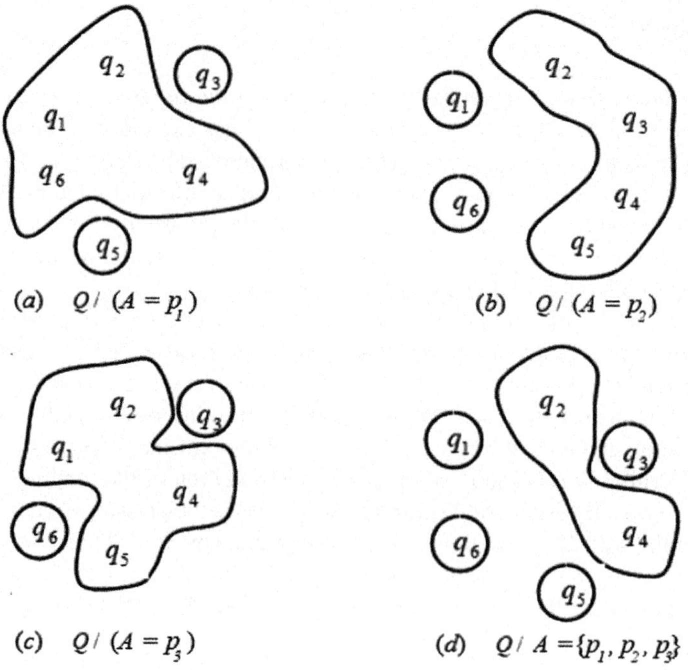

Fig. 1 Partition of information system according to conditions

Let us consider the target set X. The set X is characterized by a couple of approximations, such as upper approximation and lower approximation. It is referred to as $\underline{R}X$, as the R-lower approximation and $\overline{R}X$, as the R-upper approximation of X. These are defined below:

$$\underline{R}X = \cup\{Y \in Q/R : Y \subseteq X\} \tag{2}$$

$$\overline{R}X = \cup\{Y \in Q/R : Y \cap X \neq \phi\} \tag{3}$$

The difference between $\overline{R}X$ and $\underline{R}X$ is referred to as R-boundary of X, $BN_R(X)$. Thus, $BN_R(X) = \overline{R}X - \underline{R}X$. Hence, there arises two cases. In the first case, $\overline{R}X = \underline{R}X$ and X as said to be R-definable. Similarly in the second case, $\overline{R}X \neq \underline{R}X$ and X is termed as R-undefinable or rough set. Hence, X is a rough set concerning R iff it is not R-definable. Consider $X = \{q_2, q_3, q_6\}$ of Table 1. Let us assume the feature set $A = \{p_1, p_2, p_3\}$. Thus the upper and lower approximation is computed as $\overline{A}X = \{q_2, q_3, q_4, q_6\}$ and $\underline{A}X = \{q_3, q_6\}$, respectively. Hence, $BN_A(X) = \overline{A}X - \underline{A}X = \{q_2, q_4\}$ is considered as A-boundary of X. Figure 2 illustrates these concepts graphically.

The Concept of Reduct and Core: One of the key concepts in the rough set is the reduct and core. Reduct speeds up calculation by reducing the number of features

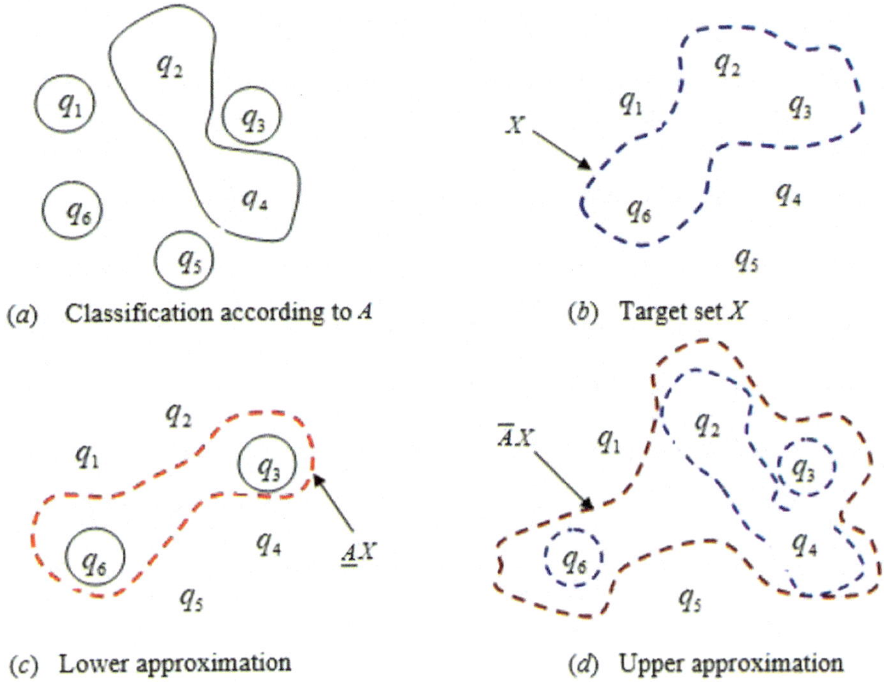

Fig. 2 Representation of approximations

in an information system. It categorizes the information system and removes the superfluous attributes from it without loss of any knowledge. Additionally, it aids in the creation of decision rules. An information system may have multiple reducts, but the core of it is present in all of them. These decision-making guidelines in turn aid decision-makers in reaching the right choices. The reduced information system consists of all minimal subsets of parameters that are both possible and have the same quality of classification. The dependent attributes are determined based on the dependency properties of the attributes [14]. Some supplementary notations are required to alternatively visualize the concepts of dependency. Let, $A \subseteq P$ and $p \in A$. The parameter p is dispensable in A if it satisfies Eq. 4. In the other case, p is said to be indispensable in A.

$$Q/A = Q/(A - \{p\}) \tag{4}$$

If all the parameters of A are indispensable, then A is said to be independent. Let $A' \subseteq A$ is said to be a reduct if both Q/A and Q/A' induce the same partition, i.e., $Q/A = Q/A'$. A reduct of the information table is denoted as $Red(A)$. The core of the information table, $Core(A)$, refers to the intersection of such $Red(A)$. It is defined as in Eq. 5.

$$Core(A) = \cap Red(A) \tag{5}$$

The most important feature of an information system is the core. The features belonging to core cannot be redundant. Consider the information system presented in Table 1 for an illustration.

1. Foremost, calculate the equivalence classes relating to the combination of features. Thus, we have $Q/\{p_1, p_2\} = \{\{q_2, q_4\}, \{q_1\}, \{q_5\}, \{q_3\}, \{q_6\}\}$; $Q/\{p_2, p_3\} = \{\{q_1\}, \{q_2, q_4, q_5\}, \{q_3\}, \{q_6\}\}$; $Q/\{p_1, p_3\} = \{\{q_1, q_2, q_4\}, \{q_3\}, \{q_5\}, \{q_6\}\}$.
2. It is seen from overhead analysis that $Q/\{p_1, p_2\} = Q/\{p_1, p_2, p_3\}$. Therefore, the attribute p_3 is dispensable. Similarly, $Q/\{p_2, p_3\} \neq Q/\{p_1, p_2, p_3\}$ and hence the attribute p_1 is indispensable. Similarly, $Q/\{p_1, p_3\} \neq Q/\{p_1, p_2, p_3\}$ and hence p_2 is not dispensable.
3. Considering individual parameters, it is obtained as $Q/\{p_1, p_2\} \neq Q/\{p_1\}$ and $Q/\{p_1, p_2\} \neq Q/\{p_2\}$. Therefore, the only reduct is given as $Red(A) = \{p_1, p_2\}$.
4. Eventually, the core as $Core(A) = \{p_1, p_2\}$.

The features needed to display the information system in the example are $\{p_1, p_2\} = \{design, dimension\}$. Only these qualities are present in the reduct subset. However, different information systems may give one access to multiple reducts.

3.2 Fuzzy Rough Set

The rough set has particular restrictions. For instance, rough set generates too many rules and cannot be applied directly to an information system with real values. To get around these restrictions, the idea of a fuzzy rough set (FRS) is proposed [15]. This model's introduction was primarily done to analyze various real-world issues. This idea has a fuzzy indiscernible relation as its analytical perspective point. Further, it results in fuzzy indiscernible classes rather than indiscernible ones.

Consider $Q = \{q_1, q_2, \ldots, q_n\}$ is the finite collection of objects, $A \subseteq P = \{p_1, p_2, \ldots, p_m\}$ is the set of features. Let $p_i \in A$ and $Q/Ind\{p_i\}$ be the fuzzy indistinguishable class generated by the feature p_i. Assume, $Q/Ind\{p_i\} = \{X_{p_i}, Y_{p_i}\}$. It states that the feature p_i offers to two fuzzy sets X_{p_i} and Y_{p_i}. If $p_i, p_j \in P$, and $Q/Ind\{p_j\} = \{X_{p_j}, Y_{p_j}\}$, then $Q/Ind\{p_i, p_j\} = \{X_{p_i} \cap X_{p_j}, X_{p_i} \cap Y_{p_j}, Y_{p_i} \cap X_{p_j}, Y_{p_i} \cap Y_{p_j}\}$ [16].

Let $Q/A = \{C_1, C_2, C_3, \ldots, C_k\}$. Assume U be a fuzzy set. Thus, A-lower $(\mu_{\underline{A}U}(q))$ and A-upper $(\mu_{\overline{A}U}(q))$ approximation relating to fuzzy rough set is elucidated in Eqs. 6 and 7 respectively, in which $q \in Q$ [17].

$$\mu_{\underline{A}U}(q) = sup_{C \in Q/A}\left(\mu_C(q) \wedge inf_{r \in Q}(\{1 - \mu_C(r)\} \vee \mu_U(r))\right) \tag{6}$$

$$\mu_{\overline{A}U}(q) = sup_{C \in Q/A}\left(\mu_C(q) \wedge sup_{r \in Q}(\mu_C(r) \vee \mu_U(r))\right) \tag{7}$$

The order pair $(\underline{A}U, \overline{A}U)$ is called as a fuzzy rough set. Equations 6 and 7 compute the A-lower and A-upper approximation of every object of U. Moreover, the membership of an object $q \in Q$ relating to such an indistinguishable category $C_i \in Q/A$, $i = 1, 2, \ldots, k$ is calculated using Eq. 8.

$$\mu_{\{C_1 \cap C_2 \cap \ldots \cap C_k\}}(q) = \mu_{C_1}(q) \wedge \mu_{C_2}(q) \wedge \cdots \wedge \mu_{C_k}(q) \quad (8)$$

The key idea behind this theory is that even if the information table lacks any uncertainty, the degree of dependency in FRS is approximately equal to 1 [16]. Besides, a comparison of FRS with other types of the rough set is found in the literature [18].

3.3 Rough Set on Fuzzy Approximation Space

Pawlak's fundamental theory of rough sets is based on the idea of equivalence relations that are specified across a whole universe. Equivalence relations in practical settings are, nevertheless, quite uncommon. As a result, initiatives are being taken to reduce the significance of the relations, and the idea of fuzzy proximity relations on Q is presented. A generalization of the knowledge base concept is the idea of fuzzy approximation space, which is dependent on a fuzzy proximity relation specified on a set Q. As a result, the idea of rough sets on knowledge bases, as developed by Acharjya and Tripathy, is extended to Rough Set on Fuzzy Approximation Spaces (RSFAS) [19]. In this section, we give the core RSFAS principles, notations, and outcomes in order to reveal the article's basis.

Consider Q be a universe. A fuzzy relation on Q as a fuzzy subset of $(Q \times Q)$. If $\mu_R(q_i, q_i) = 1$ and $\mu_R(q_i, q_j) = \mu_R(q_j, q_i)$ for $q_i, q_j \in Q$, then the fuzzy relation R on Q is said to be a fuzzy proximity relation. In such cases, two elements q_i and q_j are α-similar, $\alpha \in [0, 1]$, with respect to R if $(q_i, q_j) \in R_\alpha$ or $\mu_R(q_i, q_j) \geq \alpha$. It is written as $q_i R_\alpha q_j$. If either q_i is α-similar to q_j or q_i is transitively α-similar to q_j concerning R, then the two elements q_i and q_j are said to be α-identical concerning R. It means that, there exists a sequence $q_1, q_2, \ldots, q_n \in Q$ such that $q_i R q_1$, $q_1 R q_2$, $q_2 R q_3$, ..., $q_n R q_j$. If q_i and q_j are α-identical concerning R, then it is denoted as $q_i R(\alpha) q_j$, where the relation R for each fixed $\alpha \in [0, 1]$ is an equivalence relation on Q.

The pair (Q, R) is called a fuzzy approximation space. For any $\alpha \in [0, 1]$, we denote by R^*_α, the set of all almost equivalence classes of $R(\alpha)$. Also $(Q, R(\alpha))$ is the generated approximation space concerning R and α. Consider (Q, R) be a fuzzy approximation space and say $X \subseteq Q$ be the target set. Then the RSFAS of X in $(Q, R(\alpha))$ is defined as $(\underline{R}_\alpha X, \overline{R}_\alpha X)$. The α-lower approximation of X denoted as $\underline{R}_\alpha X$ is defined using Eqn. 9 whereas $\overline{R}_\alpha X$ is the α-upper approximation of X defined as in Eqn. 10.

$$\underline{R}_{\alpha} X = \cup\{Y \in R_{\alpha}^{*} : Y \subseteq X\} \tag{9}$$

$$\overline{R}_{\alpha} X = \cup\{Y \in R_{\alpha}^{*} : Y \cap X \neq \phi\} \tag{10}$$

The objects belonging to equivalence classes obtained through this way are almost identical or α-identical. On considering α as 100%, i.e., $\alpha = 1$; rough set on fuzzy approximation space reduces to rough set. The target set X is said to be α-crisp if $\underline{R}_{\alpha} X = \overline{R}_{\alpha} X$. Similarly, the target set X is said to be α-rough if $\underline{R}_{\alpha} X \neq \overline{R}_{\alpha} X$. The α-boundary region of X is given as $BL_{R_{\alpha}}(X) = \overline{R}_{\alpha} X - \underline{R}_{\alpha} X$. The objects belonging to $\underline{R}_{\alpha} X$ are α-certain whereas objects belonging to $BL_{R_{\alpha}}(X)$ are uncertain. Objects belonging to $NG_{R_{\alpha}}(X) = (Q - \overline{R}_{\alpha} X)$ are known as negative region.

3.4 Rough Set on Intuitionistic Fuzzy Approximation Space

The fuzzy set was developed to handle ambiguity and uncertainty [20]. Due to the presence of hesitation, it has been further extended to the intuitionistic fuzzy set [21] and is quite helpful in many real-life applications. When an element q in a fuzzy set has degree μ of membership, the degree of non-membership of q is determined using an algebraic equation under the premise that the entire degree of membership is deterministic and the indeterministic part is equal to zero. In addition, if the indeterministic component is zero, the intuitionistic fuzzy set simplifies to a fuzzy set. As a result, the intuitionistic fuzzy set model is a more comprehensive and superior model than the fuzzy set model. Thus Rough Sets on Intuitionistic Fuzzy Approximation Spaces (RSIFAS) is a generalized model over RSFAS. In RSIFAS, the fuzzy proximity relation is generalized to intuitionistic fuzzy proximity relation. Here, the definitions, notations, and results of RSIFAS are briefly presented [22]. The fundamental ideas of RSIFAS use the conventional notations μ for membership and ν for non-membership.

Consider Q be a universal set of objects. An intuitionistic fuzzy relation R on a universal set Q is an intuitionistic fuzzy set defined on $(Q \times Q)$. An intuitionistic fuzzy relation R on Q is said to be an intuitionistic fuzzy proximity relation if and only if $\mu_R(q_i, q_i) = 1$, $\nu_R(q_i, q_i) = 0$ for all $q_i \in Q$; and $\mu_R(q_i, q_j) = \mu_R(q_j, q_i)$, $\nu_R(q_i, q_j) = \nu_R(q_j, q_i)$ for all $q_i, q_j \in Q$. Let R be an intuitionistic fuzzy proximity relation on Q. Then for any $(\alpha, \beta) \in [0, 1]$ and $0 \leq (\alpha + \beta) \leq 1$, the (α, β)-cut of R is defined as in Eqn. 11.

$$R_{\alpha,\beta} = \{(q_i, q_j) \in (Q \times Q) : \mu_R(q_i, q_j) \geq \alpha, \nu_R(q_i, q_j) \leq \beta\} \tag{11}$$

Assume R be an intuitionistic proximity relation on Q. If $(q_i, q_j) \in R_{\alpha,\beta}$, then the two elements q_i and q_j are (α, β)-similar concerning R and it is written as $q_i R_{\alpha,\beta} q_j$. Two elements q_i and q_j are (α, β)-identical with respect to R for $(\alpha, \beta) \in [0, 1]$, written as $q_i R(\alpha, \beta) q_j$ if and only if $q_i R_{\alpha,\beta} q_j$ or there exists a sequence of elements

$q_1, q_2, q_3, \ldots, q_n$ in Q such that $q_i R_{\alpha,\beta} q_1, q_1 R_{\alpha,\beta} q_2, q_2 R_{\alpha,\beta} q_3, \ldots, q_n R_{\alpha,\beta} q_j$. In the second case, q_i is transitively (α, β)-similar to q_j concerning R.

Assume $R^*_{\alpha,\beta}$ be the set of equivalence classes generated by the intuitionistic fuzzy proximity relation $R(\alpha, \beta)$ for each fixed $(\alpha, \beta) \in [0, 1]$. The pair (Q, R) is an intuitionistic fuzzy approximation space. Let $X \subseteq Q$ be the target set. The lower and upper approximation of RSIFAS in the generalized approximation space $(Q, R(\alpha, \beta))$ is defined in Eqs. 12 and 13, respectively.

$$\underline{R}_{\alpha,\beta} X = \cup\{Y \in R^*_{\alpha,\beta} : Y \subseteq X\} \tag{12}$$

$$\overline{R}_{\alpha,\beta} X = \cup\{Y \in R^*_{\alpha,\beta} : Y \cap X \neq \phi\} \tag{13}$$

This method produces (α, β) equivalence classes as equivalence classes. Such a class contains (α, β)-identical objects. When β is taken to be zero, RSIFAS becomes RSFAS. Similar to this, RSIFAS becomes a rough set when $\alpha = 1$ and $\beta = 0$ [23]. If $\underline{R}_{\alpha,\beta} X \neq \overline{R}_{\alpha,\beta} X$, the target set X is considered to be (α, β)-rough. The (α, β)-boundary region of X is defined as $BL_{R_{\alpha,\beta}}(X) = \overline{R}_{\alpha,\beta} X - \underline{R}_{\alpha,\beta} X$. The objects belonging to lower approximation are (α, β)-certain whereas objects belonging to $BL_{R_{\alpha,\beta}}(X)$ are uncertain.

Rough set has been extended to many variations. Additionally, clear discussion on all these variations is beyond the scope of this chapter. However, Fig. 3 will provide

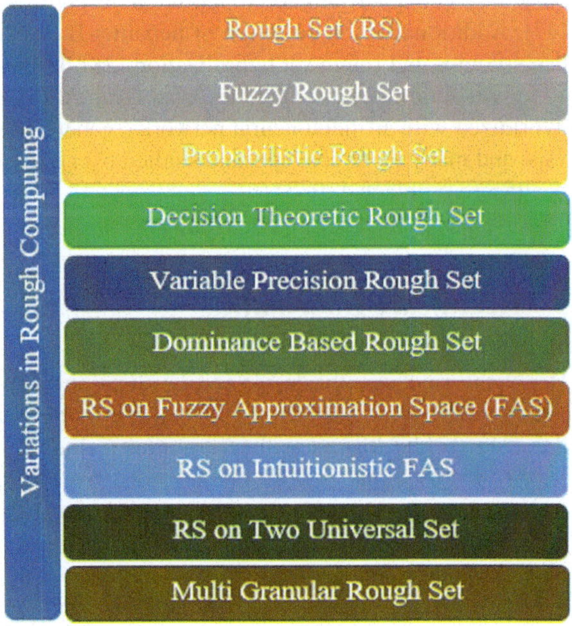

Fig. 3 Overview of variations of rough computing

an overview of variations of rough computing. More details on these variations can be found in the literature [12].

4 Hybridized Rough Computing

Increasing computing processes have also had a remarkable impact on the healthcare consortium. To include a programmed solution for disease diagnostics, researchers have employed numerous hybridized computing techniques. Additionally, bio-inspired computing combined with intelligent computing is becoming more and more common in healthcare data analytics. The numerous bio-inspired and rough computing hybridizations used in perception with healthcare applications are briefly covered in this section.

The analysis of the healthcare issue through research design has become increasingly important in recent years. Bio-inspired computing is integrated into the overall architecture, along with a rough set of tools for disease diagnostics. This hybridization's primary goal is to increase disease diagnostic accuracy while taking fewer features into account. Similar to this, a shuffling frog leaping algorithm is combined with a rough set to diagnose lung cancer [24]. As an alternative, the diagnosis of hepatitis-B disease is examined using a rough set coupled with a fish swarm method [25]. Similar to this, a hybrid algorithm inspired by bats and a rough set is being researched for the identification of chronic liver disease [26]. Likewise, the cuckoo search algorithm and rough set are used to diagnose heart disease diagnosis [27]. Additionally, the detection of hepatitis-B disease is examined using artificial bee colonies and rough set hybridization [28]. In the literature, a precise definition of hybridization and its range are also covered [9]. Besides, the rough set is also hybridized with artificial neural networks, genetic algorithms, formal concept analysis, and many more concepts and studied over numerous engineering, science, and healthcare problems [29–31].

5 Healthcare Informatics

Health informatics is the design and development of software that keeps patient information where it can be accessed by people providing care for the patient. It combines the science of health care with information technology. As a result, better health care is provided by clinicians. Additionally, it strives to create techniques and tools for gathering, analyzing, and researching patient data, which might come from a variety of sources and modalities. Health informatics is generally an interdisciplinary discipline that studies the design, development, and use of computational technologies to enhance health care [32]. The involved disciplines integrate the fields of computing and medicine. Computing in health care includes computer science and engineering, information system processing, bio-inspired computing, data science, information technology, and behavior informatics. In research, medical informatics primarily focuses on applications of artificial intelligence in health care and designing medical devices based on embedded systems [33].

Artificial intelligence and machine-learning algorithms are primarily used in healthcare informatics to simulate human cognition in the analysis, interpretation, and comprehension of challenging medical and healthcare data [34]. Artificial intelligence programs are used in processes including disease classification, clustering, disease feature selection, and diagnosis procedures. Additionally, it deals with patient monitoring and care, drug research, treatment, and personalized medicine. However, a significant portion of the healthcare consortium is focused on medical decision support systems. Machine-learning algorithms evolve as more data is gathered and make it possible for responses and solutions to be more reliable. The key artificial intelligence elements utilized in healthcare informatics are discussed in this section.

5.1 Feature Selection

An object and its features are contained in the information system. The parameters or attributes are alternate names for the features. It can be difficult to identify the key features of deducing knowledge. The machine-learning process known as "feature selection" identifies the fewest number of features, and these features are then used to categorize objects in an information system. Additionally, knowledge extraction can be performed using these minimized features. As a result, the scientific model has a simplified design with no loss of predicted accuracy. A collection of features is selected as part of the preprocessing operating method for data analysis to provide roughly high predicted accuracy [35]. Numerous feature selection techniques have been proposed by various academics [36–38].

The research survey states that Naive Bayes, artificial neural networks, and decision trees are the initial feature selection strategies used [39]. The Naive Bayes classifier using gain ratio, the decision tree using information gain, and the multilevel perceptron utilizing Chi-square are found to produce great results for feature selection. An introduction to knowledge discovery in databases using feature selection is provided by a recent survey [40]. The wrapper, filter, and embedding methods are the various feature selection techniques examined in this survey. The literature provides a review of feature selection that includes assessment metrics like correlation, mutual information, symmetric uncertainty, Euclidean distance, Laplacian score, Fisher score, and dependency index [41].

There are three different kinds of feature selection processes available in the literature including wrapper, filter, and embedding techniques, depending on the search methodology. The filter technique conducts the selection process independently, without working with any research algorithms. However, in the wrapper approach, the algorithms for feature selection and those for exploring are tied together.

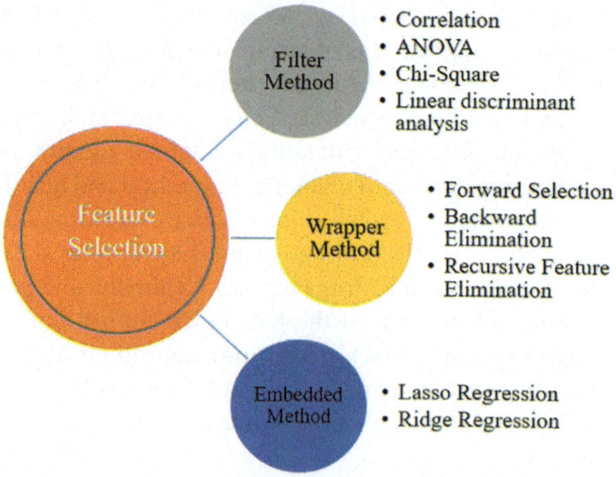

Fig. 4 Representation of feature selection process

The selection process is built into the classifier model in the embedded technique, which is comparable to the wrapper approach. Figure 4 shows an overview of feature selection methods.

For feature selection, a rough set-based reduct is employed in many research [11]. Without reducing the classification power of the information system, it eliminates superfluous features. It has been used to get rid of redundant features in many kinds of information systems. The biggest benefit is that data dependencies can be found without the need for additional knowledge. A thorough analysis of feature selection methods based on rough sets is available in the literature [42]. Similar to that, a statistical solution using rough sets is provided for pattern identification and feature reduction [43]. Likewise, a backpropagation neural network and a rough set equivalence relation are utilized to create a knowledge mining model using clinical datasets [44]. In addition, a method for feature selection called feature space decomposition utilizing a rough set is created and it is analyzed over a hybrid information system [45]. The technique is protracted and hasn't been properly evaluated on a hybridized information system, which is the main drawback. Another technique that chooses features based on boundary regions is also suggested in the literature. In this method, significant measures are constructed using the boundary region [46].

5.2 *Classification*

Classification is a task that requires the use of machine-learning procedures that learn how to assign a decision label to instances from the problem area. A simple illustration is the classification of hepatitis diseases as "live" or "die". Machine learning uses a variety of classification problems, each of which can be modeled using a different specialized approach. Predictive modeling classification, binary classification, multi-class classification, multi-label classification, and imbalanced classification are among the several types of classification [47].

In a classification problem using predictive modeling, a class label is predicted for a specific set of input attribute values. An example would be the computation of a decision class label from a set of input feature values. Determining the recognized character of a particular handwritten character is an example of classification. For the machine to learn, a training dataset with several examples of inputs and decisions is needed for this classification modeling. This model will determine the best way to match a hypothetical instance of input data to a given set of decision class labels using the training dataset. Due to this, the training dataset needs to have a wide representation of the problem as well as several instances for each class label [48].

Predicting one of two classes is known as binary classification. Binary classification problems often require two classes, one representing the normal state and the other representing the abnormal state. Typically, normal condition is represented by 0 and abnormal state is represented by 1. One example of a binary classification is the detection of email spam [49]. Logistic regression, k-nearest neighbors, decision trees, support vector machines, and naive Bayes algorithms are examples of popular algorithms that come under this category.

Predicting one of more than two class labels is the goal of multi-class classification. Multi-class classification does not have the idea of normal and abnormal outcomes, in contrast to binary classification. Instead, instances are grouped into one of several well-known classes [50]. For instance, in a system for forecasting heart disease, the diagnosis might be hypertensive heart disease, coronary heart disease, heart failure, or cardiomyopathy. K-nearest neighbors, decision trees, naive Bayes, random forests, and gradient-boosting techniques are examples of popular multi-class algorithms. Multi-class problems can also be solved using algorithms created for binary classification.

In multi-label classification, each example is assigned to one or more classes. For instance, a certain photo could be classified as a bike, banana, animal, etc. Multi-label decision trees, multi-label random forests, and multi-label gradient-boosting procedures are classification algorithms that fall under this category [51]. The term "imbalanced classification" describes classification problems where there is an unequal distribution of examples among the classes. For instance, fraud detection and outlier identification are examples of imbalanced classification. Cost-sensitive logistic regression, cost-sensitive decision trees, and cost-sensitive support vector machines are the techniques used for imbalanced classification [52].

It is typically advised that a practitioner undertakes controlled tests to determine which algorithm and algorithm configuration produces the greatest performance for a certain classification task. This is because there is no strong theory on how to map algorithms onto classification types available in the literature. However, rough computing is widely used for healthcare classification [9]. A diagrammatic representation of classification is depicted in Fig. 5.

5.3 *Clustering*

This section explains clustering, another key idea in healthcare informatics. Healthcare data analysis employs the concepts of classification and clustering. The main distinction between classification and clustering is that classification analyzes data already marked with decisions. Additionally, when the data are not decision-labeled, clustering groups similar data occurrences. Put another way, clustering is grouping similar things using a machine-learning algorithm. For artificial intelligence models, grouping related objects into clusters is advantageous. It is extensively utilized in data analytics, including image processing, data discovery, unsupervised learning, and numerous healthcare applications [53].

Hard clustering and soft clustering are two different types of clustering. One data instance can only be a part of one cluster when using hard clustering. In contrast, the probability likelihood of a data instance belonging to each of the pre-defined numbers of clusters is the output that is given in soft clustering. In contrast to supervised learning, clustering is an unsupervised learning method that employs datasets where the relationship between the data instances is unknown. The clustering techniques applied in data analytics include hierarchical, centroid-based, distribution-based, density-based, model-based, and grid-based methods. These techniques help cluster data points together. Hierarchical clustering is a cluster analysis technique used in data analytics to create a hierarchy of groups. Each data point is initially treated as an independent cluster. Further, the algorithm iteratively aggregates the nearest clusters until a stopping requirement is met. The different strategies used for hierarchical clustering are agglomerative and divisive. Divisive is a top-down strategy, whereas agglomerative is a bottom-up strategy. The merges and splits are greedily chosen in these methods. The clear benefit of hierarchical clustering is that any legitimate distance metric can be applied [54].

In contrast to hierarchical clustering, centroid-based clustering groups the data into non-hierarchical clusters. The most popular centroid-based clustering algorithm

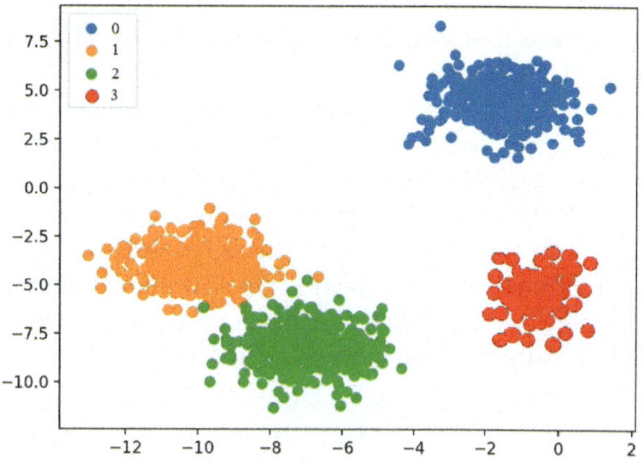

Fig. 5 A general view of classification process

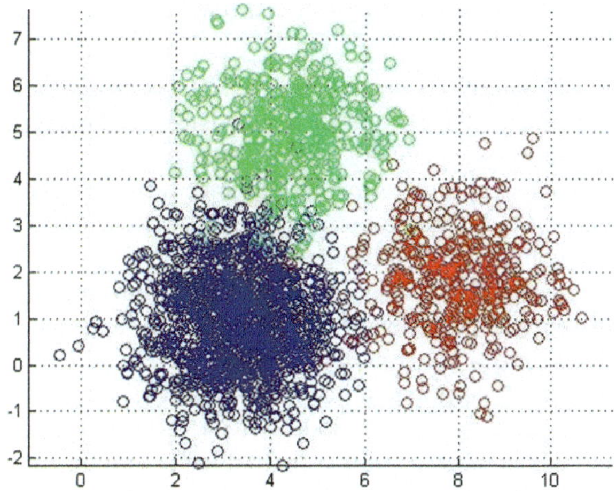

Fig. 6 A general view of clustering process

is the k-means algorithm. Further, the k-means algorithm has been extended to the fuzzy k-means algorithm, rough k-means algorithm, and many more. Although efficient, centroid-based algorithms are sensitive to beginning conditions and outliers. Distribution-based clustering organizes the data by grouping data instances likely to belong to the same probability distribution. A probability-based distribution groups the data items based on statistical distributions [55].

Density-based clustering in a data space is a contiguous region of high point density, separated from other clusters by sparse regions. It is an unsupervised machine-learning technique that detects distinctive groups in the data [56]. A statistical method for clustering data is known as model-based clustering. It is presumed that a finite mixture of component models produced the observed data. A probability distribution, often a parametric multivariate distribution, makes up each component of a component model. Similarly, grid-based clustering uses multi-resolution grid data structure techniques. In such cases, the object areas are quantized into a limited number of grid-like cells, on which all clustering processes are carried out [55].

The various clustering algorithms widely used in machine learning are the k-means clustering algorithm, where k is the number of clusters, Density-Based Spatial Clustering of Applications with Noise (DBSCAN) algorithm, Gaussian mixture model algorithm, Balance Iterative Reducing and Clustering using Hierarchies (BIRCH) algorithm, affinity propagation clustering algorithm, mean-shift clustering algorithm, Ordering Points to Identify the Clustering Structure (OPTICS) algorithm, and agglomerative hierarchy clustering algorithm. These algorithms are further extended to many extents to handle various real-life problems [54]. A diagrammatic overview of clustering is depicted in Fig. 6.

5.4 Decision Support System

A computerized system known as a decision support system collects, evaluates, and synthesizes data to create detailed information reports. Four components constitute a decision support system: data management, model management, knowledge management, and user interface management. The various types of decision support systems are data-driven, model-driven, knowledge-driven, document-driven, and communication-driven [57].

Decisions are made based on data from internal or external databases in data-driven decision support systems. Data mining techniques are used to make these decisions for potential future events. These decision support systems are frequently utilized in corporate processes, inventory, sales, and healthcare consortiums. Decisions are tailored in a model-driven decision support system following a predetermined list of user needs. It is frequently employed to support the creation of financial statements or schedules. Data is stored in a constantly updated knowledge base managed by a knowledge management system in a knowledge-driven decision support system. Users then have access to information consistent with the enterprise's operational procedures and knowledge base. An information management system called a document-driven decision support system uses documents to obtain data. Users of this decision support system can search databases containing information on business records, policies, and processes. Multiple people can work on the same activity at once with the help of a communication-driven decision support system. A communication-driven decision support system improves collaboration between users and the system [58].

The functions of decision support systems include model building, risk analysis, graphical analysis, what-if analysis, and goal-oriented analysis. In model building, decision support systems aid in the decision-making process by identifying the most appropriate model for solving problems. Risk analysis is beneficial while making medium- or high-risk decisions. Graphical analysis is helpful to visualize the findings before and after implementation. What-if analysis provides the advantages and limitations of the undertaking decisions [59]. In recent years, artificial intelligence has been associated with decision support systems, called intelligent decision support systems. The various phases of the healthcare decision support system are intelligence, design, choice, implementation, and monitoring [60]. An overview of decision support system is depicted in Fig. 7.

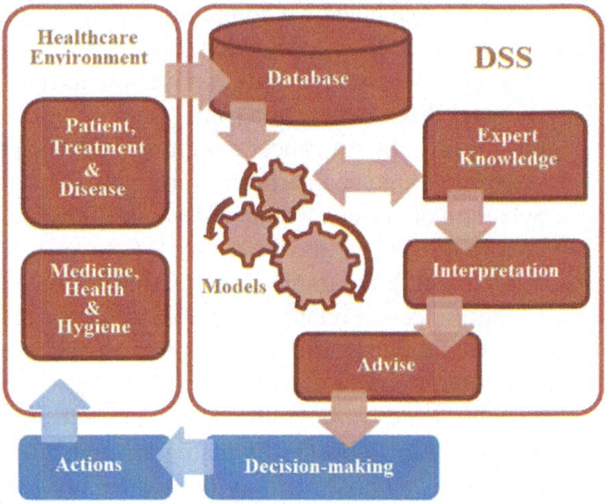

Fig. 7 An overview of decision support system

6 Healthcare Applications

The economy and the daily lives of the people are significantly impacted by a nation's effective healthcare management system. These healthcare data are laborious to manage and analyze, necessitating the use of rough and intelligent techniques. Clustering, classifying diseases, early diagnosis, and choosing the most important disease-related parameters are some of the difficult tasks. The previous section emphasized the intelligent approaches used in the aforementioned issues in the preceding part, including rough set, fuzzy rough set, rough set on fuzzy approximation space, and rough set on intuitionistic fuzzy approximation space. In addition, the hybridized and rough computing methods applied to the aforementioned issues are emphasized. Applications of rough computing in healthcare informatics are primarily discussed in this section. For the aforementioned reason, a number of research publications are examined and briefly highlighted for better visualization.

Bio-inspired algorithms are meta-heuristic and essential for solving time-sensitive challenges that are highly important. When there is a variance in the constraints and a lack of knowledge with a constrained amount of computations, these techniques become crucial. The hybridization of rough and bio-inspired computing in healthcare applications includes feature selection, disease classification, diagnosis, and decision support system. These applications are almost always fraught with uncertainty and incomplete data. When analyzing healthcare information systems, classification, clustering, prediction, feature selection, attribute reduction, and rule mining are difficult tasks. These problems have been partially addressed through the development of hybridized rough computing. This section briefly states few applications of hybridized rough computing concerning healthcare informatics.

1. Hybridized rough and bat computing in healthcare informatics:
   – Chronic liver disease detection [26].
   – ECG analysis for myocardial infarction detection [61].
   – Diabetes disease detection [62].
   – Diagnosis of breast cancer mammographic mass [63].
2. Hybridized rough and particle swarm optimization in healthcare informatics:
   – Selection of various features of medical diseases [64].
   – Diagnosis of hepatitis disease [65].
   – Detection of brain tumor diseases [66].
   – Feature selection relating to kidney and liver diseases [67].
3. Hybridized rough and firefly algorithm in healthcare informatics:
   – Brain tumor image classification [68].
   – Rough set-firefly integration for feature selection for MRI brain tumor [69].
   – Rough firefly algorithm for histogram-based fuzzy image clustering [70].
   – Rough set and firefly fusion for medical image segmentation [71].
   – Firefly and rough set integration for breast cancer classification [72].
   – Integrated rough set-firefly algorithm for heart disease prediction [73].
4. Hybridized rough and artificial bee colony algorithm in healthcare informatics
   : – Artificial bee colony and rough set integration for hepatitis-B diagnosis [28]
   . – Rough set bee colony reduction for medical domain datasets [74]
   . – Rough set and bee colony optimization for attribute reduction [75].
5. Hybridized rough and cuckoo search algorithm in healthcare informatics:
   – Diagnosis and decision rule generation relating to heart disease [27].
   – Cuckoo search with rough set algorithm for feature selection [75].
   – Hybrid binary cuckoo search and rough set for classification [76].
   – Binary cuckoo search with rough set for feature selection [77, 78].
6. Hybridized rough and fish swarm algorithm in healthcare informatics:
   – Rough set and fish swarm algorithm for attribute reduction [79].
   – Rough set and fish swarm algorithm for hepatitis diagnosis [25].
   – Rough set and fish swarm algorithm for rule mining [80]
7. Hybridized rough and ant colony optimization in healthcare informatic:
   – Ant colony rough set hybridization for attribute reduction [81].
   – Ant colony rough set hybridization for feature selection [82].

7 Conclusion

This chapter summarized several studies on healthcare informatics that include feature selection, disease classification, clustering, disease diagnosis, and decision support system conducted by diverse research communities. The foundations of rough computing and hybridized rough computing used in healthcare informatics are important in artificial intelligence which are covered in this chapter, which serves as the basis for the further investigation. This chapter has provided information on healthcare applications pertaining to the rough computing as an introduction to research.

Additionally, it examines several applications that make use of this rough and biologically inspired computing. Additionally, it aids readers in comprehending the theories upon which this study is built.

References

1. Gheyas, I.A., Smith, L.S.: Feature subset selection in large dimensionality domains. Pattern Recognit. **43**(1), 5–13 (2010)
2. Foithong, S., Pinngern, O., Attachoo, B.: Feature subset selection wrapper based on mutual information and rough sets. Expert Syst. Appl. **39**(1), 574–584 (2012)
3. Arji, G., Ahmadi, H., Nilashi, M., Rashid, T.A., Ahmed, O.H., Aljojo, N., Zainol, A.: Fuzzy logic approach for infectious disease diagnosis: a methodical evaluation, literature and classification. Biocybernet. Biomed. Eng. **39**(4), 937–955 (2019)
4. Pawlak, Z.: Rough classification. Int. J. Human-Comput. Stud. **51**(2), 369–383 (1999)
5. Sarkar, M.: Rough-fuzzy functions in classification. Fuzzy Sets Syst. **132**(3), 353–369 (2002)
6. Maji, P., Paul, S.: Rough-fuzzy clustering for grouping functionally similar genes from microarray data. ACM Trans. Comput. Biol. Bioinform. **10**(2), 286–299 (2012)
7. Rorissa, A., Yuan, X.: Visualizing and mapping the intellectual structure of information retrieval. Inf. Process. Manag. **48**(1), 120–135 (2012)
8. Pawlak, Z.: Rough set approach to knowledge-based decision support. Eur. J. Oper. Res. **99**(1), 48–57 (1997)
9. Kumari, N., Acharjya, D.P.: Data classification using rough set and bioinspired computing in healthcare applications-an extensive review. Multimedia Tools Appl. **82**(9), 13479–13505 (2023)
10. Acharjya, D.P., Geetha, M.A.: Privacy preservation in information system. In: Censorship, Surveillance, and Privacy: Concepts, Methodologies, Tools, and Applications, IGI Global, USA, pp. 1695–1720 (2019)
11. Pawlak, Z.: Rough sets. Int. J. Comput. Inf. Sci. **11**, 341–356 (1982)
12. Acharjya, D.P., Abraham, A.: Rough computing-A review of abstraction, hybridization and extent of applications. Eng. Appl. Artif. Intell. **96**, 103924 (2020)
13. Pawlak, Z., Skowron, A.: Rudiments of rough sets. Inf. Sci. **177**(1), 3–27 (2007)
14. Pawlak, Z.: Rough set theory and its applications to data analysis. Cybernet. Syst. **29**(7), 661–688 (1998)
15. Dubois, D., Prade, H.: Rough fuzzy sets and fuzzy rough sets. Int. J. General Syst. **17**(2–3), 191–209 (1990)
16. Jensen, R., Shen, Q.: Fuzzy-rough sets assisted attribute selection. IEEE Trans. Fuzzy Syst. **15**(1), 73–89 (2007)
17. Pei, D.: A generalized model of fuzzy rough sets. Int. J. Gen Syst **34**(5), 603–613 (2005)
18. Radzikowska, A.M., Kerre, E.E.: A comparative study of fuzzy rough sets. Fuzzy Sets Syst. **126**(2), 137–155 (2002)
19. Acharjya, D.P., Tripathy, B.K.: Rough sets on fuzzy approximation spaces and applications to distributed knowledge systems. Int. J. Artif. Intell. Soft Comput. **1**(1), 1–14 (2008)
20. Zadeh, L.A.: Fuzzy sets. Inf. Control **8**(3), 338–353 (1965)
21. Atanassov, K.T.: Intuitionistic fuzzy sets. Fuzzy Sets Syst. **20**(1), 87–96 (1986)
22. Tripathy, B.K.: Rough sets on intuitionistic fuzzy approximation spaces. Notes Intuit. Fuzzy Sets **12**(1), 45–54 (2006)
23. Acharjya, D.P.: Comparative study of rough sets on fuzzy approximation spaces and intuitionistic fuzzy approximation spaces. Int. J. Comput. Appl. Math. **4**(2), 95–106 (2009)
24. Kumari, N., Acharjya, D.P.: A hybrid rough set shuffled frog leaping knowledge inference system for diagnosis of lung cancer disease. Comput. Biol. Med. **155**(3), 106662 (2023)

25. Kumari, N., Acharjya, D.P.: A decision support system for diagnosis of hepatitis disease using an integrated rough set and fish swarm algorithm. Concurr. Comput.: Pract. Exp. **34**(21), e7107 (2022)
26. Acharjya, D.P., Ahmed, P.K.: A hybridized rough set and bat-inspired algorithm for knowledge inferencing in the diagnosis of chronic liver disease. Multimed. Tools Appl. **81**(10), 13489–13512 (2022)
27. Ahmed, P.K., Acharjya, D.P.: A hybrid scheme for heart disease diagnosis using rough set and cuckoo search technique. J. Med. Syst. **44**(1), 1–16 (2020)
28. Ahmed, P.K., Acharjya, D.P.: Knowledge inferencing using artificial bee colony and rough set for diagnosis of hepatitis disease. Int. J. Healthcare Inf. Syst. Inf. **16**(2), 49–72 (2021)
29. Anitha, A., Acharjya, D.P.: Neural network and rough set hybrid scheme for prediction of missing associations. Int. J. Bioinform. Res. Appl. **11**(6), 503–524 (2015)
30. Rathi, R., Acharjya, D.P.: A rule based classification for vegetable production using rough set and genetic algorithm. Int. J. Fuzzy Syst. Appl. **7**(1), 74–100 (2018)
31. Tripathy, B.K., Acharjya, D.P., Cynthya, V.: A framework for intelligent medical diagnosis using rough set with formal concept analysis. Int. J. Artif. Intell. Appl. **2**(2), 45–66 (2011)
32. Nadri, H., Rahimi, B., Timpka, T., Sedghi, S.: The top 100 articles in the medical informatics: a bibliometric analysis. J. Med. Syst. **41**, 1–12 (2017)
33. Shortliffe, H.E., Cimino, J.J.: Biomedical Informatics: Computer Applications in Health Care and Biomedicine. Springer, London (2014)
34. Quan, X.I., Sanderson, J.: Understanding the artificial intelligence business ecosystem. IEEE Eng. Manag. Rev. **46**(4), 22–25 (2018)
35. Schiezaro, M., Pedrini, H.: Data feature selection based on artificial bee colony algorithm. EURASIP J. Image Video Process. **2013**, 1–8 (2013)
36. Jain, A., Zongker, D.: Feature selection: evaluation, application, and small sample performance. IEEE Trans. Pattern Anal. Mach. Intell. **19**(2), 153–158 (1997)
37. Kohavi, R., John, G.H.: Wrappers for feature subset selection. Artif. Intell. **97**(1–2), 273–324 (1997)
38. Guyon, I., Elisseeff, A.: An introduction to variable and feature selection. J. Mach. Learn. Res. **3**(3), 1157–1182 (2003)
39. Karabulut, E.M., Özel, S.A., Ibrikci, T.: A comparative study on the effect of feature selection on classification accuracy. Proc. Technol. **1**, 323–327 (2012)
40. Chandrashekar, G., Sahin, F.: A survey on feature selection methods. Comput. Electr. Eng. **40**(1), 16–28 (2014)
41. Cai, J., Luo, J., Wang, S., Yang, S.: Feature selection in machine learning: a new perspective. Neurocomputing **300**, 70–79 (2018)
42. Jensen, R.: Rough set-based feature selection: a review. In: Rough Computing: Theories, Technologies and Applications, IGI Global, USA, pp. 70–107 (2008)
43. Świniarski, R.W.: Rough sets methods in feature reduction and classification. Int. J. Appl. Math. Comput. Sci. **11**, 565–582 (2001)
44. Nahato, K.B., Harichandran, K.N., Arputharaj, K.: Knowledge mining from clinical datasets using rough sets and backpropagation neural network. Comput. Math. Methods Med. **2015**, 1–14 (2015)
45. Kim, K.J., Jun, C.H.: Rough set model based feature selection for mixed-type data with feature space decomposition. Expert Syst. Appl. **103**, 196–205 (2018)
46. Lu, Z., Qin, Z., Zhang, Y., Fang, J.: A fast feature selection approach based on rough set boundary regions. Pattern Recognit. Lett. **36**, 81–88 (2014)
47. Ganda, D., Buch, R.: A survey on multi label classification. Recent Trends Program. Lang. **5**(1), 19–23 (2018)
48. Olson, D.L., Delen, D., Olson, D.L., Delen, D.: Performance evaluation for predictive modeling. In: Advanced Data Mining Techniques, Springer, USA, pp. 137–147 (2008)
49. Kumari, R., Srivastava, S.K.: Machine learning: a review on binary classification. Int. J. Comput. Appl. **160**(7), 11–15 (2017)

50. Sahare, M., Gupta, H.: A review of multi-class classification for imbalanced data. Int. J. Adv. Comput. Res. **2**(3), 160–164 (2012)
51. Godbole, S., Sarawagi, S.: Discriminative methods for multi-labeled classification. In: Proceedings of the Pacific-Asia Conference on Knowledge Discovery and Data Mining. Springer, Berlin, Heidelberg, pp. 22–30 (2004)
52. Zou, Q., Xie, S., Lin, Z., Wu, M., Ju, Y.: Finding the best classification threshold in imbalanced classification. Big Data Res. **5**, 2–8 (2016)
53. Ezugwu, A.E., Ikotun, A.M., Oyelade, O.O., Abualigah, L., Agushaka, J.O., Eke, C.I., Akinyelu, A.A.: A comprehensive survey of clustering algorithms: state-of-the-art machine learning applications, taxonomy, challenges, and future research prospects. Eng. Appl. Artif. Intell. **110**, 104743 (2022)
54. Xu, R., Wunsch, D.: Survey of clustering algorithms. IEEE Trans. Neural Netw. **16**(3), 645–678 (2005)
55. Madhulatha, T.S.: An overview on clustering methods. IOSR J. Eng. **2**(4), 719–725 (2012)
56. Bhattacharjee, P., Mitra, P.: A survey of density based clustering algorithms. Front. Comput. Sci. **15**, 1–27 (2021)
57. Er, M.C.: Decision support systems: a summary, problems, and future trends. Decis. Support Syst. **4**(3), 355–363 (1988)
58. Eom, S., Kim, E.: A survey of decision support system applications (1995–2001). J. Oper. Res. Soc. **57**, 1264–1278 (2006)
59. Moskowitz, H., Kim, K.J.: QFD optimizer: a novice friendly quality function deployment decision support system for optimizing product designs. Comput. Ind. Eng. **32**(3), 641–655 (1997)
60. Gottinger, H.W., Weimann, P.: Intelligent decision support systems. Decis. Support Syst. **8**(4), 317–332 (1992)
61. Kora, P., Kalva, S.R.: Improved Bat algorithm for the detection of myocardial infarction. Springerplus **4**, 1–18 (2015)
62. Cheruku, R., Edla, D.R., Kuppili, V., Dharavath, R.: Rst-batminer: a fuzzy rule miner integrating rough set feature selection and bat optimization for detection of diabetes disease. Appl. Soft Comput. **67**, 764–780 (2018)
63. Africa, A.D.M., Cabatuan, M.K.: A rough set based data model for breast cancer mammographic mass diagnostics. Int. J. Biomed. Eng. Technol. **18**(4), 359–369 (2015)
64. Inbarani, H.H., Azar, A.T., Jothi, G.: Supervised hybrid feature selection based on PSO and rough sets for medical diagnosis. Comput. Methods Programs Biomed. **113**(1), 175–185 (2014)
65. He, F., Wang, G., Yang, H. M.: A novel method for hepatitis disease diagnosis based RS and PSO. In: Proceedings of the 4th Electronic System-Integration Technology Conference, pp. 1289–1292, (2012)
66. Sharif, M., Amin, J., Raza, M., Yasmin, M., Satapathy, S.C.: An integrated design of particle swarm optimization (PSO) with fusion of features for detection of brain tumor. Pattern Recognit. Lett. **129**, 150–157 (2020)
67. Gunasundari, S., Janakiraman, S., Meenambal, S.: Velocity bounded boolean particle swarm optimization for improved feature selection in liver and kidney disease diagnosis. Expert Syst. Appl. **56**, 28–47 (2016)
68. Jothi, G.: Hybrid Tolerance Rough Set-Firefly based supervised feature selection for MRI brain tumor image classification. Appl. Soft Comput. **46**, 639–651 (2016)
69. Jothi, G.: Hybrid tolerance rough set-firefly based supervised feature selection for MRI brain tumor image classification. Appl. Soft Comput. **46**, 639–651 (2016)
70. Dhal, K.G., Das, A., Ray, S., Gálvez, J.: Randomly attracted rough firefly algorithm for histogram based fuzzy image clustering. Knowl.-Based Syst. **216**, 106814 (2021)
71. Chinta, S.S.: Kernelised rough sets based clustering algorithms fused with firefly algorithm for image segmentation. Int. J. Fuzzy Syst. Appl. **8**(4), 25–38 (2019)
72. Farouk, R.M., Mustafa, H.I., Ali, A.E.: Hybrid firefly and swarm algorithms for breast cancer mammograms classification based on rough set theory features selection. In: Proceedings of the Future Technologies Conference, Springer International Publishing, USA, vol. 3, pp. 849–867 (2022)

73. Long, N.C., Meesad, P., Unger, H.: A highly accurate firefly based algorithm for heart disease prediction. Expert Syst. Appl. **42**(21), 8221–8231 (2015)
74. Suguna, N., Thanushkodi, K.: A novel rough set reduct algorithm for medical domain based on bee colony optimization. J. Comput. **2**(6), 49–54 (2010)
75. Chebrolu, S., Sanjeevi, S.G.: Attribute reduction on real-valued data in rough set theory using hybrid artificial bee colony: extended FTSBPSD algorithm. Soft. Comput. **21**, 7543–7569 (2017)
76. Aziz, M.A.E., Hassanien, A.E.: Modified cuckoo search algorithm with rough sets for feature selection. Neural Comput. Appl. **29**, 925–934 (2018)
77. Alia, A.F., Taweel, A.: Feature selection based on hybrid binary cuckoo search and rough set theory in classification for nominal datasets. Algorithms **14**(21), 65 (2017)
78. Alia, A., Taweel, A.: Enhanced binary cuckoo search with frequent values and rough set theory for feature selection. IEEE Access **9**, 119430–119453 (2021)
79. Luan, X.Y., Li, Z.P., Liu, T.Z.: A novel attribute reduction algorithm based on rough set and improved artificial fish swarm algorithm. Neurocomputing **174**, 522–529 (2016)
80. Huang, Y., Fu, B., Cai, X., Xing, X., Yuan, X., Yu, L.: Rules extraction by clustering artificial fish-swarm and rough set. Res. J. Appl. Sci. Eng. Technol. **4**(2), 127–130 (2012)
81. Ke, L., Feng, Z., Ren, Z.: An efficient ant colony optimization approach to attribute reduction in rough set theory. Pattern Recognit. Lett. **29**(9), 1351–1357 (2008)
82. Chen, Y., Miao, D., Wang, R.: A rough set approach to feature selection based on ant colony optimization. Pattern Recognit. Lett. **31**(3), 226–233 (2010)

Computational Intelligence in Ethical Issues in Healthcare

Ethical Issues on Drug Delivery and Its Impact in Healthcare

Afsana Zannat Ahmed and Kedar Nath Das

Abstract Medical ethics is all about study on a specific clinical problem, facts, and finding reasoning to determine the best medical practices. In the healthcare sector, it mostly assists the patients, communities, and healthcare providers in making a decision on treatment options, procedures, and other issues that emerge in the health sectors. Drug discovery is the process of discovering novel candidate treatments in biotechnology and pharmacology. Nowadays, chronic illnesses such as cancer is a major concern. Every year, around 1.5 million Americans are diagnosed with cancer. Inappropriate proportion of drug delivery or stopping a medication regimen in cancer treatment may lead to several side effects. Cancer chemotherapy involves use of anti-cancer drugs which are not free from toxic side effects. A suitable medication schedule is required for the drug scheduling that might help in making decision on which drug is to be delivered in time. It, in turn, reduces the size of the cancer cell under minimizing adverse effects. In this chapter, analytical studies have been conducted in order to plan such medication dose regimens to achieve the desired results. In the earlier study, the problem is modeled as an optimization problem that has been solved by Particle Swarm Optimization (PSO) and Population-Based Incremental Learning (PBIL). In this chapter, a Complete Elitist Genetic Algorithm (CEGA) has been proposed to solve the problem. Later, both efficiency and efficacy of CEGA are better demonstrated in terms of cancer reduction and minimal adverse effects. Further, CEGA has been proven to be better than the conventional treatment methods and is also concluded superior to both PSO and PBIL.

A. Zannat Ahmed
IIT Kharagpur, Kharagpur, West Bengal, India
e-mail: zannat989@gmail.com

K. Nath Das (✉)
NIT Silchar, Silchar, Assam, India
e-mail: kedarnath@math.nits.ac.in

D. P. Acharjya and K. Ma (eds.), *Computational Intelligence in Healthcare Informatics*, Studies in Computational Intelligence 1132,
https://doi.org/10.1007/978-981-99-8853-2_19

1 Introduction

The use of mathematical algorithms to replicate human cognitive capacities is known as Artificial Intelligence (AI). The exponential expansion of AI in the recent decade has been demonstrated to be a viable platform for super intelligence decision-making, where the human mind is inadequate to deal out big quantities of data in a short amount of time. The findings are consistent independent of contextual conditions since a mathematical algorithm simulates the best of human intellect. The chief goal of health-related AI applications is to scrutinize the relations between preventive or treatment strategies and patient outcomes. Diagnoses, treatment schedule, medication research, customized medicine, and patient monitoring are all areas where AI systems are being used. These AI procedures have been adopted by medical organizations such as The Mayo Clinic, Memorial Sloan Kettering Cancer center, and the British National Health Service. Huge companies such as IBM and Google have also invested in AI for healthcare.

AI has made significant contributions to the treatment of a variety of medical issues, including cancer, during the last decade. With a large population of cancer patients diagnosed every year around the world, AI applications in oncology are gaining attention. Increased processing power, algorithmic advancements, and data encoding strategies have created a plethora of new data sources for AI to use. Treatment choices and patient monitoring can be informed by personalized estimates of treatment response to alternative medicines, as well as their possible side effects.

AI is primarily concerned with the interaction between drugs and patients in cancer therapy. Managing chemotherapy drug usage, prediction of chemotherapy drug tolerance, and optimization of chemotherapy are among of AI's most notable contributions. Personalized or precision medicine is a field of medicine that uses a person's genetic information to help direct decisions on prevention, diagnosis, and treatment of disease. Drug-related disorders in patients cost $177 billion in the United States alone every year. These astronomical expenses reflect how difficult it has become for healthcare practitioners to prescribe medications. The accuracy of medicine that can be dosed using traditional methods has a limit. To date, the most appealing strategy to tackling this critical problem has been to use artificial intelligence to enable precise dosage.

Cancer is a bodily condition in which uncontrolled cell division leads to an overabundance of cells that invade other tissues. There are two forms of cancer: benign and malignant. Breast cancer is one of the most frequent malignancies in women. The chemotherapeutic optimum medication schedule in the case of breast cancer is designed in this research. From [1], clinical data on anti-cancer medicines used to treat breast cancer has been gathered. Cancer chemotherapy refers to the intake of anti-cancer drugs which are delivered via the bloodstream. The drugs being extremely toxic directly damage the normal tissues elsewhere in the body.

The objective is to reduce the tumor size to quite an extent with the application of Genetic Algorithm (GA) and tuning of the GA operators to get a better dosing scheme. The chapter is structured as follows: Sect. 2 throws light on review of literature. Foundations of genetic algorithm are presented in Sect. 3. The problem statement, mathematical model, and its discussion are presented in Sect. 4 followed by methodology in Sect. 5. Results and discussions are presented in Sect. 6. Section 7 concludes the chapter and presents the future scope.

2 Review of Literature

In a research work, optimizing techniques are explored in order to optimize cancer chemotherapy and showed the application of GAs to optimize single-drug and multi-drug chemotherapy treatment leading to findings of feasible treatment strategies for both curative and palliative treatments authenticating the aptness and efficacy of GAs for cancer chemotherapy optimization [1].

An optimal parameter selection model of cancer chemotherapy in order to describe the cancer treatment for a fixed duration by the recurrent application of a single drug is conferred in the literature. The model's selected dosage regime is at equal intervals of time is delivered. Through numerical solutions it was shown that an optimal regimen suppressing the bulk of the doses until the end of the treatment period outperforms the conventional regimen that delivers the entire drug at the onset of the treatment [2].

Further, Particle Swarm Optimization (PSO) is used to ease the finding of optimal chemotherapeutic treatments, and compared its performance with that of GAs. It illustrates that PSO achieves the same optimization objective as that of GA, but in a new and faster way. This means PSO is proven to be more efficient than GA [3]. The GA is applied taking two chromosomes for elitism.

Designing of efficient chemotherapeutic treatments was done satisfying the constraints and investigating a larger solution space of possible treatment schedules with the help of two computational heuristic algorithms such as Estimation of Distribution Algorithm (in the form of PBIL) and GA. The experimental results show that PBIL outperforms GA in both the speed of finding a feasible treatment schedule and in the quality of the final solution. It is also to be noted that the GA used here has been designed to use two chromosomes only for elitism [4].

The application of two directed intervention crossover approaches to the problem is demonstrated. The Calculated Expanding Bin (CalEB) method and Targeted Intervention with Stochastic Selection (TInSSel) approaches are computed and it indicates that these approaches lead to significant improvements over uniform crossover when applied to cancer chemotherapy drug schedule designing [5]. Additionally, a novel approach with the aim of optimizing chemotherapy with regard to contradictory treatment objectives is presented. The approach is made on the PSO algorithm that decomposes a multi-objective optimization problem into quite a few scalar aggregation problems, thus reducing its complications and enabling an efficient appliance of computational intelligence techniques. The uniqueness of the algorithm lies in presenting particles in the swarm with information from a set of defined neighbors as well as leaders which help in achieving resourceful chemotherapeutic treatments [6].

Similarly, a novel approach has been made toward computationally feasible, speedy, and self-governing algorithm known as Deterministic Oscillatory Search (DOS) algorithm with the aim of optimizing chemotherapy. The model is tested with Fixed Interval Variable Dose (FIVD) and Variable Interval Variable Dose (VIVD) interval schemes under a period of 52 weeks. It shows the superiority of FIVD over VIVD [7]. Further, the limitations of the existing theoretical research and several directions are provided to improve research in optimizing chemotherapy treatment. It plans using real protocol treatments defined by the oncologist [8].

Likewise, the variables prioritizing precision dosing are discussed while highlighting key examples of precision dosing that have been successfully used to improve patient care [9]. Besides, a sophisticated automating drug scheduling approach based on evolutionary computation and computer modeling with Adaptive Elitist population-based Genetic Algorithm (AEGA) is presented to solve and discuss the situation of multiple optimal solutions under different parameter settings [10].

3 Rudiments of Genetic Algorithm

Genetic algorithm is a search-based evolutionary algorithm used for near-optimal solution to complex problems in machine learning and AI. It is significant as it solves difficult problems that would acquire a long time to solve. It has various real-life applications such as in data centers, electronic circuit design, code-breaking, image processing, and artificial creativity. GA is a heuristic technique following Darwin's theory of evolution known as survival of the fittest. It comes from the concepts of genes and chromosome where terms like mating, crossover, mutation, parents, offspring, etc. are the keys that take hold of the whole algorithm. It is a probabilistic process. GA biases in the selection of good solutions and passes onto the valuable information achieved so far to the next generation. Mostly, it is applied to a larger scale due to its robustness making it a more versatile method.

Chromosome: Just like how a chromosome is the building blocks of the characteristics of an individual, an encoded chromosome is the building block of GA. A chromosome symbolizes a solution and thus its build up plays the most vital part in the performance of the algorithm. Chromosomes may be designed as binary, real, or integer coded. In binary coding, a chromosome is encoded as a string of binary digits 0 and 1, where each bit in the string will refer to a gene. The chromosome length is dependent on the problem's necessity and the data available to manipulate the string. The genes or the bits in the strings display the characteristic of a chromosome as good or bad on the basis of fitness function evaluation, like DNA brings out the characteristics of a human being. The chromosome having maximum fitness is considered as the better chromosome. A bit string chromosome is presented in Fig. 1.

Fig. 1 A binary coded string of length 8

1	0	1	1	0	1	0	0

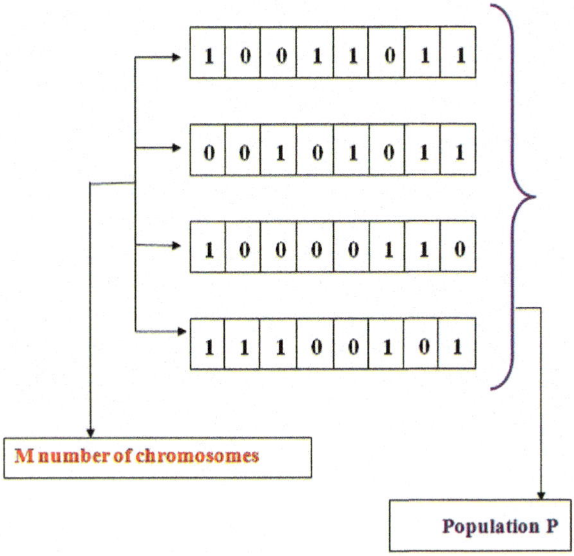

Fig. 2 A population of M chromosomes

Initialization of Population: A set of chromosomes is collectively known as population. The first step in GA is the initialization of population. The initialization of population P is executed by random generation of M number of individuals referred to as chromosomes that serve as a predefined set of solutions. To initialize a population, a set of chromosomes are randomly encoded as binary or real or integer string. After forming the population, the chromosomes are decoded by means of some decoding function and then undergo the step where a fitness function allocates a numerical value to all the M chromosomes of the population P. One such example is depicted in Fig. 2.

3.1 Fitness Function

A fitness function which determines the fitness of the chromosome is the key to the optimal solution. The fitness function will take the chromosome as an input and assign a numerical value to the chromosome as the output. The fitness criteria depends upon the problem undertaken and the researcher. For example, if the cost of a commodity is considered as the fitness function, the chromosome with lesser fitness value will be considered more fit than the one with a greater fitness value. The fittest chromosomes are carried on to the subsequent generation thus following the survival

Table 1 Descriptions of various parameters under study

Chromosomes	Decoded value	Fitness function $f(x) = \log x$
1001	9	0.9542
0010	2	0.3010
1100	12	1.0790
0101	5	0.6989

of the fittest. The fitness function evaluation reflects the quality of a chromosome as a solution. A very common factor in the designing of fitness function is the use of penalties, which are applied upon the violation of constraints. This helps to keep in check that the better the fitness value, the more the number of constraints the candidate solution is satisfying. For example, if the fitness function is $f(x) = \log x$ and it is to be maximized, then maximum fitness value is 1.079, hence 1100 is fitter chromosome than the rest of the chromosomes as it has higher fitness value as shown in Table 1.

3.2 Selection

After the fitness function evaluation, the fitness values are scaled or sorted according to their ranks. On the basis of this scaling or ranking, the fittest chromosomes of the population P undergo the selection process whereby parents are selected from them and introduced in the mating pool where they are allowed to breed. Selection emphasizes the presence of better solutions in the existing population. Based on the fitness both good and bad solutions exist and hence the purpose of selection is for the better solution to hold more possibility of being selected for reproduction. The selection methods vary as tournament selection, roulette wheel selection. In tournament selection k chromosomes are chosen at random from the population and whichever has better fitness value is copied into the mating pool for reproduction. In this case, there is a chance of getting repeated individuals. The tournament selection is depicted in Fig. 3.

3.3 Crossover

The selected chromosomes are introduced in the mating pool for breeding. In this way, the characteristics of the parents to serve as good solutions will be passed on to the next generation, i.e., the off springs. The offsprings may result in better or worse solutions than their predecessors. The basic idea is to offer the offspring a mixture of genetic materials of their parents. Crossover may be varied as uniform crossover,

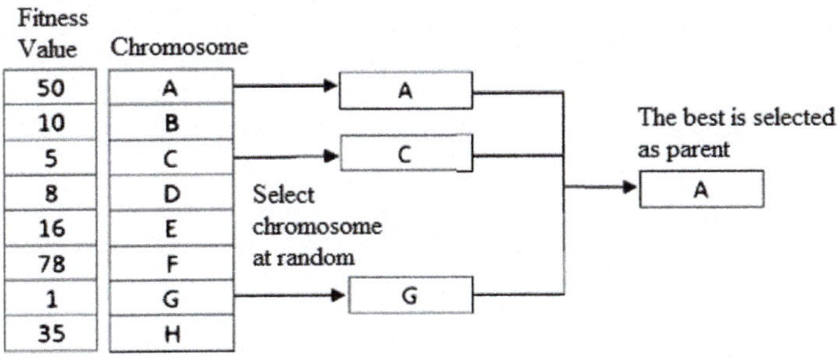

Fig. 3 An illustration of tournament selection

Fig. 4 An illustration of one-point crossover

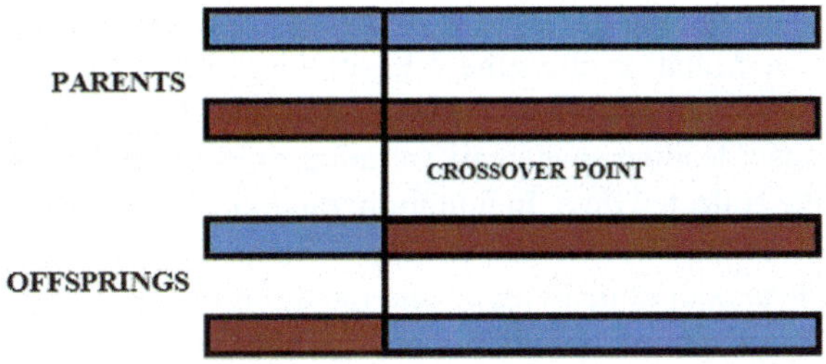

Chromosome 1	1101 01101000101
Chromosome 2	0010 10110001011
Offspring 1	0010 01101000101
Offspring 2	1101 10110001011

one-point crossover, two-point crossover, and so on. In each of these variations, the common parameter used is the crossover probability, P_c. The crossover will take place only when the generated random number between 0 and 1 is less than the crossover probability P_c. Otherwise there will be no crossover between the parents. Depending on this probability the exchange of genes between the parents takes place. Further, the randomly generated crossover points are selected and the genetic materials are exchanged in the mating pool for breeding. In this way, the characteristics of the parents to serve as good solutions will be passed on to the next generation, i.e., the offsprings. An illustration of one-point crossover is depicted in Fig. 4.

Table 2 An example of mutation with $P_m = 0.1$

Random number	Before mutation	After mutation
0.05	1	0
0.20	0	0
0.60	0	0
0.50	1	1

3.4 Mutation

There is a chance that the offspring produced from the crossover will undergo mutation. Mutation alters the value of the genes or the genetic material of a chromosome. The idea is to change a small number of genes in the offsprings keeping the rest intact leading to a small variation on the population with very little disturbance in the collected data. In mutation, random alteration in an allele takes place whereby it is replaced by another. In case of binary string, mutation causes the bits to get flipped at random depending on the mutation operator known as the mutation probability P_m. A random number is generated between 0 and 1, if the number is less than P_m then mutation will take place otherwise not, i.e., 1 will become 0 or 0 will become 1. For better understanding, an example is shown in Table 2.

4 Problem Formulation

The present study aims at designing a robust evolutionary algorithm to reduce the tumor size at the end of one treatment cycle and at the same time keeping the overall tumor burden minimum while minimizing the toxic side effects. The algorithm that is going to be applied is GA with complete elitism.

4.1 Modeling of the Problem

Modeling the problem requires to control the side effects. Toxicity constraints must be taken into account on the concentration of drug fed at any time interval, on the cumulative drug dosage over the treatment period. The damage to different sensitive tissues and organs is also to be taken into consideration. Besides toxicity constraints, the size of the tumor has to be kept lower than the fatal level throughout the treatment phase. Usually delivery of anti-cancer drugs is done by means of a discrete dose strategy wherein number of doses is to be delivered at times $t_1, t_2, \cdots, t_s$. Here, every single dosage is a mixture of d drugs symbolized as the concentration $C_{ij}, i \in (1, s), j \in (1, d)$ in the blood plasma. These variables serve as the decision vectors for the present problem. The search space I is the set of these decision vectors $c = (C_{ij})$ [4]. The general form of constraints considered by [4] is given below:

1. **Constraint on maximum instantaneous dose**: Any drug's concentration C_{ij} must never exceed the maximum dose C_{maxj} that may be tolerated at any point during therapy which may otherwise give rise to temporary side effects as vomiting, pain, etc.

$$g_1(c) = \{C_{maxj} - C_{ij} : \forall i \in \{1, s\}, j \in \{1, d\}\} \tag{1}$$

2. **Constraint on maximum cumulative dose**: To avoid the cells becoming resistant to the medicine and long-term harm, the cumulative dosage concentration of any drug C_{ij} must never exceed the maximum cumulative dose C_{cumj} throughout the treatment period, otherwise the patient might suffer from long-term side effects.

$$g_2(c) = \{C_{cumj} - \sum_{i=1}^{s} C_{ij} : \forall i \in \{1, s\}, j \in \{1, d\}\} \tag{2}$$

3. **Constraint on maximum tumor size**: At any given period t, the number of cancer cells $N(t)$ must never surpass the maximal tumor size N_{max}. The ratio of resistant cells to non-resistant cells increases as the tumor size $N(t)$ grows [1]. As a result, tumor size must be limited during treatment, as assessed at each point in time. Furthermore, a higher tumor burden raises the likelihood of a patient's mortality, as the increased quantity of tumor cells prevents key organs from functioning correctly.

$$g_3(c) = \{N_{max} - N(t_i) : \forall i \in \{1, s\}\} \tag{3}$$

4. **Constraint on the drug toxicity level**: There is a toxic side effect limit that must not be surpassed since doing so would be deadly to the patient's life. The greatest side effects that a patient can tolerate are C_{s-effk}, with η_{kj} being the risk factor that represents the damage caused to the kth organ, such as the heart, liver, or lungs when the drug j is used where ω is the number of organs taken into account [3].

$$g_4(c) = \{C_{s-effk} - \sum_{j=1}^{d} \eta_{kj} C_{ij} \geq 0 : \forall i \in \{1, s\}, j \in \{1, d\}, k \in \{1, \omega\}\} \tag{4}$$

Mathematical modeling can be used to replicate the tumor's response to treatment. To simulate tumor growth under chemotherapy, a variety of models are available, the most frequent of which being the Gompertz growth model, which also contains linear cell-kill factors [11]. The initial specific rate of exponential growth and the rate of exponential decrease in the initial growth rate are the two factors that define Eq. 5 [8].

$$\frac{dN}{dt} = N(t).\left[\lambda \cdot \ln(\frac{\theta}{N(t)}) - \sum_{j=1}^{d} \kappa_j \sum_{i=1}^{s} C_{ij}\{H(t - t_i) - H(t - t_{i+1})\}\right] \quad (5)$$

Here, $N(t)$ denotes the number of cancers cells at any time t; $\frac{dN}{dt}$ is rate of change of number of cancer cells with respect to time t; λ is the doubling time of the tumor, i.e., the time required for the tumor to become the double of its initial size without treatment; θ is the plateau size, i.e., the maximum size of the tumor which is taken to be 10^{12}. This is that point where the tumor cannot be cured and the patient may not survive [12]; $H(t)$ is the heavy side function where the terminologies have been well explained in [1]; and κ_j is the drug efficacy for the jth drug.

The first term in the right-hand side of the equation depicts the number of cancer cells at any time t and the second term depicts the number of cancer cells being killed as a result of the exposure to the chemo drugs. The number of cancer cells being killed is directly proportional to the concentration of the drug j at time i and the efficacy of the drug j. The analytical solution of the differential solution has been obtained by substituting $y(t)$ as

$$y(t) = \ln\left(\frac{\theta}{N(t)}\right) \quad (6)$$

where $y(t)$ is inversely proportional to $N(t)$. The solution is given as

$$y(t_p) = y_0 \cdot e^{-\lambda t_o} + \left(\frac{1}{\lambda}\right) \cdot \left(e^{\lambda \Delta t} - 1\right) \cdot \sum_{j=1}^{d} \kappa_j \cdot \sum_{i=1}^{j} C_{ij} \cdot \left(e^{\lambda(t_{i-1} - t_p)}\right); \; p = 0, 1, \cdots n$$

Here, p denotes the instances throughout the treatment intervals, and Δt is the time difference between two consecutive treatment instances. The objective function is given as

$$Minimize\ F(c) = \sum_{i=1}^{s} N(t_i)$$

Therefore, the optimization problem thus constructed is to minimize the objective function, i.e., the overall tumor burden, upon following up the treatment regime c, subject to all the constraints explained above.

5 Methodology

In the initial stage, representation of chromosome for the problem is carried out. The goal of this issue is to find a solution that is a schedule for 10 drugs and 10 treatment intervals, i.e., the values of $s = 10$ and $d = 10$. The schedule will specify the time

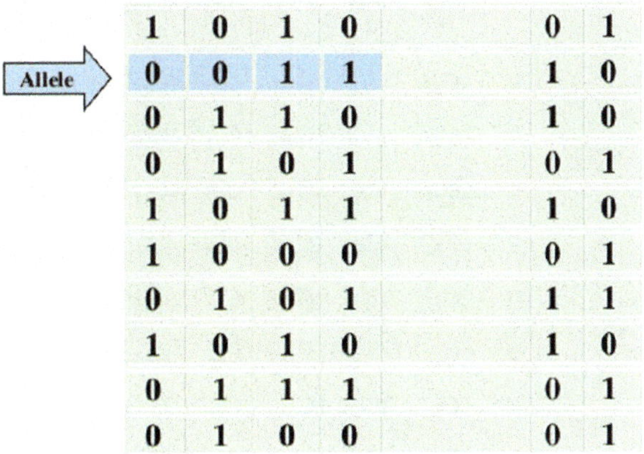

Fig. 5 A binary chromosome which is represented as 10×40 matrix

and drugs or pharmacological combinations that will be given. The decision vectors $c = (C_{ij})$, $i \in \{1, 10\}$, $j \in \{1, 10\}$, will refer to the chemotherapy drug schedules. The encoding of these decision vectors for the present work is done in the form of binary strings. Chromosomes are represented by these binary strings. From this point onward, the term binary string will eventually be used to mean a chromosome. The solution space I is thus represented as the Cartesian product of allele sets A_i^j, where each A_i^j is a set of 4 consecutive bits which when decoded under a decoding scheme gives a value between 0 and 15. The representation of I is given as

$$I = A_1^1 \times A_1^2 \times \cdots \times A_1^{10} \times A_2^1 \times A_2^2 \times \cdots \times A_2^{10} \times \cdots \times A_{10}^1 \times A_{10}^2 \times \cdots \times A_{10}^{10}$$

where $A_i^j = \{x_1x_2x_3x_4 : x_k \in \{0, 1\} \forall k \in \{1, 4\}\}$. Thus, a chromosome $x \in I$ as per [12] can be expressed as a binary string of 400 bits defined as $x = \{x_1x_2x_3 \cdots x_{400} : x_k \in \{0, 1\}, \forall k \in \{1, 400\}\}$ and it will look like as shown in Fig. 5. The chromosome string is a string of length 400 which may be taken as a 40×10 matrix where each set of four alleles will give out one integer this converting to a 10×10 matrix after decoding.

The decoding function from the set of 4 bit string to the set of integer from 0 to 15 is represented as given below [12]:

$$C^*(i, j) = \sum_{k=1}^{4} 2^{4-k} x_{40(i-1)+4(j-1)+k}; \forall i \in \{1, 10\}, j \in \{1, 10\} \quad (7)$$

The transformation maps of the solution space I to the set of decision vectors are defined as $C_{ij} = \Delta C_j \times C^*(i, j)$, where ΔC_j means unit concentration of drug j. The standard concentration unit ΔC^* can be used to compute this number. Adriamycin is taken as the standard for calculating this value as defined in Eq. 8.

	drug 1	drug 2	drug 3	drug 4	drug 5	drug 6	drug 7	drug 8	drug 9	drug 10
t_1	C_{11}	C_{12}	C_{13}	C_{14}	C_{15}	C_{16}	C_{17}	C_{18}	C_{19}	C_{110}
t_2	C_{21}	C_{22}	C_{23}	C_{24}	C_{25}	C_{26}	C_{27}	C_{28}	C_{29}	C_{110}
t_3	C_{31}	C_{32}	C_{33}	C_{34}	C_{35}	C_{36}	C_{37}	C_{38}	C_{39}	C_{110}
t_4	C_{41}	C_{42}	C_{43}	C_{44}	C_{45}	C_{46}	C_{47}	C_{48}	C_{49}	C_{110}
t_5	C_{51}	C_{52}	C_{53}	C_{54}	C_{55}	C_{56}	C_{57}	C_{58}	C_{59}	C_{110}
t_6	C_{61}	C_{62}	C_{63}	C_{64}	C_{65}	C_{66}	C_{67}	C_{68}	C_{69}	C_{110}
t_7	C_{71}	C_{72}	C_{73}	C_{74}	C_{75}	C_{76}	C_{77}	C_{78}	C_{79}	C_{110}
t_8	C_{81}	C_{82}	C_{83}	C_{84}	C_{85}	C_{86}	C_{87}	C_{88}	C_{89}	C_{110}
t_9	C_{91}	C_{92}	C_{93}	C_{94}	C_{95}	C_{96}	C_{97}	C_{98}	C_{99}	C_{110}
t_{10}	C_{101}	C_{102}	C_{103}	C_{104}	C_{105}	C_{106}	C_{107}	C_{108}	C_{109}	C_{1010}

Fig. 6 Chemotherapy drug regimen with 10 time intervals and 10 drugs

$$\Delta C^* = \frac{\text{Maximum tolerable dose for Adriamycin}}{2^4 - 1} \tag{8}$$

Each drug is assigned a potency factor ρ_j, which is derived using the highest acceptable dose for the jth drug. A list of these factors is defined in [1]. Finally, a population's chromosomes will resemble Fig. 6.

It is noted that $\Delta C_j = \frac{\Delta C^*}{\rho_j}$. Figure 6 represents a chromosome where each C_{ij} is a drug concentration at specified time. For instance, the concentration C_{32} will refer to the concentration of drug 2 given in time t_3.

Fitness Function: The fitness function designing is one of the most important jobs in GA. The fitness function will determine how good or bad a chromosome is. The value of the fitness function is given by Eq. 9.

$$F(c) = \sum_{p=1}^{10} \sum_{j=1}^{10} \kappa_j \sum_{i=1}^{p} C_{ij} e^{\lambda(t_{i-1} - t_p)} - \sum_{s=1}^{4} P_s d_s; \quad p = 0, 1, \cdots 10 \tag{9}$$

where $F(c)$ is a maximizing function, d_s are values of constraint violation. That means if a constraint is violated the corresponding distance returns a non-zero number upon which penalties P_s are applied. The idea is that more the violation of constraints, lesser is the fitness value. The first term signifies the efficacy of the schedule, i.e., more the value of the first term, more the number of cancer cells is killed. It is referred to as the cell-kill term.

Selection: For the present work, tournaments selection has been applied. Here, a population size of 30 is taken and initialization of population takes place by random

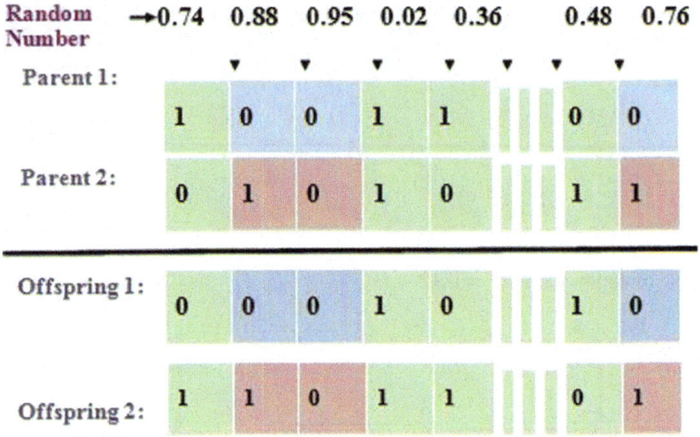

Fig. 7 Uniform crossover with $P_c = 0.75$

generation of chromosomes. Following that, the fitness function evaluation is done for each of the individuals. The winner of each match moves to the next round. In this way, a population of 30 individuals is obtained for the next step which will be considered for manipulation. A point to remember is one or more chromosomes may be repeatedly present in the population after selection.

Crossover: In the present work, after selection, uniform crossover is applied. The crossover probability taken is $P_c = 0.75$. The process of crossover is defined earlier. Take two parent chromosomes and introduce them into the mating pool. For every bit of the chromosomes, a random number is generated. If this random number is less than or equal to P_c, then crossover will take place and genes between the parents will get swapped, otherwise not. For each bit, the probability of the genes getting swapped is 0.75. After the swapping of the genes, the newly formed chromosomes are known as offsprings. Figure 7 depicts the uniform crossover.

Mutation: The mutation probability, $P_m = 0.12$. The mutation process is defined earlier. Take each of the individuals from the population of offspring and apply mutation separately. Just like in crossover, for each single bit, a random number is generated. If this random number is less than or equal to P_m then mutation will take place and the genes will get altered, i.e., 0 will change to 1 and 1 will change to 0. For each bit, the probability of the genes getting altered is 0.12. Figure 8 depicts the mutation.

Complete Elitism: After mutation, the next step applied in the present work is complete elitism. Here, the new population P' is combined with the population P from the previous generation. The main advantage of complete elitism is that the good solutions of the previous generation are retained for the next iteration. Sometimes worse chromosomes may be present in the present generation than the previous one. Hence, the need of including all the good chromosomes urges the need for use of complete elitism. Now the working principle of the complete elitism is as follows:

Random Number
0.15 0.08 0.2 0.45 0.1 0.33 0.65
Before mutation
1 1 0 1 0 0 1
After mutation
1 0 0 1 1 0 1

Fig. 8 Bitwise mutation with $P_m = 0.12$

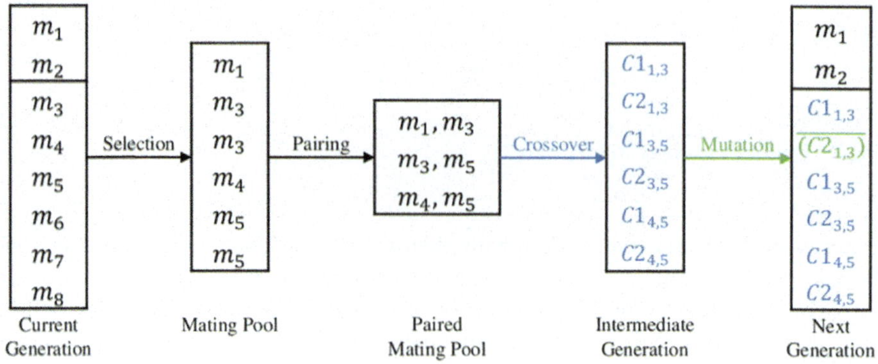

Fig. 9 Two-chromosome elitism

1. Combine both populations ($P \cup P'$) to get a double size population.
2. Sort the individuals according to the descending order of their fitness.
3. Take the best (upper) half and reject the rest.

Figure 9 depicts two-chromosome elitism process where only two best chromosomes from previous generations are included in the subsequent generations.

5.1 Complete Elitist Genetic Algorithm

As a result of the facts presented in Sect. 3, an attempt is made to propose a new GA that utilizes a specific set of GA operators, as explained above. The robustness of the proposed GA is significantly predicated on absolute elitism. The proposed GA is referred to as the Complete Elitist Genetic Algorithm (CEGA) since in each iteration complete elitism is conducted at the end of the GA cycle. The pseudocode describes the CGEA functioning mechanism and the flowchart is depicted in Fig. 10.

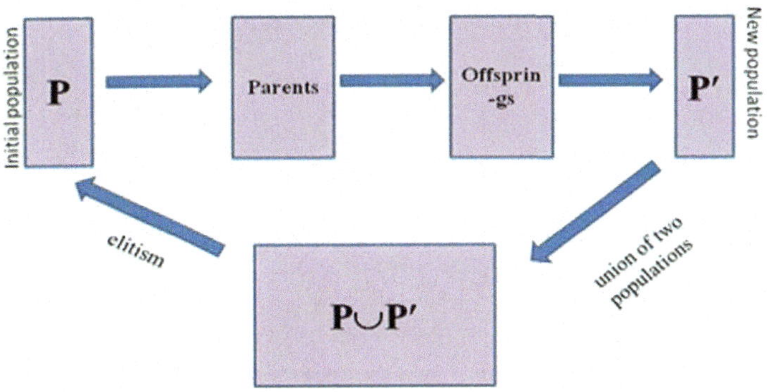

Fig. 10 Methodology flowchart of CEGA

Algorithm 8 (Complete Elitist Genetic Algorithm)

1. *Initialize population P.*
2. *Evaluate fitness value.*
3. *While (number of generation not equal to Gmax) do*
4. *Apply tournament selection.*
5. *Apply probability-based crossover.*
6. *Apply mutation.*
7. *New population P'.*
8. *Merge P and P'.*
9. *Sort in decreasing order of fitness value.*
10. *Keep the best half and input this as P.*
11. *End while.*

6 Results and Discussions

In order to design a chemotherapy drug schedule, CEGA has been applied. The computational task that takes the time is the evaluation of fitness functions. More the number of fitness function evaluations, more is the time taken to find the solution. The quality of solution in this study is the final tumor size and the overall tumor burden which indicates how successful the schedule is reducing the tumor size on the whole. The final tumor size received is approximately of the order of less than or equal to 10^6. The population size has been taken to be 30 and the number of generation is taken to be 300.

CEGA has been encoded as a MATLAB program that is used to create a chemotherapy treatment schedule. After a thorough examination of the parameters, the following experimental setup was chosen. CEGA uses a random population

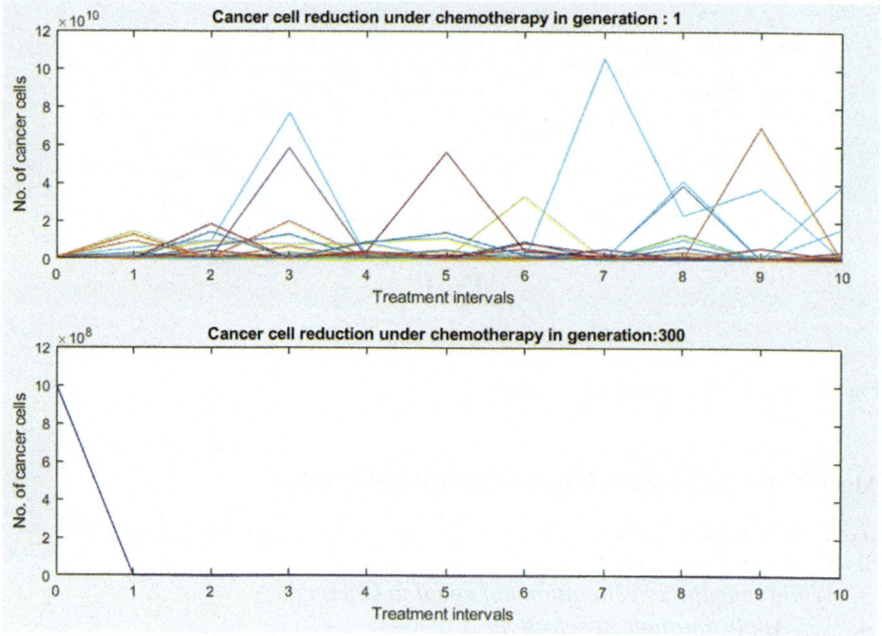

Fig. 11 Number of cancer cells throughout the treatment interval

initialization method with a population size of 30 people. In the CEGA cycle, tournament selection, uniform crossover, bit-wise mutation, and complete elitism are applied in that order. The likelihood of crossover is set at 0.75, while the probability of mutation is set to 0.12. The suggested CEGA has been compared to the results of [4] to assess its efficacy. Furthermore, the same set of factors as in [4] are reviewed in this study to guarantee a fair comparison. The efficiency in terms of computing time and the quality solution in terms of the consequent optimal objective function value are the two most important aspects. The ultimate tumor size and total tumor burden arriving at the end of a run determine the quality of solutions among comparable algorithms. The more fitness function evaluations there are, the longer it takes to discover a solution. Figure 11 shows the number of cancer cell during the treatment period and how it reduces to a considerable amount toward the end of the treatment cycle. From Fig. 11, it is clearly visible how the number of cancer cells is decreasing with the schedule designed by the CEGA.

Figure 12 shows the variation of fitness function value with the number of generations. The algorithm was run for 500 generations and it is seen that the efficacy of the algorithm is reflected in Fig. 12 and the fitness function value keeps on increasing sharply and gains stability within a range of 300 generations. For the present work, the initial number of cancer cells, N_o, has been taken as the maximum number of cancer N_{max} which is 10^9.

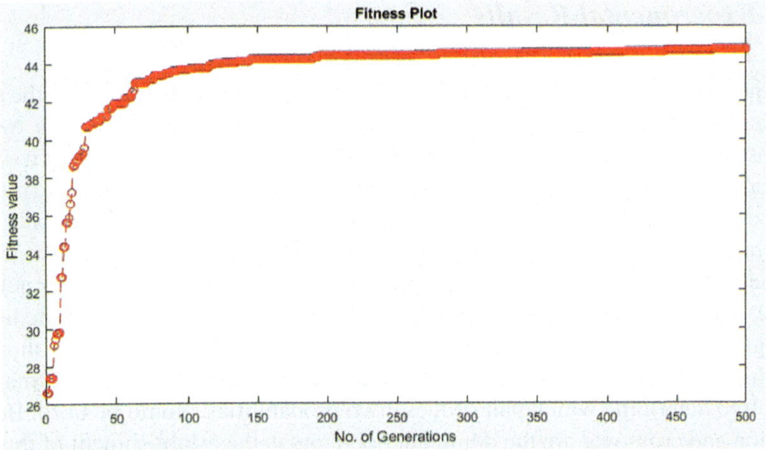

Fig. 12 Fitness function evaluation with each generation

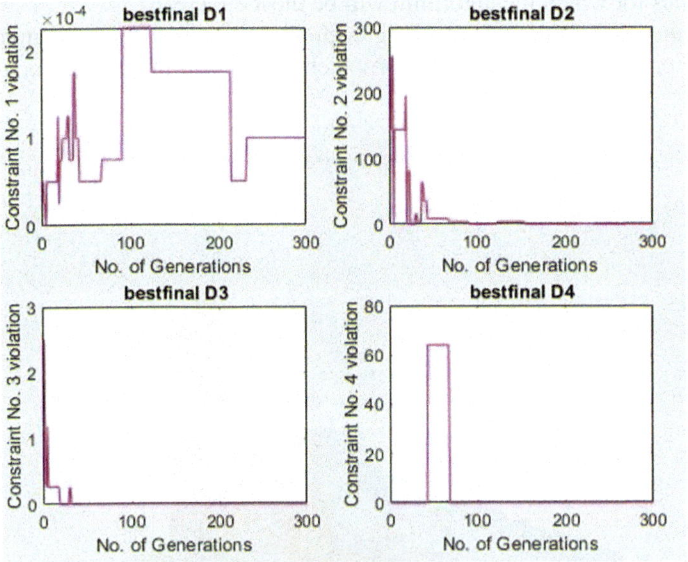

Fig. 13 The constraint violations with each generation

In Fig. 13, it is seen how the final constraint violation by the 300th generation is zero thus giving a feasible solution and moving toward an optimal solution. Constraint violation D_1 is considered negligible as satisfying D_2 makes the first condition to be satisfied as well.

6.1 Experimental Results

An experiment was carried out for the current paper to demonstrate how the results vary based on the mutation probability and crossover probability. The eventual decrease of cancer cells differs for each distinct pair of P_m and P_c, making it difficult to determine which pair of these two probabilities should be used. Because mutation and crossover are the determining elements in the establishment of the genes of the offspring, it is necessary to identify the ideal couple for the greatest outcomes. The parameter was tuned by utilizing a population size of 50 and a number of generations of 500. An experiment was carried out for the current paper to demonstrate how the findings vary based on the mutation probability and probable crossover threshold. The ultimate number of cancer cells changes for each pair of P_c and P_m, making it difficult to determine which pair of these two probabilities should be used. Because mutation and crossover are the dominant operators in the establishment of the genes of the offspring, it is necessary to identify the ideal match for the greatest outcomes. The experiment, together with Fig. 14, yields the ideal pair of crossover and mutation probabilities for which the algorithm will be most effective.

From the above experiment along with Fig. 14, a few pairs of mutation and crossover probability are obtained for which the algorithm has the maximum effi-

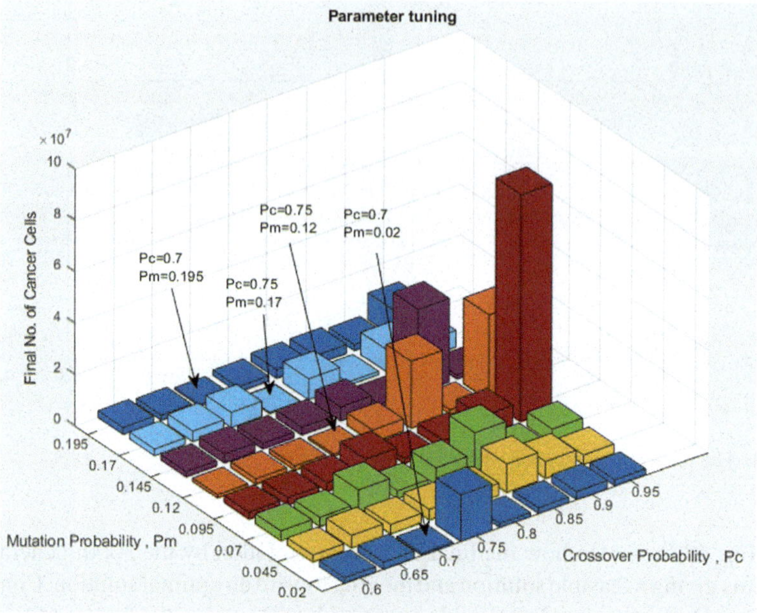

Fig. 14 Tuning of parameter for crossover and mutation probability

cacy. The algorithm is run for each pair 10 times and finally it has been seen that the pair of P_c and P_m as 0.75 and 0.12 has been giving better results consistently and thus has been selected for the algorithm.

6.2 Analysis of the Findings

CEGA is permitted 50 independent runs to demonstrate its higher order efficiency. The fitness function value for each iteration of a certain run is shown in Table 3. According to Table 3, CEGA is capable of capturing the greatest fitness value with an exponentially enhanced fitness graph and a greater rate of stability. It achieves a bigger height after only a few iterations. In contrast, it depicts the number of cancer cells over time. The graph clearly shows that as the iterations keep on going, there is a dramatic decline in the number of cancer cells. The CEGA regimen received is depicted in Table 3 and the number of cancer cells for each time interval has been received as depicted in Table 4.

6.3 Comparative Study

The CEGA regimen has been compared with the two-chromosome elitist GA regimen and the clinical application of CAF (Cyclophosphamide, Adriamycin, and Fluorouracil) and CMF (Cyclophosphamide, Methotrexate, and Fluorouracil) and has been found to be superior to the later three regimens. The comparative study is depicted in Fig. 15.

Table 3 CEGA treatment regimen

Time interval	Drug 1	Drug 2	Drug 3	Drug 4	Drug 5	Drug 6	Drug 7	Drug 8	Drug 9	Drug 10
t_1	50	40	100.005	1.46×10^3	1800	80	9.33×10^3	4	80.0040	0.0750
t_2	60	30	33.3350	399.9990	2600	8	9.33×10^5	3	53.3360	0.2750
t_3	75	25	53.3360	1.46×10^3	800	88	8.67×10^3	9	73.3370	0.2250
t_4	75	5	80.0040	133.3330	3000	112	1.00×10^4	0	93.3380	0.3250
t_5	5	0	33.3350	1.07×10^3	3000	48	7.33×10^3	11	93.3380	0.2250
t_6	65	70	73.3370	1.47×10^3	2400	104	9.33×10^3	3	0	0.3750
t_7	0	75	20.0010	533.3370	3000	32	8.67×10^3	4	40.0020	0.2750
t_8	70	70	6.6670	933.3310	2800	96	1.00×10^4	3	13.3340	0.0750
t_9	75	25	20.0010	1.86×10^3	3000	32	9.33×10^3	1	0	0.2750
t_{10}	60	65	93.3380	533.3320	3000	0	1.00×10^4	1	46.6500	0.0250

Table 4 Number of cancer cells for 10 time intervals

Time interval	Number of cancer cells
t_1	3.3390×10^6
t_2	6.2016×10^6
t_3	7.9381×10^6
t_4	3.2455×10^6
t_5	6.3285×10^6
t_6	9.3416×10^5
t_7	4.8829×10^6
t_8	1.1451×10^6
t_9	3.0545×10^5
t_{10}	1.1900×10^6

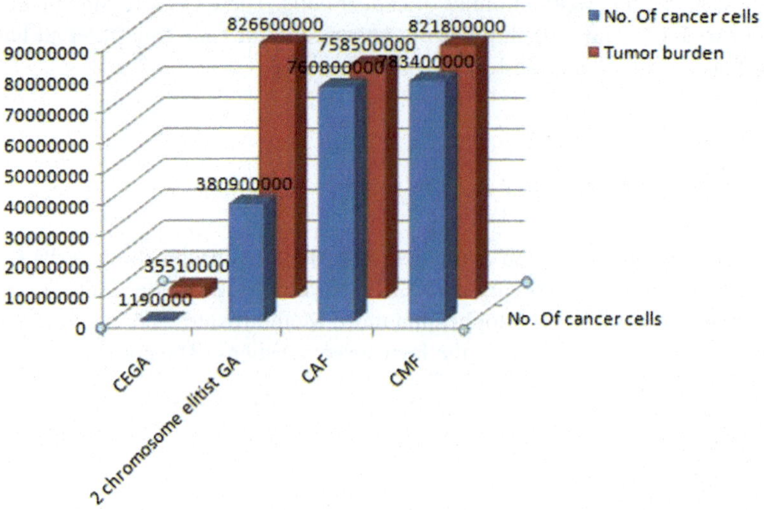

Fig. 15 Comparative plot between CEGA, two-chromosome elitist GA, CAF, and CMF

7 Conclusion and Future Extensions

Cancer chemotherapy is used when metastasis has occurred. Chemotherapy is implemented with the help of chemo drugs which are highly toxic in nature. Therefore, there is a need to feed the drugs in such a way that the patient suffers least toxicity while the cancel cell count is being reduced too. In the present study to avoid the traditional methods of finding a chemotherapy drug schedule, an approach has been made using GA with complete elitism. This attempt has been made in order to explore the vast solution space of this chemotherapeutic drug scheduling problem with the help of CEGA. In the experiment, it has been shown how a perfect pair of mutation and crossover probability can aid to find best results and thus emphasizing its need to make the GA more robust. Complete elitism has been approached with the idea so as not exclude any good chromosome from the previous generation. From experimental results, it is seen that the number of cancer cells after using a regime founded by CEGA is of the order of 10^6, where the original number of cancer cells was 10^9. It has also been demonstrated how CEGA discovers a nearly optimal situation in a very few number of generations, beyond which the improvement in fitness score is essentially inconsequential.

Figure 13 also demonstrates how the constraints were violated for the treatment regimes in the initial generations. By the 300th generation, however, all constraint violations became zero, with the exception of the first constraint, where the violation was minor and ignored because it was indirectly included in the second constraint. The results showed that CEGA outperformed all of them when compared to the conventional clinical applications of CAF and CMF regimens and two-chromosome elitist GA.

There are scopes of working on the encoding of the chromosome. Since binary encoding has been used, there is a scope to use floating point numbers for encoding giving a larger area of solution space for this multi-drug chemotherapy problem. Also, discrete time intervals have been taken, but for future work it can use continuous time interval for the varying effects of the drugs on the body alongside the natural processes taking place inside the body. Further, a hybridized GA can be modeled which may improve the accuracy of the machine learning algorithms.

References

1. Petrovski, A.: An application of genetic algorithms to chemotherapy treatment, Robert Gordon University, Ph.D. Thesis (1998)
2. Martin, R., Teo, K.L.: Optimal Control of Drug Administration in Cancer Chemotherapy. World Scientific, New Jersey, USA (1994)
3. Petrovski, A., Sudha, B., McCall, J.: Optimising cancer chemotherapy using particle swarm optimisation and genetic algorithms. In: Proceedings of the International Conference on Parallel Problem Solving from Nature, pp. 633–641 (2004)
4. Petrovski, A., Shakya, S., McCall, J.: Optimising cancer chemotherapy using an estimation of distribution algorithm and genetic algorithms. In: Proceedings of the 8th Annual Conference on Genetic and Evolutionary Computation, pp. 413–418 (2006)
5. Godley, P., Cowie, J., Cairns, D., McCall, J., Howie, C.: Optimisation of cancer chemotherapy schedules using directed intervention crossover approaches. In: Proceedings of the IEEE Congress on Evolutionary Computation, pp. 2532–2537 (2008)
6. Al Moubayed, N., Petrovski, A., McCall, J.: Multi-objective optimisation of cancer chemotherapy using smart pso with decomposition. In: Proceedings of the IEEE Symposium on Computational Intelligence in Multicriteria Decision-Making, pp. 81–88 (2011)
7. Archana, N., Benedict, A.M.F., Niresh, J.: Chemotherapy drug regimen optimization using deterministic oscillatory search algorithm. Trop. J. Pharmaceut. Res. **17**(6), 1135–1143 (2018)
8. Sbeity, H., Younes, R.: Review of optimization methods for cancer chemotherapy treatment planning. J. Comput. Sci. Syst. Biol. **8**(2), 74–95 (2015)

9. Tyson, R.J., Park, C.C., Powell, J.R., Patterson, J.H., Weiner, D., Watkins, P.B., Gonzalez, D.: Precision dosing priority criteria: drug, disease, and patient population variables. Front. Pharmacol. **11**, 420 (2020)
10. Liang, G., Fan, W., Luo, H., Zhu, X.: The emerging roles of artificial intelligence in cancer drug development and precision therapy. Biomed. Pharmacother. **128**(8), 110255 (2020)
11. McCall, J., Petrovski, A.: A decision support system for cancer chemotherapy using genetic algorithms. In: Proceedings of the International Conference on Computational Intelligence for Modeling, Control and Automation, pp. 65–70 (1999)
12. McCall, J., Petrovski, A., Shakya, S.: Evolutionary algorithms for cancer chemotherapy optimization. Comput. Intell. Bioinform. **7**, 263–296 (2007)

Privacy-Preserving Deep Learning Models for Analysis of Patient Data in Cloud Environment

Sandhya Avasthi and Ritu Chauhan

Abstract A substantial amount of patient data is being generated every second by the healthcare sector. The medical data especially patient data could be used for analysis through advanced deep learning models, but the private nature of patient data limits the use. Massive volumes of diverse data must be collected, which is often only possible through multi-institutional collaborations. One way to create large central repositories is through multi-institutional studies. This method is limited to privacy issues, intellectual property, data identification, standards, and data storage when data sharing is done. As a result of these challenges, cloud data storage has become increasingly viable. The various models for exchanging medical records on the cloud while protecting privacy are discussed in this chapter. Furthermore, vertical partitioning of medical datasets that exploits attribute categories in health records is explained and analyzed in order to examine distinct areas of medical data with varying privacy issues. These methods can ease the strain on communication costs while minimizing the need to communicate sensitive patient information.

1 Introduction

Medical institutions and research agencies have been collecting medical records in the forms of Electronic Health Records (EHR) and Personal Health Records (PHR) of patients due to the use and availability of all kinds of health applications by patients and healthcare providers. Such medical data in the forms of patient records, clinical trials, and other medical reports are massive that is stored on third-party storage solution providers because such data and its maintenance are very expensive and time-consuming. In addition to the primary use of medical data like diagnosis,

S. Avasthi
Department of Computer Science and Engineering, ABES Engineering College, Ghaziabad, India
e-mail: sandhya_avasthi@yahoo.com

R. Chauhan (✉)
Center for Computational Biology and Bioinformatics, Amity University, Noida, India
e-mail: rituchauha@gmail.com

D. P. Acharjya and K. Ma (eds.), *Computational Intelligence in Healthcare Informatics*, Studies in Computational Intelligence 1132,
https://doi.org/10.1007/978-981-99-8853-2_20

treatment, and prediction of patient disease, secondary use of such data has become quite common due to data analytics by medical institutions. Medical institutions have collected massive genetics data, medical, clinical, and biological. Smartphones and many other wearable devices have enabled third-party corporations to mHealth services through these services they can collect health data of patients. Secondary use of health data is defined by the American Medical Informatics Association as "any use of Personal Health Information (PHI) for purposes other than direct care, including but not limited to analysis, research, quality measurement, public health, payment, provider certification or accreditation, marketing, and other business activities, including strictly commercial activities [1, 2]". Secondary use of data analytics in conjunction with advanced deep learning models aids in clinical decision-making, the extraction of useful patterns and information about medicine and diseases, and the improvement of patient care while lowering healthcare costs and benefiting public health policy [3–5].

Health Insurance Portability and Accountability Act (HIPAA) defines 18 categories of protected health information to preserve the privacy of patient information [6], introduced in the United States. Privacy issues constrain the use of health big data for secondary purposes. To make a balance between privacy protection and the secondary use of all kinds of patient information, organizations are incorporating technical and legal solutions [7]. COVID-19, which was just announced, may highlight the challenge of preserving health information while ensuring its accessibility to address the challenges posed by a significant worldwide epidemic. China, South Korea, and other countries are following the guidelines of the suit [8] to have a mandate for the usage of data from contact surveillance devices. To be able to get useful insights and results, deep learning technologies and cloud infrastructure as well as collaborative model training can be suitable. Because user devices have limited resources, transferring resource-intensive operations to external infrastructure, such as the cloud, which has high-power computing and huge storage, is the solution. Collaborative learning, on the other hand, improves learning accuracy by large-scale diverse datasets originating from disparate sources like patient devices, hospitals, and medical institutions. However, there have been privacy concerns expressed for private medical data usage for model training and inference in deep learning.

Since the advent of the latest information technology tools, healthcare applications have become very adaptable and scalable, to achieve universal scalability and accessibility cloud-based healthcare solutions are in demand. Private cloud infrastructures enhance information flow between healthcare centers and end-users. Heterogeneous communication standards that support various applications and services promote accessibility and information transmission. The cloud platform service provides centralized and decentralized storage as well as ubiquitous access to information across several terminals. Access and retrieval are key storage design capabilities that enable the flexible distribution of information via networked systems [9, 10]. The end-users and intelligent processing systems have access to the same information store but with distinct control and constraint mechanisms. Many intelligent systems or applications process raw patient data to identify patterns or even provide diagnostic information. Further, many end-user applications provide statistical reports and analysis for which they depend on health data [11]. Effective indexing and storage retrieval qualities are required for information storage and retrieval, notably in healthcare applications. Among many other requirements, flexibility, concurrency, and efficient data retrieval are required from a cloud-based healthcare system [12].

This chapter reviews various methods and strategies to store massive health data and the secondary use of such data in deep learning to gain insights. Since medical data is huge, the cloud platform has become important to store medical data. Privacy is safeguarded purely by technical measures in the cloud where medical data is stored, and users have to trust the cloud platform service providers. Authentication, anonymous communication, anonymous yet approved transactions, data deidentification, and database pseudonymization are some technical measures for privacy-preserving systems. The rest of the chapter is organized as follows. Section 2 discusses foundations of medical data, deep learning, and cloud computing. Electronic health records and its categories are discussed in Sect. 3 followed by protected health information and regulations in Sect. 4. Section 5 discusses deep learning approaches for privacy-preservation. Further, in Sect. 6, cloud environment and privacy-preserving framework are discussed. The vertical partitioning approach is discussed in Sect. 7 followed by conclusion in Sect. 8.

2 Medical Data, Deep Learning, and Cloud Computing

Consistency of clinical data from different time frames, organizations, and research sites is crucial for secondary data analytics utilization. To use current state-of-the-art technologies, deep learning models, and other implementations, standard terminology and ontologies need to be applied to biomedical data.

2.1 Medical Data and Secondary Usage

Any use of patient data beyond its regular intended purposes such as diagnosis, identifying symptoms, and generating reports is called secondary use. Analysis, research, measuring safety and quality, payments, and marketing purpose are some activities that organizations do which are categorized as secondary use. This entails a taxonomy to identify to clarify all kinds of technical and legal issues that might arise due to secondary usage. These medical data mainly contain a database of administrative information, claims, and patient details. The purpose of secondary use is research and applications to improve the quality of treatment and personal care with the help of the latest technologies [13, 14]. The difficulties associated with secondary usage is refactoring, management, aggregation of variables, and maintaining the quality of data (missing data). The other very critical concern associated with secondary use is the security and privacy of patient information. The typical flow of medical data and secondary usage to improve clinical care are shown in Fig. 1.

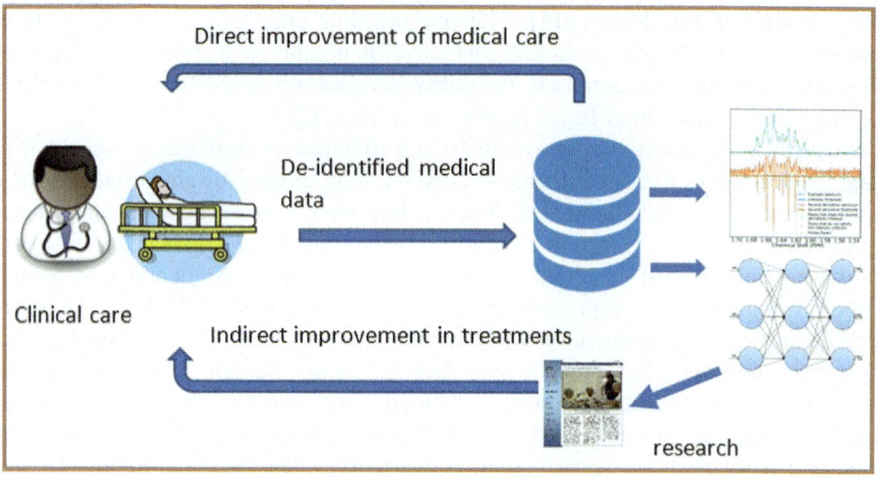

Fig. 1 Clinical and medical data flow in research and analysis

2.2 *Deep Learning*

A multi-layer computational approach for learning data representations at various levels of abstraction is called deep learning. The model begins with raw data, and each level can employ non-linear transformations to translate the previous level's representation into a higher level representation. Compiling a significant number of these changes enables the learning of complex functions. Deep learning frameworks are applied in a varied range of applications such as AI, image identification [15], object identification, speech recognition, biometric systems, face detection [16], and expert systems [17]. Deep learning is typically divided into two phases: a training phase used to enhance the model's accuracy and an inference phase used to classify the system.

Deep learning is used in medicine to seek patterns in patient histories and spot anomalies in medical imaging that assist in illness diagnosis and prognosis. The application of machine learning in health care can threaten patient privacy by identifying genetic markers [18]. Deep learning is also commonly utilized in finance for a variety of purposes, including price prediction and portfolio creation. In these instances, an entity normally trains its model, and the model parameters are kept private. It is deemed an invasion of privacy to be able to locate or infer them [19]. The aforementioned advancements have been made possible by the ease of access to big datasets and high processing capacity (GPUs and TPUs). These datasets are frequently gathered from the public and may contain sensitive information. Because neural networks are employed in several facets of our life [20–22], they raise severe privacy problems.

2.3 *Cloud as a Platform to Store Health Data*

Cloud computing platform combines distributed, grid, and utility computing to provide infrastructure, platforms, and applications over the Internet. Cloud platforms and services intend to provide the hosting of patient data, installation, or software services and also provide the necessary infrastructure to manage healthcare operations. It can also help with health record transfer, availability, and retrieval, as well as huge data volume sharing and Electronic Medical Record (EMR) exchange between hospitals and healthcare institutions. Furthermore, it allows medical providers to monitor patients remotely. On the other hand, there are limitations to cloud-based health data exchange. Because it includes personal information, medical history, treatment, linked diseases, symptoms, and even family health history, patient health information is extremely sensitive. Moving sensitive health data or health records to a cloud controlled by third parties exposes them to illegal access and, as a result, poses serious privacy concerns. For example, a trustworthy and approved doctor in a hospital has complete access to all health information, including personal information. Patients, on the other hand, do not want their personal and sensitive information to be shared with someone who is not trustworthy or approved.

3 Electronic Health Records and Categories

Individuals can access and manage their health information via the Internet, which is particularly relevant for our continued consideration of PHRs. Individuals can also share this information with others and have access to the information of others to whom they have permission. Apart from the main use of medical data, Health Records (HRs) also have a feature known as secondary data use. It is concerned with the use of PHI for uses other than direct treatment, such as analysis, research, quality and safety measurement, public health, payment, provider certification or accreditation, marketing, and other commercial operations. In contrast to the benefits listed above, HRs are not without risk. This is simply because the central availability of health-related data increases the risk of these data being misused. Data protection officials and representatives from medical institutions frequently address this issue, with the so-called transparent patient and physician being a prominent concern. The former implies that a person's health status be completely transparent to anybody who has access to their HRs.

Any information on a user's health could be deemed health information in general. Clinical data, particularly Electronic Medical Records (EMR), is the most important sort of health data. It is created by various levels of hospitals. Many other types of health data are being recorded like diet or exercise or heart rate data, and IoT devices generate huge amounts of such data. Technological advancement has made people dependent on wearable devices and other mobile applications making this situation even more data-centric. In general, health-related data can be categorized into four categories [23]. This research focuses on the first two types of data, which are directly related to users' health and privacy.

Health data are generated by the healthcare system referred to as Category 1. When a patient receives healthcare services in a hospital or clinic, clinical personnel collect clinical data. Clinical data sources include the EMR, prescriptions, test results, pathology images, radiography, and payor claim data. Patients' prior and current conditions are documented to determine therapy needs. It is vital to collect and exchange clinical data with several healthcare professionals over time to improve patient care. It was proposed that patients' clinical data from several institutions and across their lifetime be combined into a Personal Health Record (PHR). This sort of health data is created and gathered regularly as part of the healthcare process to analyze and enhance therapy. Clinical data has a high level of health-related privacy due to the nature of clinical treatment and the high amount of trust consumers have in healthcare experts and organizations. As a result, clinical data privacy is the focus of the health privacy legislation. Thus, a significant proportion of clinical data has been designated for internal use solely by medical institutions. Meanwhile, clinical data is especially valuable for secondary use because it is provided by professionals and provides an up-to-date picture of customers' health status. The delicate balance between utility and privacy associated with this type of health data has been one of the most perplexing issues in the age of big data.

The consumer health and wellness industry provides health data that comes under Category 2. This type of health information complements clinical data well. Consumer attitudes around health have evolved substantially away from passive treatment and toward active health as a result of the widespread adoption of next generation information technologies such as the Internet of Things, mobile health, smartphones, and wearable gadgets. Consumer health data can be generated through wearable fitness trackers, medical wearables such as insulin pumps and pacemakers, health monitoring apps, and online health services. Examples of health data include breathing, heart rate, blood pressure, blood glucose, walking, weight, diet preference, location, and online health consultation. These goods and services, as well as health data, are key components of consumers' daily health management, especially for those with chronic diseases. Industry and academia are increasingly concentrating their efforts on this subject. Consumer health informatics is a representative field [23, 24]. Although this type of unusual health-related data is frequently as revealing of health status as conventional data, it is typically less accessible to physicians, patients, and public health officials to improve individual and community health. To protect the privacy of patients, these massive amounts of health data are divided among organizations. Apart from the utility-privacy trade-off, integration and connectivity of this type of health data at the individual level present other challenges. Table 1 summarizes the two aspects of health data and their distinctions.

Table 1 Clinical data and health data of patient summary

Categories	Category 1	Category 2
Generated/recorded by	Medical equipment, Clinical professional, Healthcare system	Wearable device in treatments, IoT devices
Data detail	Name, age, id, phone, medical history, family history, conditions, medicine use, therapy, narratives, prescriptions, test results	Name, id, type, phone, address, position, age, weight, heart rate, breathing pattern, test, blood pressure, blood glucose, exercise data, diet preference, online health consultation
Data features	Discrete but more professional, more clinical information and more privacy, stored in the healthcare system, passive	Less standardized, more health information, privacy tends to be ignored, stored by different providers, active, vast amounts

4 Protected Health Information and Regulations

Due to the sensitive nature of patient personal and health data, several privacy protection regulations have been established to govern the secondary use of clinical and personal health data. The Fair Information Practices Principles System (FIPPS) serves as the bedrock of contemporary data protection regulation [25]. One of these rules is the Health Insurance Portability and Accountability Act (HIPAA), which was enacted to oversee and regulate the use of medical information [26]. The act protects healthcare systems and insurance firms from all sorts of fraud, theft, and misuse. The HIPAA Safe Harbor (SH) rule requires the destruction of 18 types of expressly or potentially identifying characteristics, collectively referred to as protected health information, before the distribution of health data to a third party. HIPAA also applies to electronically protected health information. This includes, but is not limited to, medical imaging and EHRs. Table 2 is primarily composed of PHI components that pertain to identity data and do not contain any sensitive aspects. That is, HIPAA does not provide instructions on how to safeguard sensitive attribute data; rather, the primary goal of the HIPAA SH rule is to safeguard privacy by preventing identity exposure. On the other hand, other sensitive characteristics can be uniquely combined into a quasi-identifier, enabling data users to reidentify the individuals to whom the data pertains. As a result, strict adherence to the SH rule may not be adequate to guarantee data quality or privacy to whom the data relates [27].

Table 2 Patient information protected by HIPAA

Categories	Description	Categories	Description
1	Date, time	10	Certificate/license number
2	Location	11	Health plan beneficiary numbers
3	Patient Names	12	Vehicle identifier and serial number
4	Phone numbers	13	Device identifier and serial numbers
5	e-mail address	14	URLs of website, domain name
6	Social security numbers	15	Speech, fingerprints, and other biometric markers
7	Medical record numbers	16	Face images and other images
8	IP address details	17	Unique identifying number, characteristics or code
9	Account numbers		

5 Deep Learning Approaches for Privacy-Preservation

Deep Learning (DL) approaches based on artificial neural models can be applied in various applications, including image classification, autonomous driving, natural language processing, medical diagnosis, intrusion detection, and credit risk assessment. Reverse engineering and stealing of data through DL are very common, and it is possible to recreate images of a patient, and sensitive training data could be inferred. This section reviews the research on strategies and techniques of DL to protect the privacy of sensitive medical data belonging to a patient. The different privacy-preserving mechanism is summarized and classified in Fig. 2. Mainly three types of strategies can be found. The first method is data aggregation which collects data and aggregates data from various sources into one dataset keeping contributor anonymity [28, 29]. The second mechanism encompasses a substantial amount of research that focuses on finding strategies to keep the model training process private so that sensitive information about the training dataset's participants is not revealed. Finally, the third mechanism focuses on the inference phase of deep learning.

Data Aggregation: Here are some of the most well-known data privacy-preserving techniques. Although not all of these tactics apply to deep learning, this chapter will go through them quickly for completeness. Context-free and context-aware privacy approaches are the two types of strategies available. Differential privacy and other context-free privacy solutions have no way of knowing what context or purpose the data will be used for. Context-aware privacy solutions, such as information-theoretic privacy, consider how the data will be used and can improve the privacy-utility trade-off.

Naive Data Anonymization: This is the process of removing identifiers from data, such as participants' names, addresses, and full postcodes, to protect privacy. This

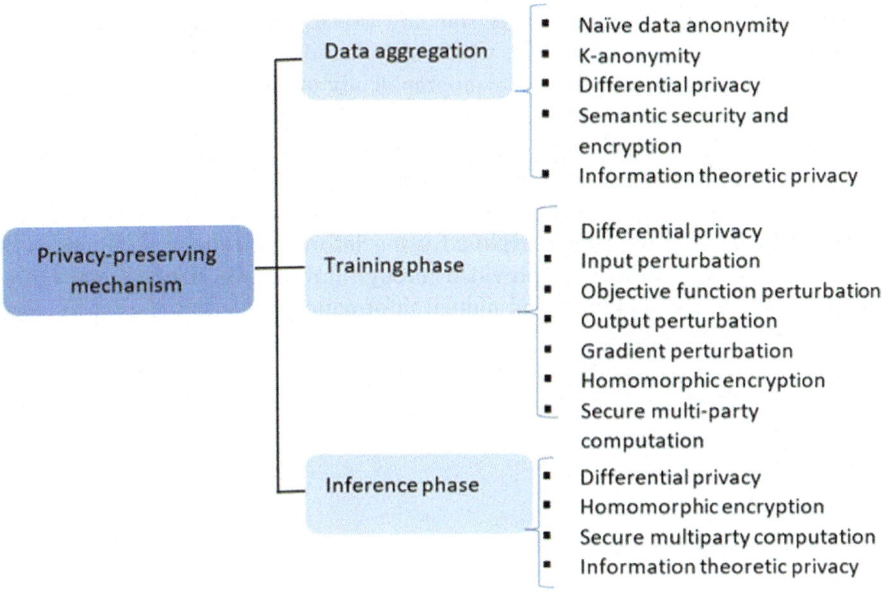

Fig. 2 Different privacy-preserving paradigms in the DL training phase

technique was implemented to protect patients while processing medical data, but it has repeatedly been shown to be useless. Perhaps the most egregious failure is the Netflix Prize instance, in which Narayanan and Shmatikov apply their de-anonymization technique on the Netflix Prize dataset. This dataset includes the anonymous movie ratings of 500,000 Netflix subscribers. They showed that an attacker with additional information about a user (from publicly available Internet Movie Database entries) can quickly identify the person and extract potentially sensitive information [30].

K-Anonymity: If each participant's information cannot be discriminated from the information of at least $(k - 1)$ other participants in the dataset, the dataset has the k-anonymity property. K-anonymity states that there are at least k rows with an identical set of attributes available to the adversary for any given combination of attributes [31, 32]. k-anonymization, on the other hand, has been demonstrated to perform poorly when it comes to anonymizing high-dimensional datasets.

Differential Privacy: When two mechanisms with Privacy Budgets 1 and 2 are applied to the same datasets, together they use a privacy budget 1+2. As such, composing various differential private mechanism consumes a privacy budget that increases linearly [33, 34]. Without relying on a centralized server, differential privacy can be achieved by having each participant apply differentially private randomization to their data before sharing it. The approach "randomized response" is demonstrated to be locally differentially private [35], and the model is called the local model of differential privacy.

Semantic Security Encryption: A standard privacy requirement of encryption methods is semantic security [36], which specifies that an opponent given background information should have a cryptographically minimal advantage, a measure of how successfully an adversary can attack a cryptographic algorithm.

Information Theoretic Privacy: Information-theoretic privacy is a context-aware privacy solution. Context-aware approaches model the dataset statistics, as opposed to context-free solutions, which assume worst-case dataset statistics and adversaries. Privacy and fairness have been explored using information-theoretic methods, in which privacy and fairness are provided through information degradation, obfuscation, or adversarial learning, and mutual information reduction is used to verify them. Further, Generative Adversarial Privacy (GAP), a context-aware privacy system that generates private datasets using Generative Adversarial Networks (GANs) is discussed. An attacker tries to deduce secret attributes, whereas a sanitizer strives to eradicate them [37].

5.1 Learning Phase

The private training in deep learning and research articles on them can be categorized based on the guarantee that these methods provide differential privacy or semantic security and encryption. Privacy using encryption can be achieved by doing computation over encrypted data. Homomorphic Encryption (HE) and Secure Multi-party Computation (SMC) are two common methods for encryption of data.

HE is a program that allows one to compute encrypted data [38]. A client can transmit their data to a server in an encrypted format, and the server can compute over it without decrypting it, then send the client a ciphertext for decryption. Because HE is so computationally intensive, it has yet to be used in many production systems [39, 40]with confidentiality. SMC tries to create a network of computing parties that carry out a particular calculation while ensuring that no data leaks. Only an encrypted portion of the data is accessible to each party in this network.

Figure 3 describes a deep learning aggregated model for multi-institutional learning or known as federated learning. The randomization required for differential privacy can be inserted in five places that are input, objective function, gradient updates, output, and labels [41]. Input perturbations can be considered equivalent to using a sanitized dataset for training. Objective function perturbation and output perturbation are explored for machine learning tasks with convex objective functions. For instance, in the case of logistic regression, it is proved that objective perturbation requires sampling noise on the scale of $\frac{2}{n\epsilon}$ [42], and output perturbation required sampling noise on a scale of $\frac{2}{n\lambda\epsilon}$, where n is the number of samples and λ is the regularization coefficient. The proposed work is more practical and is a general objective perturbation approach that works for high-dimensional real-world data.

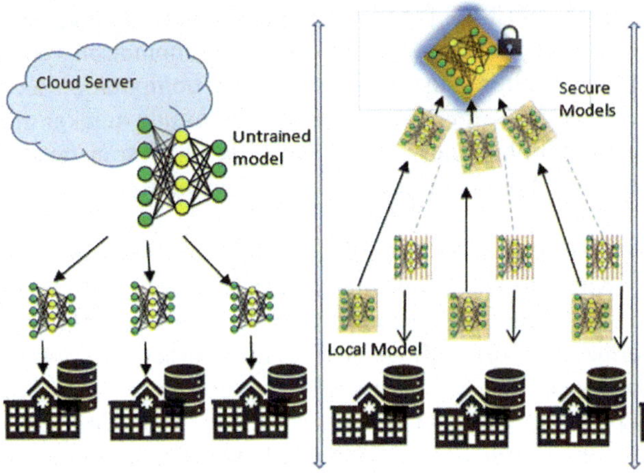

Fig. 3 The framework of deep learning on the cloud

5.2 *Inference Privacy*

Inference privacy, as opposed to training, aims to provide Inference-as-a-Service through a cloud-based health inferencing system. This cloud-based system can efficiently analyze data on cloud storage and can learn hidden patterns underlying medical data. The situation is only expected to carry out the inference task that has been assigned to it. The classification of literature for inference privacy is similar to that for training, except for one additional area called Information-Theoretic (IT) privacy. The context-aware mechanism is intelligent and sends only information sufficient enough for analysis purposes removing much information content about the patient.

Differential Privacy: The major reason in differential privacy is that it provides a worst-case guarantee, which necessitates applying high-intensity noise to all input segments. On pre-trained networks, this automatically degrades performance. Arden is a data nullification and differentially private noise injection strategy used for inference [43]. Arden divides the DNN between the edge and the cloud. The mobile device performs a simple data transformation, while the cloud data center handles the computationally intensive and complicated inference. Simultaneously, it employs data nullification and noise injection to make various requests indistinguishable, protecting clients' anonymity. The suggested method necessitates noisy retraining of the entire network, with noise inserted at various layers.

Homomorphic Encryption: CryptoNets is one of the first efforts in the field of HE inferences [44]. CryptoNet is a method for transforming a learned neural network into an encrypted one. This allows inference service clients to provide their data in an encrypted format and obtain the result without having to decode it. CryptoNets enables Single Instruction Multiple Data (SIMD) operations, which boost the deployed system's throughput. However, the latency of this technique is still considerable for single queries. GAZELLE [45] is a system for secure and private neural network inference with shorter latency. It is a two-party computation system that combines homomorphic encryption and regular two-party computation techniques such as garbled circuits. Besides, it is three orders of magnitude quicker than CryptoNets because of its homomorphic linear algebra kernels, which transfer neural network operations to efficient homomorphic matrix vector multiplication and convolutions.

Secure Multi-party Computation: Unlike GAZELLE, which utilizes Additively Homomorphic Encryption (AHE) to speed up linear algebra directly, MiniONN [46] uses AHE in a preprocessing stage. When compared to CryptoNets, MiniONN shows a considerable performance boost without sacrificing accuracy. However, it is a two-party computation system that does not support multi-party calculation. Further, Chameleon, a two-party computation system whose vector dot product of signed fixed-point integers increase prediction performance in heavy matrix multiplication classification algorithms. Chameleon improves on MiniONN by 4.2 seconds in latency. The majority of work in the field of SMC for deep learning is focused on speeding up computation.

Information Theoretic Privacy: Privacy-preserving strategies that rely on information-theoretic approaches often assume a nonsensitive task, the task that the service is supposed to perform, and strive to degrade any unnecessary information in the input data [47–49]. Anonymization techniques for securing temporal sensory data through obfuscation are proposed [50] which offer a multi-objective loss function for training deep autoencoders to extract and conceal user identity-related data while maintaining the sensor data's value. The training method instructs the encoder to ignore user identifiable patterns and tunes the decoder to shape the output without regard for the training set's users.

6 Cloud Environment and Privacy-Preserving Framework

Cloud computing incorporates many techniques like grid, utility, and distributed computing to provide infrastructure to host data and provide software services. Recent healthcare trends, which emphasize accessing information at any time and from any location, favor sending healthcare data to the cloud. Although the cloud has many advantages, it also poses new privacy and security risks to health data [51].

6.1 Cloud Privacy Threats

To protect patient privacy, various measures need to be taken to safeguard medical data on the cloud. Four primary components that can violate the privacy of patients are data representation, data transmission, data distribution and processing, and data storage.

Data representation refers to any user that processes medical data as input, retrieve, and visualization. This could generally be a client browser. Data transmission refers to the transmission of medical data from the client machine to the health record system on the cloud and vice versa. Data distribution and processing refers to the system that handles medical data, and gives a representation of health records on the cloud. This part of the system also manages the efficient storage of data in a distributed manner. In data storage, all kinds of medical data related to the patient are permanently stored on the cloud in a distributed manner. This is based on a database management system and other storage solutions to provide query processing.

6.2 Privacy-Preserving Framework

The word "privacy-preservation" refers to a lot more than just keeping data secret. Spoofing identification, tampering with data, repudiation, and information exposure are all challenges to data privacy in the cloud [52]. In a spoofing identity attack, the attacker impersonates a legitimate user, whereas data tampering entails harmful content change and modification. Users who deny executing an action with the data are considered repudiation threats. The exposing of information to entities with no permission to access it is known as information disclosure [53]. An EMR is a record of patient medical records information. Some other type of medical records can be EHR. The Health Information and Management System Society (HIMSS) and ISO/TS 18308 standards describe these types of medical records [54]. A healthcare institution, such as a hospital, creates and manages EMR. These records are proof of overall patient health, treatment, and other results to track the patient's treatment. The EHR is created and maintained inside a single institution or community. It's a digital record that can be shared throughout a community, region, or state's many institutions. Data from EMRs can be fed into EHRs. When EMR data is shared with other institutions, EHRs are established. Figure 4 depicts the conceptual framework for sharing medical data in the cloud while maintaining privacy.

Vertical Partitioning of Medical Data: This component in any system preserves the privacy of the medical data by dividing data and then storing it at various locations in the cloud. The EMR 'T' is partitioned into three different tables T_p, T_a, and T_e. The T_p is a plaintext table; T_a is an anonymized table where many of the personal identifiers of the personal has been removed. The third table T_e is a table of explicit identifiers and quasi-identifier. After the portioning, these three tables are moved to cloud storage in separate locations.

Merging Process: All the stored medical data in the cloud can be accessed through the merging process. The recipient has to merge partitioned medical data into one dataset. The data recipient can access T_p directly for medical data research or analysis with the approval of the data owner. When the plaintext table T_p is unable to meet the information needs of data consumers, this component is utilized to integrate T_p with T_a and T_e, resulting in two additional medical dataset access paradigms. Similarly,

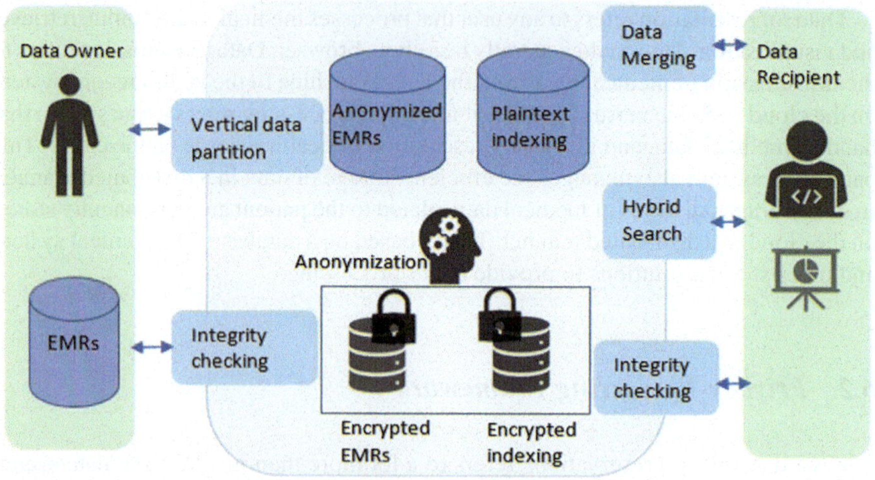

Fig. 4 Conceptual framework of a cloud environment for medical data

the data recipient can access anonymized medical dataset and full version medical dataset access through original EMR table T.

Integrity Checking This component is often used by both the data owner and the data recipient to confirm that the data saved in the cloud is equivalent to what was originally recorded. Two integrity checking systems are given due to the varied requirements of data owners and data recipients.

Hybrid Search This component is used by the data recipient to achieve record-level medical data access, i.e., to locate one or more interested EMRs in the shared medical dataset. For the implementation of information retrieval over remote medical data storage, a hybrid search technique is proposed that combines encrypted and plaintext search methods.

In any cloud-based healthcare system, the entities are mainly patients, physicians, pharmacists, pathologists, nurses, laboratory staff, insurance companies, reports, and cloud service providers. Various ways to maintain e-Health cloud privacy are based on adversarial models. One model considers cloud servers to be untrustworthy entities that might potentially reveal sensitive health information. Furthermore, untrusted cloud servers are vulnerable to both internal and external attacks. The adversary may not only use forged credentials to access the encrypted health data, but they may also obtain access to the data as privileged users. Inside enemies may pose a danger to health data kept on trusted cloud servers in the second scenario. Parts of the data, for example, may be stored by a doctor, who may then share the data with unauthorized parties, resulting in information leakage [55].

Additionally, the entity identities must be protected. The third strategy entails the usage of semi-trusted cloud servers, while semi-trusted cloud services are considered trustworthy that collect information about health data and collaborate with unauthorized users [55]. The intruder in these cases might steal or can manipulate the patient's personal information; even in worse cases might even sell it to third parties for monetary gains. For instance, a physician's prescription medicine may be disclosed to representatives of pharmaceutical companies, or insurance company spending information may be falsified. To deal with such cases, the healthcare system should provide a mechanism and guarantee to protect private sensitive information belonging to a patient.

7 Vertical Partitioning Approach

If the data volume is high, the data is growing at a high rate and it is of varied kinds; the healthcare system needs partitioning of data to store them because they cannot be stored in one repository. Many large-scale solutions split data so that it can be maintained and retrieved separately. Scalability, conflict reduction, and performance optimization can all be aided by partitioning. It can also be used to categorize data according to usage trends.

There are three aspects to the EMR conceptual model such as patient data (name, birth date, address); profile of patient (medical history and reports); and data such as symptoms, diagnoses, and treatments, for each patient's hospital visit [56, 57]. Overall data like patient data is in proper format and structure, but patient profiles and clinical reports are semi-structured that include a lot of text information. Consider that 'T' is a table that holds EMR data, to maintain data privacy; T's properties can be categorized into three distinct groups like Quasi-Identifiers (QID), Explicit Identifiers (EID), and Medical Information (MI).

All collected information that together identifies a patient or individual is called QID attributes. EID refers to information such as name, social security number, and phone number that can uniquely identify individual patient records. MI refers to information containing clinical, reports, and medical data about patients and is called medical information. Knowing that MI contains all the sensitive patient information, data analysis of such data might reveal the identity of the person and so vertical partitioning use can be done to hide the details by keeping the table in a partitioned manner.

Three vertical partitioned tables T_p, T_a, and T_e are created from the original EMR table T. T_p contains attributes from MI, T_e contains attributes from EID, and T_a has information from QID. Their precise values are encrypted and saved in ciphertext. Allow A to signify the set of all attributes $\{A_1, A_2, \cdots, A_m\}$ and $t[A_j]$ to denote the value of attribute A_i for tuple t. The attributes in A are categorized as EID, QID, and MI, with EID equal to $\{A_1, A_2, \cdots, A_{|EID|}\}$, QID equal to $\{A_1, A_2, \cdots, A_{|QID|}\}$, and MI equal to $\{A_1, A_2, \cdots, A_{|MI|}\}$.

8 Conclusion

In light of the developing cloud computing realm and deep learning, models use in bioinformatics; protecting patient information has become a serious issue. EHRs and personal health records keep all kinds of personal and clinical information. Privacy considerations are critical when dealing with such sensitive medical data; yet, with today's sophisticated cloud apps, privacy concerns are a distant second. The necessary technical steps have not been widely addressed and are not included in standard solutions. The chapter introduces the notion of deep learning as well as several ways for protecting patient privacy from insider threats. The chapter's primary objective is to provide a detailed overview of the secondary use of medical data, deep learning for data analytics, and storage solutions on cloud platforms. A cloud provider can host an effective health record system even if it can't link people and their health information. In addition, the chapter goes through four essential parts of cloud storage: horizontal and vertical partitioning, integrity, and privacy-preserving query processing. Data splitting makes it possible to store information straightforwardly and effectively. It also enables more flexible access and lowers the cost of data storage by integrating cryptographic techniques and statistical analysis.

References

1. Solares, J.R.A., Raimondi, F.E.D., Zhu, Y., Rahimian, F., Canoy, D., Tran, J., Gomes, A.C.P., Payberah, A.H., Zottoli, M., Nazarzadeh, M., Conrad, N., Rahimi, K., Salimi-Khorshidi, G.: Deep learning for electronic health records: a comparative review of multiple deep neural architectures. J. Biomed. Inform. **101**(1), 103337 (2020)
2. Azencott, C.A.: Machine learning and genomics: precision medicine versus patient privacy. Philos. Trans. R. Soc. A: Math. Phys. Engin. Sci. **376**(2128), 20170350 (2018)
3. Si, Y., Du, J., Li, Z., Jiang, X., Miller, T., Wang, F., Zheng, W.J., Roberts, K.: Deep representation learning of patient data from Electronic Health Records (EHR): a systematic review. J. Biomed. Inform. **115**(3), 103671 (2021)
4. Jensen, P.B., Jensen, L.J., Brunak, S.: Mining electronic health records: towards better research applications and clinical care. Nat. Rev. Genet. **13**(6), 395–405 (2012)
5. Jiang, M., Chen, Y., Liu, M., Rosenbloom, S.T., Mani, S., Denny, J.C., Xu, H.: A study of machine-learning-based approaches to extract clinical entities and their assertions from discharge summaries. J. Am. Med. Inform. Assoc. **18**(5), 601–606 (2011)
6. Office for Civil Rights, H.H.S.: Standards for privacy of individually identifiable health information. Final rule. Feder. Regist. **67(157)**, 53181–53273 (2002)
7. McGraw, D., Mandl, K.D.: Privacy protections to encourage use of health-relevant digital data in a learning health system. NPJ Digital Med. **4**(1), 2 (2021)
8. Liu, Y., Zhang, L., Yang, Y., Zhou, L., Ren, L., Wang, F., Liu, R., Pang, Z., Deen, M.J.: A novel cloud-based framework for the elderly healthcare services using digital twin. IEEE Access **7**, 49088–49101 (2019)
9. Jungkunz, M., Köngeter, A., Mehlis, K., Winkler, E.C., Schickhardt, C.: Secondary use of clinical data in data-gathering, non-interventional research or learning activities: definition, types, and a framework for risk assessment. J. Med. Internet Res. **23**(6), e26631 (2021)
10. Xue, J., Xu, C., Bai, L.: DStore: A distributed system for outsourced data storage and retrieval. Futur. Gener. Comput. Syst. **99**, 106–114 (2019)

11. Manogaran, G., Shakeel, P.M., Fouad, H., Nam, Y., Baskar, S., Chilamkurti, N., Sundarasekar, R.: Wearable IoT smart-log patch: an edge computing-based Bayesian deep learning network system for multi access physical monitoring system. Sensors **19**(13), 3030 (2019)
12. Li, D., Huang, L., Ye, B., Wan, F., Madden, A., Liang, X.: FSRM-STS: Cross-dataset pedestrian retrieval based on a four-stage retrieval model with Selection Translation Selection. Futur. Gener. Comput. Syst. **107**(6), 601–619 (2020)
13. Avasthi, S., Chauhan, R., Acharjya, D.P.: Processing large text corpus using N-gram language modeling and smoothing. In: Proceedings of the Second International Conference on Information Management and Machine Intelligence, Springer Singapore, pp. 21-32 (2021)
14. Hutchings, E., Loomes, M., Butow, P., Boyle, F.M.: A systematic literature review of researchers' and healthcare professionals' attitudes towards the secondary use and sharing of health administrative and clinical trial data. Syst. Control Found. Appl. **9**(1), 1–27 (2020)
15. Ozyurt, F.: Efficient deep feature selection for remote sensing image recognition with fused deep learning architectures. J. Supercomput. **76**(11), 8413–8431 (2020)
16. Santhanavijayan, A., Naresh Kumar, D., Deepak, G.: A semantic-aware strategy for automatic speech recognition incorporating deep learning models. In: Proceedings of the Intelligent System Design, pp. 247–254. Springer, Singapore (2021)
17. Jain, R., Gupta, M., Taneja, S., Hemanth, D.J.: Deep learning based detection and analysis of COVID-19 on chest X-ray images. Appl. Intell. **51**, 1690–1700 (2021)
18. Fredrikson, M., Lantz, E., Jha, S., Lin, S., Page, D., Ristenpart, T.: Privacy in pharmacogenetics: An End-to-End case study of personalized warfarin dosing. In: Proceedings of the 23rd USENIX Security Symposium, pp. 17–32 (2014)
19. Bolton, R.J., Hand, D.J.: Statistical fraud detection: a review. Stat. Sci. **17**(3), 235–255 (2002)
20. Thompson, S.A., Warzel, C.: Twelve million phones, one dataset, zero privacy. In: Ethics of Data and Analytics, pp. 161–169. Auerbach Publications (2022)
21. Schiff, J., Meingast, M., Mulligan, D. K., Sastry, S., Goldberg, K.: Respectful cameras: detecting visual markers in real-time to address privacy concerns. In: Protecting Privacy in Video Surveillance, pp. 65–89 (2009)
22. Senior, A., Pankanti, S., Hampapur, A., Brown, L., Tian, Y.L., Ekin, A., Connell, J., Shu, C.F., Lu, M.: Enabling video privacy through computer vision. IEEE Secur. Privacy **3**(3), 50–57 (2005)
23. Geetha Mary, A., Acharjya, D.P., Iyengar, N.C.S.: Improved anonymization algorithms for hiding sensitive information in hybrid information system. Int. J. Comput. Netw. Inf. Secur. **6**(6), 9–17 (2014)
24. Avasthi, S., Chauhan, R., Acharjya, D.P.: Extracting information and inferences from a large text corpus. Int. J. Inf. Technol. **15**(1), 435–445 (2023)
25. Cate, F.H.: The failure of fair information practice principles. In: Consumer Protection in the Age of the Information Economy, pp. 341–377. Routledge (2016)
26. Mendes, R., Vilela, J.P.: Privacy-preserving data mining: methods, metrics, and applications. IEEE Access **5**, 10562–10582 (2017)
27. Li, X.B., Qin, J.: Anonymizing and sharing medical text records. Inf. Syst. Res. **28**(2), 332–352 (2017)
28. Sweeney, L.: k-anonymity: A model for protecting privacy. Internat. J. Uncertain. Fuzziness Knowl.-Based Syst. **10**(05), 557–570 (2002)
29. Dwork, C., McSherry, F., Nissim, K., Smith, A.: Calibrating noise to sensitivity in private data analysis. J. Privacy Confident. **7**(3), 17–51 (2016)
30. Narayanan, A., Shmatikov, V.: Robust de-anonymization of large sparse datasets. In: Proceedings of the IEEE Symposium on Security and Privacy, pp. 111–125 (2008)
31. Aggarwal, C.C.: On k-anonymity and the curse of dimensionality. In: Proceedings of the VLDB Conference, Trondheim, Norway vol. 5, pp. 901–909 (2005)
32. Machanavajjhala, A., Kifer, D., Gehrke, J., Venkitasubramaniam, M.: l-diversity: Privacy beyond k-anonymity. ACM Trans. Knowl. Discov. Data **1(1)**, 3–es (2007)
33. Dwork, C., Rothblum, G.N., Vadhan, S.: Boosting and differential privacy. In: Proceedings of the IEEE 51st Annual Symposium on Foundations of Computer Science, pp. 51–60 (2010)

34. Kairouz, P., Oh, S., Viswanath, P.: The composition theorem for differential privacy. In: Proceedings of the International Conference on Machine Learning, PMLR **37**,1376–1385 (2015)
35. Kairouz, P., Oh, S., Viswanath, P.: Extremal mechanisms for local differential privacy. J. Mach. Learn. Res. **17**(1), 492–542 (2016)
36. Shafi, G., Micali, S.: Probabilistic encryption. J. Comput. Syst. Sci. **28**(2), 270–299 (1984)
37. Huang, C., Kairouz, P., Chen, X., Sankar, L., Rajagopal, R.: Context-aware generative adversarial privacy. Entropy **19**(12), 656 (2017)
38. Gentry, C.: Fully homomorphic encryption using ideal lattices. In: Proceedings of the 41st Annual ACM Symposium on Theory of Computing, pp. 169–178 (2009)
39. Riazi, M.S., Samragh, M., Chen, H., Laine, K., Lauter, K., Koushanfar, F.: XONN:XNOR-based oblivious deep neural network inference. In: Proceedings of the 28th USENIX Security Symposium, pp. 1501–1518 (2019)
40. Makri, E., Rotaru, D., Smart, N.P., Vercauteren, F.: EPIC: efficient private image classification (or: Learning from the masters). In: Proceedings of the RSA Conference, San Francisco, CA, USA, Springer International Publishing, pp. 473–492 (2019)
41. Papernot, N., Abadi, M., Erlingsson, U., Goodfellow, I., Talwar, K.: Semi-supervised knowledge transfer for deep learning from private training data. In: Proceedings of the ICLR Conference (2017)
42. Chaudhuri, K., Monteleoni, C., Sarwate, A.D.: Differentially private empirical risk minimization. J. Mach. Learn. Res. **12**(3), 1069–1109 (2011)
43. Iyengar, R., Near, J. P., Song, D., Thakkar, O., Thakurta, A., Wang, L.: Towards practical differentially private convex optimization. In: Proceedings of the IEEE Symposium on Security and Privacy, pp. 299–316 (2019)
44. Wang, J., Zhang, J., Bao, W., Zhu, X., Cao, B., Yu, P.S.: Not just privacy: improving performance of private deep learning in mobile cloud. In: Proceedings of the 24th ACM SIGKDD International Conference on Knowledge Discovery and Data Mining, pp. 2407–2416 (2018)
45. Gilad-Bachrach, R., Dowlin, N., Laine, K., Lauter, K., Naehrig, M., Wernsing, J.: Cryptonets: applying neural networks to encrypted data with high throughput and accuracy. In: Proceedings of the International Conference on Machine Learning, PMLR **48**, 201–210 (2016)
46. Juvekar, C., Vaikuntanathan, V., Chandrakasan, A.: GAZELLE: a low latency framework for secure neural network inference. In: Proceedings of the 27th USENIX Security Symposium, pp. 1651–1669 (2018)
47. Liu, J., Juuti, M., Lu, Y., Asokan, N.: Oblivious neural network predictions via minionn transformations. In: Proceedings of the ACM SIGSAC Conference on Computer and Communications Security, pp. 619–631 (2017)
48. Malekzadeh, M., Clegg, R. G., Cavallaro, A., Haddadi, H.: Mobile sensor data anonymization. In: Proceedings of the International Conference on Internet of Things Design and Implementation, pp. 49–58 (2019)
49. Malekzadeh, M., Clegg, R. G., Cavallaro, A., Haddadi, H.: Protecting sensory data against sensitive inferences. In: Proceedings of the 1st Workshop on Privacy by Design in Distributed Systems, pp. 1–6 (2018)
50. Malekzadeh, M., Clegg, R.G., Cavallaro, A., Haddadi, H.: Privacy and utility preserving sensor-data transformations. Pervas. Mob. Comput. **63**(3), 101132 (2020)
51. Avasthi, S., Chauhan, R., Acharjya, D.P.: Information Extraction and Sentiment Analysis to gain insight into the COVID-19 crisis. In: Proceedings of the International Conference on Innovative Computing and Communications, vol. 1, pp. 343–353. Springer Singapore (2022)
52. Antwi-Boasiako, E., Zhou, S., Liao, Y., Liu, Q., Wang, Y., Owusu-Agyemang, K.: Privacy preservation in distributed deep learning: a survey on distributed deep learning, privacy preservation techniques used and interesting research directions. J. Inf. Secur. Appl. **61**(9), 102949 (2021)
53. Bukowski, M., Farkas, R., Beyan, O., Moll, L., Hahn, H., Kiessling, F., Schmitz-Rode, T.: Implementation of eHealth and AI integrated diagnostics with multidisciplinary digitized data: are we ready from an international perspective? Eur. Radiol. **30**(10), 5510–5524 (2020)

54. Gomes, J., Romao, M.: Information system maturity models in healthcare. J. Med. Syst. **42**(12), 1–14 (2018)
55. Mashima, D., Ahamad, M.: Enhancing accountability of electronic health record usage via patient-centric monitoring. In: Proceedings of the 2nd ACM SIGHIT International Health Informatics Symposium, pp. 409–418 (2012)
56. Dong, N., Jonker, H., Pang, J.: Challenges in ehealth: from enabling to enforcing privacy. In: Proceedings of the First FHIES International Symposium, Johannesburg, South Africa, pp. 195–206. Springer, Berlin (2012)
57. Park, T.H., Lee, G.R., Kim, H.W.: Survey and prospective on privacy protection methods on cloud platform environment. J. Korea Inst. Inf. Secur. Cryptol. **27**(5), 1149–1155 (2017)

Computational Intelligence Ethical Issues in Health Care

Najm Us Sama, Kartinah Zen, N. Z. Jhanjhi, and Mamoona Humayun

Abstract Over the past ten years, a significant influx of multi-modality material has contributed to the rapid growth of data analytics' significance in health informatics. As a result, there is a growing demand for creating analytical, data-driven solutions in health informatics focused on deep learning approaches. Artificial neural networks are the foundation of the deep learning method of machine learning, which has recently gained prominence as a formidable tool for machine learning with the potential to revolutionize artificial intelligence. Rapid improvements in processing power, quick data storage, and parallelization, together with the technology's potential for producing optimal high-level features and semantic interpretation automatically from the incoming data, have all aided in its swift acceptance. This book chapter offers a thorough, current overview of DL research in health informatics, with a critical evaluation of the method's relative advantages, potential issues, and prospects for the future. This chapter primarily focuses on deep learning applications in translational bioinformatics, medical imaging, ubiquitous sensing, medical informatics, and public health.

N. U. Sama (✉) · K. Zen
Faculty of Computer Science and IT, Universiti Malaysia Sarawak, 94300 Kota Samarahan, Malaysia
e-mail: najmussama@gmail.com

K. Zen
e-mail: kartinah@unimas.my

N. Z. Jhanjhi
School of Computer Science and Engineering, Taylor's University, Subang Jaya, Malaysia
e-mail: noorzaman.jhanjhi@taylors.edu.my

M. Humayun
College of Computer and Information Sciences, Jouf University, Sakakah, Saudi Arabia
e-mail: mahumayun@ju.edu.sa

D. P. Acharjya and K. Ma (eds.), *Computational Intelligence in Healthcare Informatics*, Studies in Computational Intelligence 1132,
https://doi.org/10.1007/978-981-99-8853-2_21

1 Introduction

Maintaining good health is, without question, one of our highest priorities. Humans have fought deadly diseases throughout history; today, we face an unprecedented number of these threats while increasing life expectancy and general health. Due to various factors, including a lack of clinical equipment, sensors, and analytical tools for the collected medical data, medicine has historically been unable to discover the treatment for many diseases. Big data, artificial intelligence, and cloud computing have all been instrumental in dealing with these datasets. Because of the fast advancements made in nearly every area of our lives, Artificial Intelligence is already commonplace and well recognized by most individuals worldwide. The importance of artificial intelligence (AI) stems from its rapid development over the past two decades, which is now continuing, thanks to the efforts of experts from various sectors [1].

Its usefulness has been demonstrated in many contexts, including medicine, industry, and everyday life. For the classification or prediction of future or uncertain conditions, Machine Learning (ML) is a branch of artificial intelligence that requires minimal human intervention. Due to its data-driven nature, ML falls under the umbrella of symbolic AI despite its ability to extrapolate from insufficient sample data. The three datasets in ML, training, validation, and testing, are kept separate. Data characteristics are learned from the training dataset and then checked against the validation dataset to ensure accuracy. Lastly, the test dataset can be used to validate ML's reliability [2].

An ANN is a type of ML based on the human brain's same layer-based, connected node architecture. It has hidden layers between the input and output layers. In this setup, the input values are in the first layer, and the labeled output values are in the last layer. In training, learning techniques like backpropagation are used to parameterize weights, which decide the matter of individual nodes. Each node's weight is tuned in the direction to decrease losses and maximize precision. Optimized weights can be acquired through repeated backpropagation [3]. However, ANN has limitations, such as overfitting issues when training reaches a local minimum or is optimized solely for trained data. To transform ANN into Deep Neural Network (DNN), researchers have recently resorted to a deep learning strategy that involves stacking many hidden layers with connected nodes between the input and output layers. The greater the number of interconnected nodes in a network, the deeper the network is said to be. DNNs can have hundreds of layers, whereas more basic ones often have just two or three [4]. The multilayer can handle increasingly complicated issues by composing intermediate judgments across layers. It performs better than the shallow layered network in prediction tasks like classification and regression. The detailed steps of deep learning are illustrated in Fig. 1.

As a result, the Deep Learning (DL) algorithm is receiving a great deal of focus nowadays to address various challenges in the medical imaging industry. Detecting diseases or disorders in X-ray pictures and then categorizing them into different disease categories or severity levels is one such use in radiology [5]. Even though

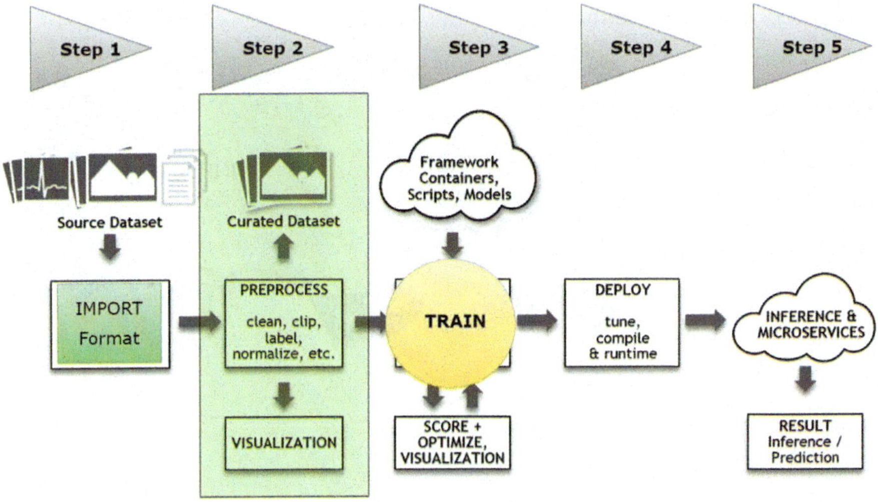

Fig. 1 Steps of deep learning

DL technology is a huge step forward, it will not replace doctors anytime soon, particularly radiologists. It aids radiologists in making more precise diagnoses.

The rest of the chapter is organized as follows. Section 2 throws light on various DL applications followed by DL in health informatics in Sect. 3. Various issues and challenges are discussed in Sect. 4 followed by future research directions in Sect. 5. The chapter is concluded in Sect. 6.

2 Deep Learning Applications

Based on a comprehensive literature review, it is clear that numerous research community members have investigated many different domains where deep learning-based algorithms have found widespread application and improved accuracy.

2.1 Translational Bioinformatics

The field of translational bioinformatics aims to utilize the findings of previous studies to build a bridge between biomedical information and clinical informatics. It reflects the growth and maturity of conventional in silico techniques [6]. Translational bioinformatics has found new uses in the medical and biological sectors. TBI has been characterized as the discovery of storage, analytics, and interpretative tools to improve the transformation of an increased proportion of biomedical data, especially

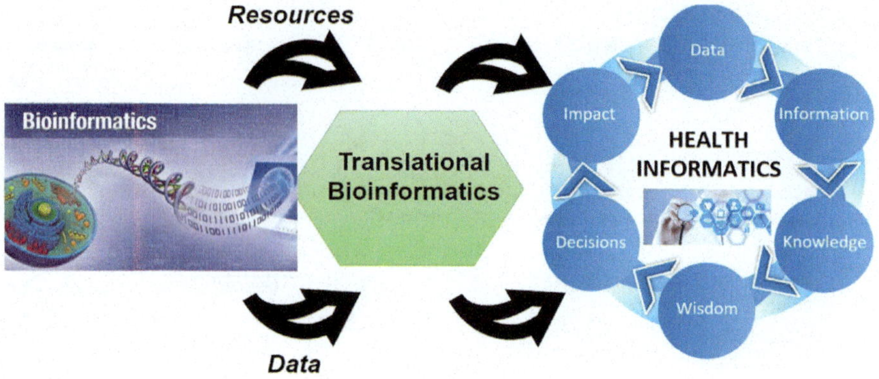

Fig. 2 Overview of translational bioinformatics

genomic data, into proactively predicted, preventative, and participatory health. The overview of translational bioinformatics is illustrated in Fig. 2.

Studies in translational bioinformatics focus on expanding the clinical informatics approach to incorporate biological observations and creating unique methods for integrating biological and clinical data. By integrating sequence-based characteristics, physicochemical property-based features, and quantitative space-derived data with knowledge gain feature extraction, the authors of this study build a DL-based technique they call Deep-Kcr for predicting Kcr (histone lysine crotonylation) locations [7]. For lysine crotonylation sites named iCrotoK-PseAAC, an enhanced predictor is proposed [8]. It incorporates various position and composition relative features and statistical moments into PseAAC. The results show 99.0% accuracy.

The online tool MusiteDeep uses a DL approach to predict and display Post Translational Modification (PTM) sites in proteins. The authors improved the existing system by using more sophisticated ensemble approaches and allowing users to directly assess potential PTM cross-talks through prognosis and visualizations for several PTMs at once [9].

2.2 Medical Imaging

Clinically, professionals rely heavily on medical images because of the wealth of information they contain. Modern tools have made it easier to fully use this data and draw insightful conclusions from it. Many answers and enhancements to analyzing these pictures by radiologists and other experts have been made possible by applying DL technologies in computer-assisted imaging environments. It is common knowledge that acquiring medical photos can be time-consuming and challenging due to the image's complexity. DL algorithms, especially convolutional networks,

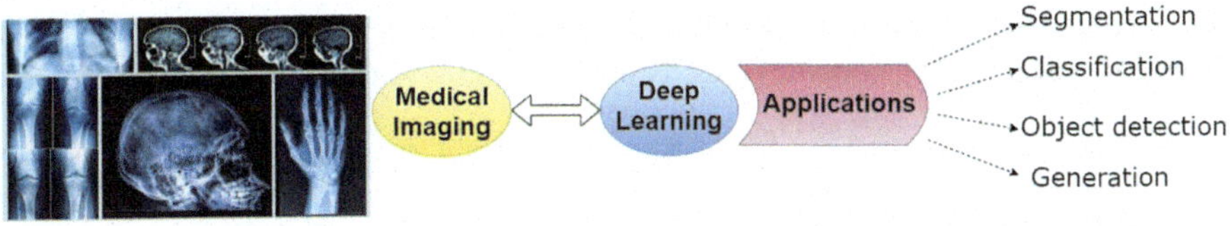

Fig. 3 Deep learning for medical imaging application

have emerged as the standard approach to a wide range of medical image analysis tasks, including segmentation, classification, object detection, and generation, as shown in Fig. 3.

Imaging modalities like MRI, CT scans, PET, and ultrasound combine deep learning to segment various anatomical components. It involves dividing a picture into smaller pieces, each of which often belongs to a specific category. Segmentation can aid in surgical planning and determining the precise edges of sub-regions (such as tumor tissues) for better guidance during immediate surgical treatment [10].

Recent advances in DNN, particularly Convolutional Neural Network (CNN), have enhanced the classification accuracy of medical images in various applications. One of the most fundamental jobs in computer vision and pattern recognition is image classification assigning one or more labels to an image. Medical picture categorization and computer assisted diagnosis are two other areas where DL techniques have been implemented [11]. Finding and classifying items is what "object detection" is all about. A detection method is also applied to biomedical pictures to determine the box coordinated regions where the patient's malignancies are situated. Despite the widespread introduction of CNN-based applications in medical imaging, finding well-balanced datasets with labels in the medical domain can be difficult. It takes a lot of time and effort to obtain accurate tags for most medical images. In addition, privacy concerns make it challenging to get medical images. Much research has used GAN to create life like synthetic images of complete X-ray or CT scans and ROIs of individual lesions to address the problems. Another study is proposed to evaluate the efficacy of the DL-based photoplethysmography (PPG) classification technique for identifying PAD from toe PPG signals. Results prove that the proposed method is 88.9% accurate [12].

DL is a rapid and precise mechanism for diagnosing skin cancer. Using the ISIC2018 dataset, researchers employed the DL technique CNN to differentiate between malignant and benign cancers [13]. First, the images were digitally altered and enhanced using ESRGAN. The images were improved, standardized, and rescaled during the preliminary processing phase. Photos of skin lesions might be categorized with a CNN approach using an average of results from numerous training iterations. The models were then fine-tuned using a variety of transfer learning methods, including Resnet50, InceptionV3, and inception Resnet. The novelty and contribution of their research are the utilization of ESRGAN as a preprocessing step, in addition to the experimentation with multiple models. The proposed strategy was validated through simulation studies on the ISIC 2018 skin lesion dataset.

When compared to Resnet50 (83.7% accuracy), InceptionV3 (85.8% accuracy), and inception Resnet (84% accuracy), CNN achieved 83.2% accuracy.

2.3 Ubiquitous Sensing

The term "Ubiquitous Sensing" (US) denotes the numerous sensor networks that are available and their capacity to investigate geographical occurrences in real time. Sensors dispersed across the surroundings provide us with the means to acquire the required data. Safety, intruder and occupancy recognition, authorization, outdoor lights, military surveillance, and industrial control and tracking systems all rely on the US [14, 15]. DL for US applications is presented in Fig. 4.

The complex nature of operating surroundings makes it difficult to interpret significant volumes of inbound High Dimensional Sensing (HDS) data feeds to improve situational awareness. To overcome it, a new data computation system called ENhanced Situational understanding with Ubiquitous-sensing REsources (ENSURE) is introduced. Then, ENSURE uses concepts from data association to "check up" on these event sequences as they happen and change over time, allowing it to "see" and "understand" its operational setting [16].

The authors categorize human motions including moving, resting, sleeping, and strolling using data collected from ubiquitous devices. In addition, they proposed a new ensemble classification technique based on the ground truth provided by a variety of machine learning and deep learning classifiers. As compared to existing algorithms, the suggested algorithm has a higher accuracy of 98% [17]. Conventional sensing networks are commonly comprised of high-priced communication machines. The high cost of these devices, in addition to the expense of their maintenance, restricts their widespread implementation and, as a result, limits the implementation of tracking. The author suggested a new self-powered sensor that is based on a Sound-Driven Tribo Electric Nano Generator (SDTENG), which is both cheap and new [18]. Materials like fluorinated ethylene propylene membranes, acrylic shells,

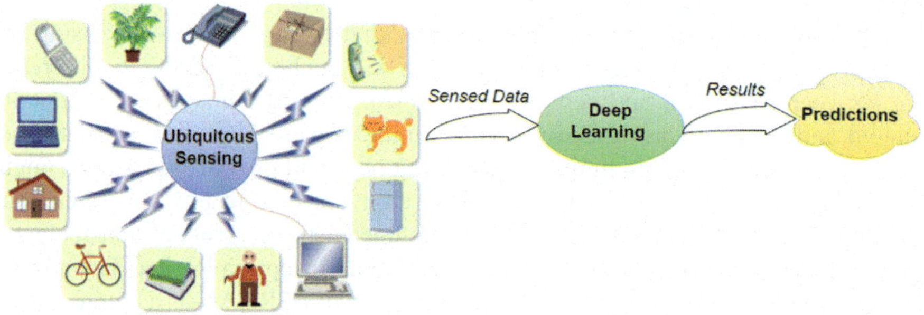

Fig. 4 Deep learning for ubiquitous sensing

conductive fabrics, and kapton spacers compose the majority of the SDTENG. The goal of this study was to develop an intelligent sound tracking and verification system by combining the SDTENG-based sensor with a DL algorithm to achieve a high rate of accuracy in classifying a range of roadway and traffic noises. The novel SDTENG-based self-powered sensor shows great promise for use in urban sound monitoring when combined with deep learning, pointing to promising future developments in the domain of ubiquitous sensing devices.

2.4 Public Health

The field of public health deals with community health for residents. The World Health Organization (WHO) defines public health as the science and art of protecting people from disease, protecting their lives, and improving their health through concerted initiatives and well-informed preferences of communities, institutions, governments, and individuals [19, 20]. Figure 5 shows public health based on deep learning.

Inciting unwarranted fear of COVID by distributing misinformation about it on social media platforms like Twitter. The purpose of this research is to examine the content of tweets posted by Indian Internet users during the COVID-19 quarantine. Data was compiled from March 23, 2020, to July 15, 2020, and tweets were classified as either expressing fear, sadness, anger, or happiness. The Bidirectional Encoder Representations from Transformers (BERT) model, a novel DL model for text analysis and performance, was used for the analysis. According to the statistics, the BERT model achieved an accuracy of 89% [21].

Re-hospitalization due to heart failure is predicted in patients with Acute Myocardial Infarction (AMI) using a DL-based model for the 6-, 12-, and 24-month follow-ups following hospital discharge is proposed [22]. The authors accessed the 13,104 patient information and 551 characteristics from the Korea Acute Myocardial Infarction National Institutes of Health (KAMIR-NIH) repository. It was found that the suggested DL-based re-hospitalization predictive algorithm performed better than other popular machine learning techniques like logistic regression, AdaBoost, sup-

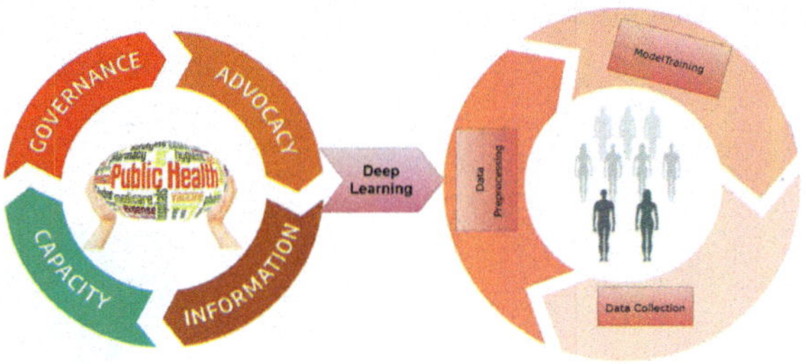

Fig. 5 Public health based on Deep Learning

port vector machine, and random forest. The suggested framework achieved an accuracy of 99.37%, an area under the curve of 99.90%, a precision of 99.90%, a recall of 99.49%, a specificity of 97.73%, and an F1 score of 98.61%.

2.5 Medical Informatics

The field of medical informatics bridges the gap between information sciences, computer technology, and medical treatment. Health informatics focuses on the tools, technologies, and practices that better the health and biomedical communities' ability to collect, store, retrieve, and apply data. Evidence-based treatment, quality improvement, and patient data security and accessibility are driving trends in the healthcare sector, which in turn is increasing demand.

Cross validation is the usual method for accurately quantifying the performance of machine learning-based algorithms, in which the algorithm is first trained on a training set and then its performance is tested on a validation set. It is recommended that both databases be devoid of human subjects so that they can more accurately mimic the conditions of a clinical trial [23]. Understanding the development of rare but serious Vaccine Adverse Event Reporting System (VAERS) reactions requires automated evaluation of post-marketing monitoring narrative reports. Extraction of events linked to nervous system disorders from vaccination safety reports was the focus of experts, and it used state-of-the-art DL techniques for named entity recognition, which were then applied and assessed [24]. Figure 6 shows the DL integration in medical informatics.

The players' EEG waves are analyzed by professionals using deep learning techniques. After that, a channel attention mechanism is proposed and designed to be linked to CNN's input layer. The suggested method learns autonomously from the EEG data across channels to identify the task's involvement. The information gathered by CNN is subsequently fed into the Recurrent Neural Network (RNN) for interpretation. The proposed technique uses the Convolutional Recurrent Neural Network (CRNN) architecture for EEG signal detection. The Stanford research project's EEG

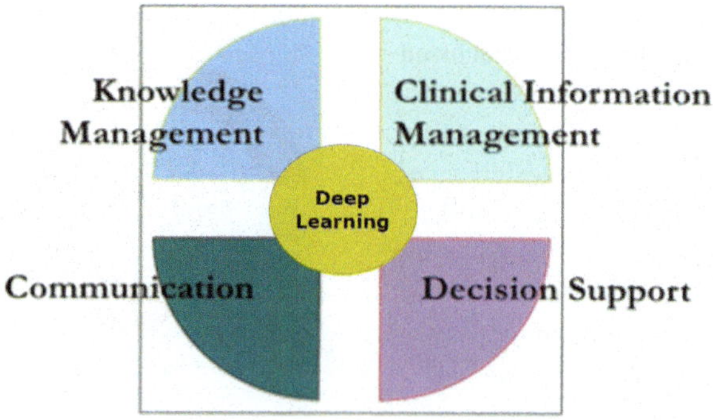

Fig. 6 Deep learning in medical informatics

dataset was used for experimental analysis in this work. The research findings show that by employing the CAMResNet13 and CRNN architecture, the suggested model was able to achieve a remarkable recognition accuracy of 91.05% [25].

To balance the level of competition while dealing with disparate datasets, the Generative Adversarial Network (GAN) model is selected [26]. In addition, they develop an ensemble method that improves the performance of the individual DL model by making use of long short-term memory (LSTM) and a GAN. The suggested GAN-LSTM model outperforms in terms of accuracy (0.992), F1-score (0.987), and area under the curve (0.984). Similarly, breast cancer can be diagnosed with the help of a technique or an early identification model of breast cancer risk based on indicators. Using a repetitive stratified K-fold cross-validation strategy, the author suggested an investigation in which Gradient Boosting Machines (GBM), extreme gradient boosting (XGBoost), and light gradient boosting (LightGBM) frameworks are used to categorize breast cancer [27].

3 Deep Learning in Health Informatics

Health informatics is a subfield of health care that makes use of IT to solve difficult problems. This is a rapidly expanding industry with promising prospects well into the future. Healthcare informatics technology will advance in parallel with the development of communication technologies. Nowadays, doctors may work from home and still provide excellent treatment to patients at walk-in clinics across the country by using videoconferencing technology. Patients can spend less time in the office overall because many of these facilities also house pharmacies. As a result, patients who previously would have had to travel hundreds of miles to visit a specialist can now have a conference with that specialist in their primary care physician's office. Professionals in health informatics may be relied upon to devise a means by which all treating physicians involved in the patient's care can access and use the same information [28].

For the automatic diagnosis of eye illnesses from color photographs of the retinal funds, the authors suggest a deep learning model integrated with a new mixture loss function. In particular, the authors present a combination of focal loss and correntropy-induced loss functions in a deep neural network framework to enhance the recognition effectiveness of classification for biomedical data due to the good generalization and reliability of these two loss functions in overcoming various datasets with imbalanced data and exceptions. A real-world ophthalmic dataset is used to test the suggested model. Meanwhile, metrics like accuracy, sensitivity, specificity, Kappa, and area under the receiver operating characteristic curve are used to evaluate the performance of our proposed loss function to that of the benchmark dataset [29]. A comparative analysis of DL models is presented in Table 1.

Table 1 Comparative analysis of deep learning models

Method	Performance	Application
iCrotoK-PseAAC [8]	99% accurate	Translational bioinformatics
MusiteDeep [9]	50% improvement in the area under the precision	Translational bioinformatics
DL-based PPG [12]	88.9% accurate	Medical imaging
ESRGAN Resnet50 [13]	83.7% accurate	Medical imaging
ESRGAN InceptionV3 [13]	85.8% accurate	Medical imaging
ESRGAN Inception Resnet [13]	84% accurate	Medical imaging
ENSURE [16]	80.03% Accurate	Ubiquitous sensing
	80.59% Precise	
Ensemble classification [17]	98% Accurate	Ubiquitous sensing
SDTENG [18]	99% accurate	Ubiquitous sensing
BERT model [21]	89% accurate	Public health
KAMIR-NIH [22]	99.37% accurate	Public health
	99.90% AUC	
	99.90% Precise	
	98.61 F1 score	
VAERS [24]	0.8078 F1 score	Medical informatics
CAMResNet13 and CRNN [25]	91.05% accurate	Medical informatics
GAN-LSTM [26]	99% accurate	Medical informatics
	0.987 F1-score	
	0.984 AUC	
GBM [27]	93.9% accurate	Medical informatics
	0.984 AUC	
	93.8% Precise	
XGBoost [27]	94.6% accurate	Medical informatics
	0.985 AUC	
	94.6% Precise	
LightGBM [27]	95.3% accurate	Medical informatics
	0.987 AUC	
	95.5% Precise	

4 Issues and Challenges

Recently, DL has opened up a new age of machine learning and pattern identification. Furthermore, it surveyed many clinical data types and health informatics applications that potentially benefit from deep learning. A meta-analysis of existing research has shown that deep learning shows promise for a variety of medical applications, including image classification, tissue classification, cancerous cell characterization, detection, and segmentation. The systematic review of existing research led to a positive conclusion, but many problems remain unsolved. Though there are many benefits, there are also some difficulties.

4.1 Data

Medical information illustrates patients' health and health care over time, but because of the complex interactions among clinical incidents, it is challenging to distinguish legitimate transmissions from the long-term perspective. The lack of thoroughly labeled data is still a problem, even though data volumes increase. Data preprocessing as well as data accuracy and reliability are thus potential areas of discussion. There are no established minimum requirements for training sets, even though more data often leads to more reliable and accurate models. However, at present, there is not enough information to make any definitive statements about the disease's cause or progression.

Also, unlike other fields of informatics, health care relies on subject-matter experts for things like labeling complex data and evaluating the model's efficacy and usability. It is expensive to buy labels, even though they usually improve the effectiveness of treatments. Due to the wide variety of equipment in use in hospitals, the information obtained from even the same CT or EHR may vary. Furthermore, there can be substantial variations in the presentation of a condition even between doctors practicing in the same hospital.

Health data is notoriously difficult to collect because it is so variable, nebulous, noisy, and incomplete, even though it is critically important to have a large quantity of clean, well-structured data. In light of this, it is crucial to establish standards for the reliability and precision of information gathered using biosensors, mobile apps, and the World Wide Web. Data collected from patients who post symptoms on websites and social media may not be profitable to use for forecasting without suitable guidelines and control measures. Data collected from patients at the time it is being generated and collected may not be reliable.

Despite these limitations, studies are needed that integrate various hospital and patient clinical data types. The analysis of the data should help patients and doctors take better care of their health. The constant detectability of patient signals presents a privacy risk. There could be a lull in people's willingness to share information.

4.2 Model

Model authenticity, understandability, and functional applicability is difficult irrespective of the data frame. The model needs to be precise and easily interpretable to persuade doctors and patients to put the results into practice. It is easy to fool a model, especially if the data used to train it is sparse, noisy, and of low quality.

So that the model could be accurate and generalizable, researchers attempted to use multi-modal learning, testing the model with both normal images and images collected from people with Parkinson's disease. Due to its importance in areas such as cost, life and death, dependability, and more, accuracy is a must when trying to sway users. At the same time, it is important to remember that interpretability is still vital even if the prediction accuracy is better than that of other algorithms. It will be very difficult for the non-specialist to understand the model if the model provider does not fully explain why and how specific people will have a specific illness with a specific probability on a specific date. Thus, in healthcare problems, it is important to give equal weight to model credibility, interpretability, and practical application.

It will be essential in the future to construct deep learning models and share them with other important research areas while protecting patient-sensitive information. The next obstacle one may face is determining how much and with whom to share data if patients consent to share information with just one clinical institution but not all. The majority of previous research used either a small number of publicly available hospital datasets or the researchers' own institution's private databases. Yet, because of gaps in data or other factors, the health conditions and statistics of patients at public hospitals may vary greatly. One need to look into how one model could be used on a global scale.

5 Future Research Directions

It is believed that health informatics coupled with deep learning has the potential to improve people's lives, despite a few limitations. Data production and reconstruction are alternatives to combining expert knowledge from medical dictionaries, online encyclopedias, and publications. For mixed-type data, a divide-and-conquer or hierarchical approach, as well as reinforcement learning, could be used to alleviate sophistication and multi-modality worries.

Research into maintaining the safety of deep learning models is something that needs to be looked into in the future. Establishing when and where the framework should be trained is yet another problem that necessitates the attention of experts for cloud-based biosensors and smartphone applications.

6 Conclusion

Recent decades have seen a growing interest in deep learning and blockchain technology from all regions of the world. Rapid progress is being made in both the theoretical foundations of DL and its real-world application in a wide range of fields and industries today. Since data analytics has become increasingly crucial, and since more data needs to be analyzed to produce effective results, many features and attributes of DL have made it a potent tool within the field of ML. Similarly, DL is utilized in clinical and health informatics settings, as well as in biomedical and medical imaging applications. This chapter therefore provides a systematic analysis of the relative merit, potential pitfalls, and prospects of DL for imaging, specifically clinical and healthcare applications. This chapter discussed some of the most important ways that DL has been put to use in the field of medicine and health, such as translational bioinformatics, medical imaging, ubiquitous sensing, medical informatics, and public health. The comparative analysis of different DL techniques is provided in the table for better understanding. This book chapter offers an overview of DL research in health informatics, its outcomes, potential issues, and future direction.

References

1. Hanif, M., Ashraf, H., Jalil, Z., Jhanjhi, N.Z., Humayun, M., Saeed, S., Almuhaideb, A.M.: AI-based wormhole attack detection techniques in wireless sensor networks. Electronics **11**(15), 2324 (2022)
2. Christopoulou, S.C.: Machine learning tools and platforms in clinical trial outputs to support evidence-based health informatics: a rapid review of the literature. BioMedInformatics **2**(3), 511–527 (2022)
3. Humayun, M., Sujatha, R., Almuayqil, S.N., Jhanjhi, N.Z.: A transfer learning approach with a convolutional neural network for the classification of lung carcinoma. Healthcare **10**(6), 1058 (2022)
4. Kwak, G.H.J., Hui, P.: DeepHealth: deep learning for health informatics reviews, challenges, and opportunities on medical imaging, electronic health records, genomics, sensing, and online communication health. ArXiv Preprint arXiv:1909.00384 (2019)
5. Ullah, A., Azeem, M., Ashraf, H., Alaboudi, A.A., Humayun, M., Jhanjhi, N.Z.: Secure healthcare data aggregation and transmission in IoT-A survey. IEEE Access **9**, 16849–16865 (2021)
6. Rahman, H.U., Mahmood, M.H., Sama, N.U., Afzal, M., Asaruddin, M.R., Khan, M.S.A.: Impact of olive oil constituents on C-reactive protein: in silico evidence. J. Oleo Sci. **71**(8), 1199–1206 (2022)
7. Khalil, M.I., Humayun, M., Jhanjhi, N.Z., Talib, M.N., Tabbakh, T.A.: Multi-class segmentation of organ at risk from abdominal ct images: a deep learning approach. In: Intelligent Computing and Innovation on Data Science: Proceedings of ICTIDS, pp. 425–434. Springer Singapore (2011)
8. Malebary, S.J., Rehman, M.S.U., Khan, Y.D.: iCrotoK-PseAAC: Identify lysine crotonylation sites by blending position relative statistical features according to the Chou 5-step rule. PloS One **14**(11), e0223993 (2019)
9. Wang, D., Liu, D., Yuchi, J., He, F., Jiang, Y., Cai, S., Li, J., Xu, D.: MusiteDeep: a deep-learning based webserver for protein post-translational modification site prediction and visualization. Nucleic Acids Res. **48**(W1), W140–W146 (2020)

10. Khan, M.Z., Gajendran, M.K., Lee, Y., Khan, M.A.: Deep neural architectures for medical image semantic segmentation. IEEE Access **9**, 83002–83024 (2021)
11. Zhao, Y.: Deep learning based medical image segmentation and classification for artificial intelligence healthcare, Doctoral Dissertation, Technische Universitt Mnchen (2021)
12. Allen, J., Liu, H., Iqbal, S., Zheng, D., Stansby, G.: Deep learning-based photoplethysmography classification for peripheral arterial disease detection: a proof-of-concept study. Physiol. Meas. **42**(5), 054002 (2021)
13. Gouda, W., Sama, N.U., Al-Waakid, G., Humayun, M., Jhanjhi, N.Z.: Detection of skin cancer based on skin lesion images using deep learning. Healthcare **10**(7), 1183 (2022)
14. Mohr, D.C., Zhang, M., Schueller, S.M.: Personal sensing: understanding mental health using ubiquitous sensors and machine learning. Annu. Rev. Clin. Psychol. **13**, 23–47 (2017)
15. Machidon, A.L., Pejovic, V.: Deep learning for compressive sensing: a ubiquitous systems perspective. Artif. Intell. Rev. **56**(4), 3619–3658 (2023)
16. Garg, V.K., Wickramarathne, T.L.: ENSURE: a deep learning approach for enhancing situational awareness in surveillance applications with ubiquitous high-dimensional sensing. IEEE J. Select. Top. Signal Process. **16**(4), 869–878 (2022)
17. Liaqat, S., Dashtipour, K., Shah, S.A., Rizwan, A., Alotaibi, A.A., Althobaiti, T., Arshad, K., Assaleh, K.,& Ramzan, N.: Novel ensemble algorithm for multiple activity recognition in elderly people exploiting ubiquitous sensing devices. IEEE Sens. J. **21**(16), 18214–18221 (2021)
18. Yao, H., Wang, Z., Wu, Y., Zhang, Y., Miao, K., Cui, M., Ao, T., Zhang, J., Ban, D., Zheng, H.: Intelligent sound monitoring and identification system combining triboelectric nanogenerator-based selfpowered sensor with deep learning technique. Adv. Funct. Mater. **32**(15), 2112155 (2022)
19. Gunasekeran, D.V., Tseng, R.M.W.W., Tham, Y.C., Wong, T.Y.: Applications of digital health for public health responses to COVID-19: a systematic scoping review of artificial intelligence, telehealth and related technologies. NPJ Digital Med. **4**(1), 40 (2021)
20. Wu, Y., Yang, Y., Nishiura, H., Saitoh, M.: Deep learning for epidemiological predictions. In: Proceedings of the 41st International ACM SIGIR Conference on Research and Development in Information Retrieval, pp. 1085–1088 (2018)
21. Chintalapudi, N., Battineni, G., Amenta, F.: Sentimental analysis of COVID-19 tweets using deep learning models. Infect. Disease Rep. **13**(2), 329–339 (2021)
22. Bat-Erdene, B.I., Zheng, H., Son, S.H., Lee, J.Y.: Deep learning-based prediction of heart failure rehospitalization during 6, 12, 24-month follow-ups in patients with acute myocardial infarction. Health Inf. J. **28**(2), 14604582221101528 (2022)
23. Tougui, I., Jilbab, A., El Mhamdi, J.: Impact of the choice of cross-validation techniques on the results of machine learning-based diagnostic applications. Healthcare Inf. Res. **27**(3), 189–199 (2021)
24. Du, J., Xiang, Y., Sankaranarayanapillai, M., Zhang, M., Wang, J., Si, Y., Pham, H.A., Xu, H., Chen, Y., Tao, C.: Extracting postmarketing adverse events from safety reports in the vaccine adverse event reporting system (VAERS) using deep learning. J. Am. Med. Inform. Assoc. **28**(7), 1393–1400 (2021)
25. Zhao, T., Zhang, J., Wang, Z., Alturki, R.: An improved deep learning mechanism for EEG recognition in sports health informatics. Neural Comput. Appl. **35**, 14577–14589 (2021)
26. Rath, A., Mishra, D., Panda, G., Satapathy, S.C.: Heart disease detection using deep learning methods from imbalanced ECG samples. Biomed. Signal Process. Control **68**, 102820 (2021)
27. Akbulut, S., Cicek, I.B., Colak, C.: Classification of breast cancer on the strength of potential risk factors with boosting models: a public health informatics application. Med. Bull. Haseki/Haseki Tip Bulteni **60**(3), 196–203 (2022)
28. Bansal, K., Batla, R.K., Kumar, Y., Shafi, J.: Artificial intelligence techniques in health informatics for oral cancer detection. In: Connected e-Health: Integrated IoT and Cloud Computing, pp. 255–279. Springer International Publishing (2022)
29. Luo, X., Li, J., Chen, M., Yang, X., Li, X.: Ophthalmic disease detection via deep learning with a novel mixture loss function. IEEE J. Biomed. Health Inform. **25**(9), 3332–3339 (2021)

Advances in Deep Learning for the Detection of Alzheimer's Disease Using MRI—A Review

S. Hariharan and Rashi Agarwal

Abstract Magnetic Resonance Imaging (MRI) has played a vital role in comprehending brain functionalities and is a clinical tool in diagnosing neuro disorders like Alzheimer's Disease (AD), Parkinson's disease, and schizophrenia. Concurrently, massive amounts of data are generated beyond the capability of traditional data processing techniques. Analyzing these complex, high-dimensional data needs intelligent algorithms. Deep Learning technology has demonstrated high capability accuracy in image processing, natural language processing, object detection, and drug discovery. It learns features from data using backpropagation and changes its internal parameters to finally segment and classify an object. Similarly, it depends on the dataset for constructing the model. Several datasets exist to cater to the neuroimaging community for research advancements. fMRI is a subset of MRI technology that holds much promise in identifying neuro disorders, and deep learning technology has assisted in solving these complicated systems. This chapter discusses the latest works in the field of deep learning-assisted MRI identification of AD.

1 Introduction

The AD is one of the most common degenerative neurological disorders affecting millions worldwide. Several factors are responsible for causing the disease. Age is one of the prime factors as it accounts for nearly 26% of the total cases [1]. One of the issues faced by the researchers is that cerebral atrophy is visible in aging. The microscopic examination reveals that there are structural changes in the cortex region of the brain. The essential pathological hallmarks of AD are the presence

S. Hariharan
University of Madras, Chennai, India
e-mail: hariharansnair77@gmail.com

R. Agarwal (✉)
Department of Computer Science and Engineering, Harcourt Butler Technical University, Kanpur, India
e-mail: dr.rashiagrawal@gmail.com

D. P. Acharjya and K. Ma (eds.), *Computational Intelligence in Healthcare Informatics*, Studies in Computational Intelligence 1132,
https://doi.org/10.1007/978-981-99-8853-2_22

of many neurofibrillary tangles and senile plaques [2]. AD tends to preferentially develop on specific neurons responsible for learning and recalling past experiences from memory. Senile plaques, another AD indicator, are polymorphous beta-amyloid proteins derived from a precursor molecule produced by the brain [3]. The senile plaques and neurofibrillary tangles increase with aging and are considerably lower than in people with AD.

Family history is the second most important risk factor for AD after aging. The mutations in amyloid protein precursor, presenilin-1 and 2, are responsible for the early onset of AD, and gene polymorphism of the apolipoprotein gene results in the late beginning of AD [4]. Environmental factors also play a role in triggering AD. AD involves mild inflammation that can lead to tissue damage in the brain. The other reasons leading to AD are head injury, poor education, and cerebrovascular disease, and specific studies show that treating hypertension and other vascular disorders coupled with a healthy diet can help prevent AD to a certain extent [5].

Timely AD diagnosis requires various clinical assessment methods like mini-mental state examination, clinical dementia rating, and imaging technologies like MRI technology and positron emission tomography [6]. In particular, MRI imaging technologies are gaining relevance as they give medical practitioners detailed pictures of the body's organs and provide 3-D images of the organs. The primary objective of this chapter is to organize the latest works in the field of deep learning-assisted MRI identification of AD.

The rest of the chapter is organized as follows. Section 2 discusses the advances in neuroimaging techniques relevant to AD detection, followed by deep learning techniques used in neuroimaging in Sect. 3. AD datasets are described in Sect. 4 followed by software details in Sect. 5. Finally, the chapter is concluded in Sect. 6.

2 Neuroimaging Techniques

Neuroimaging is a discipline that studies the structure and function of the nervous system using imaging technology and where brain images can be obtained in a non-invasive way. It explores a series of mechanisms, such as cognition, information processing, and brain changes in the pathological state. Neuroimaging has developed rapidly in recent years and become a powerful tool for medical research and diagnosis. With the increasing prevalence of neurological diseases, higher requirements have been put forward for neuroimaging technology and subsequent data analysis, and many advances have been made in this field. The nervous system of humans is imaged structurally or functionally using many different techniques. The MRI and functional MRI (fMRI) are widely used in neuroimaging [7].

The MRI is a diagnostic process that produces comprehensive images of organs and structures within the body using a considerable magnet, radio frequencies, and a computer. MRI does not employ ionizing radiation, unlike X-rays or computed tomography. In MRI, together with radio waves, the magnetic field affects the natural arrangement of hydrogen atoms in the body. A scanner's pulses of radio waves knock

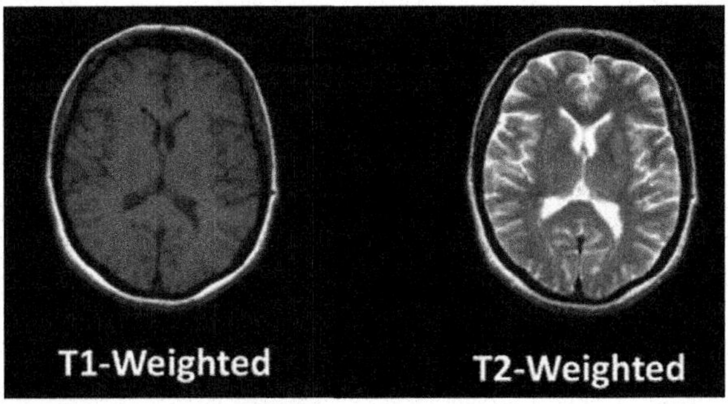

Fig. 1 T1 and T2 weighted images

the nuclei in your scraps out of their birthplace. The nuclei send radio signals when they realign back into their rightful location. A computer receives these signals, analyzes, and turns them into a 2-D image of the body organ under examination.

The magnetic characteristics of atomic nuclei are used in MRI. A high, uniform external magnetic field is used to align the protons that are ordinarily randomly orientated within the water nuclei of the tissue being studied. This alignment is perturbed or disrupted after the injection of an external Radio Frequency (RF) energy. The nuclei return to their resting alignment through numerous relaxation processes and emit RF energy. The emitted signals are measured after a given interval following the initial RF. The Fourier transform converts the frequency information in the signal from each place in the scanned plane to matching intensity levels. These are then represented as shades of gray in a pixel matrix arrangement. Different sorts of images are produced by altering the sequence of RF pulses applied and captured. The time between subsequent pulse sequences delivered to the same slice is called the Repetition Time (RT). The time between the RF pulse's delivery and the echo signal's reception is referred to as the Time to Echo (TE) [8].

Tissue is distinguished by two relaxation durations—T1 and T2. The time constant T1 controls the rate at which stimulated protons return to equilibrium. It is the time required for spinning protons to realign with the external magnetic field. T2 is the temporal constant that governs how quickly excited protons attain equilibrium or fall out of phase with one another. It is the time spinning protons take to lose phase coherence among nuclei spinning perpendicular to the main field. For example, a T1 and T2 weighted image is shown in Fig. 1. The T1 weighted imaging shows that Cerebrospinal fluid (CSF) is dark, while T2 weighted imaging shows bright.

The fMRI is used to pinpoint the exact spot of the brain where a specific function, such as speech or memory, takes place. Although the primary areas of the brain where such activities occur are known, the precise placement may differ from person to person. While having fMRI of the brain, you will be asked to execute a specific task, such as reciting the pledge of allegiance. Doctors can arrange surgery or other treatments for a particular condition of the brain by establishing the exact location of the functional center in the brain. The significant advantage is that MRI and fMRI do not employ radiation like X-rays, computed tomography, and positron emission tomography scans [9].

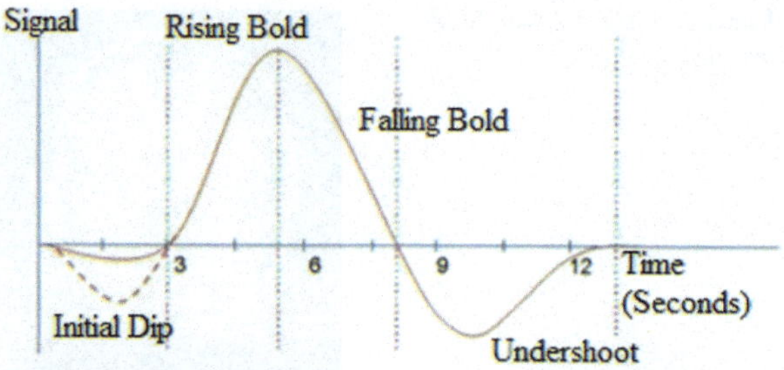

Fig. 2 Representation of hemodynamic response functions

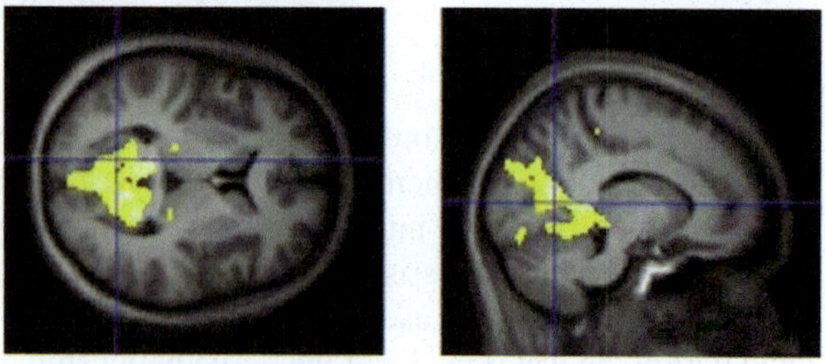

Fig. 3 A sample appearance of fMRI brain image

A person at rest has a generally accepted oxygenated to deoxygenated hemoglobin ratio. Oxygen is needed by the activated brain area, which leads to a reduction in oxygenated hemoglobin present there. Oxygenated and deoxygenated hemoglobin ratios are thereby disturbed. An MRI that employs Blood Oxygen Level Dependent (BOLD) imaging based on these principles is called fMRI.Ã¢â,¬Â¯Measuring the level of homogeneity within a magnetic field for a specified volume is called T2*. For example, when fingers are tapped, the brain is expected to become activated. Initially, blood oxygen needs to be removed from the brain for activation. MRI is affected by episodes of low oxygenation, resulting in an initial drop in the signal. As a result of activating the brain part, the blood supply system creates a blood influx, increasing BOLD signals. When the task is over, the oxygenation and MRI signal drop since the activation is removed. It slowly returns to baseline, representing the resting state of blood oxygenated by deoxygenated blood and the MRI signal.Ã¢â,¬Â¯Hemodynamic response functions are measured this way, and the graph is shown in Fig. 2.

Further, the estimation of the hemodynamic response function for that area is carried out very carefully. It is relatively well connected with local field potentials and local neuronal activity in that area. Figure 3 depicts a sample fMRI brain image.

3 Machine Learning

Machine learning is an exciting field of artificial intelligence that finds application in wide-ranging areas such as medical diagnosis [10, 11], bankruptcy prediction [12, 13], anomaly detection [14, 15], self-driving cars, and product recommendation. In

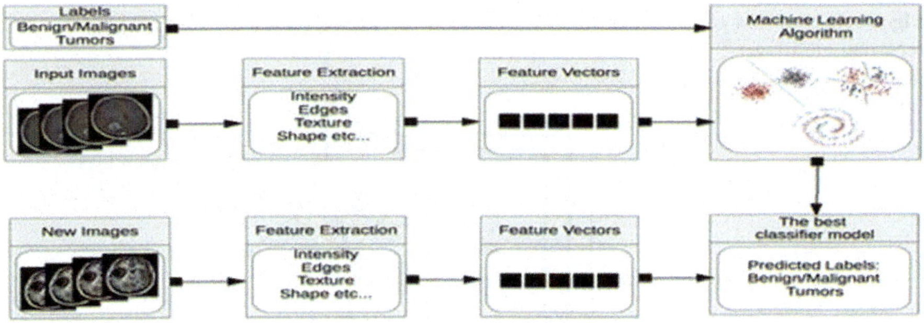

Fig. 4 An iterative machine learning model for prediction

medical diagnosis, computers using the machine learning algorithm can interpret medical images without human interpretation, thus reducing the interpretation time and improving accuracy. Machine learning involves subjecting a massive amount of data to the machine learning algorithm, from which the computer can learn the pattern and use this 'knowledge' derived to identify similar patterns in another set of instances. Figure 4 shows a typical machine learning design.

Machine learning methods include support vector machines, decision trees, networks, k-nearest neighbors, and deep learning. Of all these, deep learning methods are extensively used in the field of neuroimaging for the detection of AD. The various types of deep learning methodologies discussed here are the Recurrent Neural Network (RNN), Deep Belief Network (DBN), Probabilistic Neural Network (PNN), and the Convolutional Neural Network (CNN). Even though various deep learning architectures like RNN, DBN, PNN, and CNN are used to identify neural disorders, CNN is widely applicable.

3.1 Recurrent Neural Network

The RNN is a powerful and robust network because of its internal memory. The RNN has become more potent because of increased computer power that deals with massive data and the advent of Long Short-Term Memory (LSTM). These networks can recall critical details about their input, allowing them to accurately anticipate prediction. Compared to other algorithms, RNNs can acquire a far more profound grasp of a sequence and its context. The recurrent multilayer perceptron shown in Fig. 5 is a type of RNN design.

It consists of a linear input layer, one or more hidden layers, and an output layer. Each hidden layer's neurons are completely recurrent. It means that each neuron is fed by the outputs of every other neuron. All the connections have a weight that can be adjusted and a length that can be changed. The network's training process entails adjusting all weights. The buried layer neuron and the output neuron have

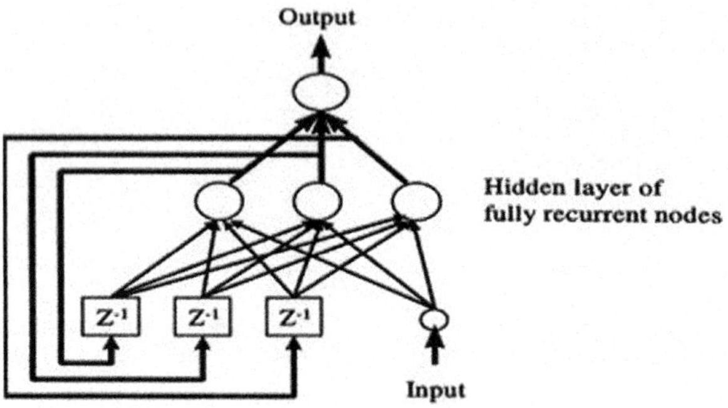

Fig. 5 An overview of recurrent multilayer perceptron

a non-linear interaction with their inputs through a bipolar sigmoid transformation. The Extended Kalman Filter (EKF)-based training is a powerful method for RNN training [16]. It adjusts the network's weights pattern by pattern, storing crucial training data in approximate error covariance matrices and providing individually modified updates for each network's weights.

The RNN can detect small linear and non-linear changes in the signal. In prior research on epileptic seizure prediction [17], remarkable signal identification abilities are displayed. Using EEG recordings and commercially available computer hardware, RNNs train raw EEG data and their wavelet-filtered sub-bands to differentiate between groups according to disease stages.

The LSTM networks are a type of RNN that can store more information. These are used as the building blocks of the layers of an RNN. It assigns weights to data, allowing RNNs to either let new lead in, forget it, or give it enough relevance to affect the output. RNNs can recall inputs for a long time because of LSTMs. This is because LSTMs store information in memory similar to that of a computer. The LSTM can read, write, and delete data from its memory. There are three gates in an LSTM: input, forget, and output. These gates are analogous in the form of sigmoids, which means they range from zero to one. Because they are analog, they can perform backpropagation.

A deep learning model based on RNNs was built to learn the informative representation and temporal dynamics of longitudinal cognitive measures from individual participants and to combine them with baseline hippocampal MRI to develop a predictive model of AD dementia progression. An LSTM autoencoder is used to learn a compact and informative representation from longitudinal cognitive tests to predict the progression from MCI to AD dementia. Without making explicit assumptions about the longitudinal process underlying the tests, these representations could encapsulate the temporal dynamics of longitudinal cognitive measures and characterize the progression trajectory of MCI participants. Cox regression assesses the chance of MCI participants progressing to AD dementia. The proposed model is applied to a large cohort obtained from the Alzheimer's Disease Neuroimaging Initiative (ADNI), and the experimental results indicate that the proposed model can achieve favorable

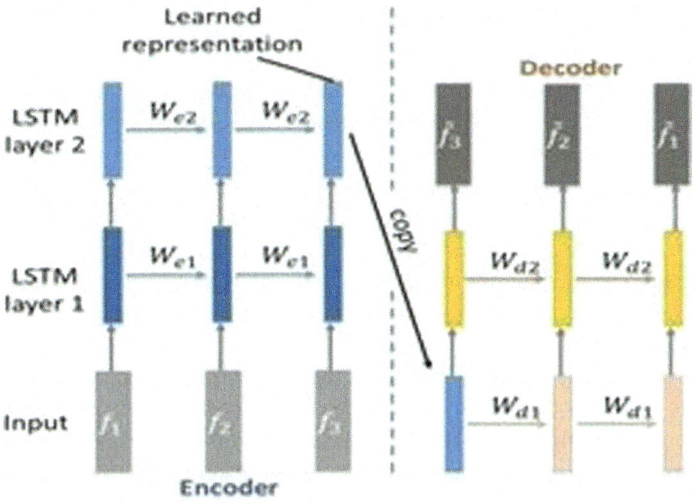

Fig. 6 LSTM autoencoder for longitudinal feature extraction

prognostic performance. At the same time, cognitive and imaging-based measures may provide complementary information for prognosis.

Figure 6 illustrates the design of the LSTM autoencoder used in an investigation. It is composed of two components: an encoder and a decoder. The encoder accepts multiple time point input data and manages the encoding of input measures and their temporal dynamics between succeeding time points. The decoder recreates the input measures at various time points in reverse order, using the encoder's learned representations. While the network is tuned to minimize the difference between the reconstructed and input measures, the encoder's learned model is expected to define the input longitudinal measurements' overall cognitive performance and dynamics.

The autoencoder comprises two LSTM layers, one for the encoder and one for the decoder. The notion f_t and $\hat{f}_t$ denote the input and reconstructed cognitive measures at time t; $t = 1, 2, 3$ for demonstration, W denotes the trainable parameters of the encoder's ith LSTM layer, and W_{ei} indicates the trainable parameters of the decoder's ith LSTM layer. Within a single LSTM layer, the trainable parameters are those associated with the forget gate, input gate, cell state, and hidden state. The Euclidean distance between the reconstructed and input measures is employed as the objective function to maximize the trainable parameters. The number of LSTM layers was set to provide generalizability with a minimum number of trainable parameters.

The autoencoder for cognitive assessments was developed in this study using longitudinal cognitive tests from the ADNI-1 cohort. After obtaining, the autoencoder was applied to all MCI patients from the ADNI1 and GO&2 cohorts to extract latent cognitive traits from longitudinal cognitive assessments, which were then employed in the subsequent predictive analysis.

3.2 Probabilistic Neural Network

The PNN is made up of four layers. The layers are the input, summation, pattern, and output layers. This is a feed-forward neural network that works on a single-pass algorithm. It is based on the statistical principles derived from the non-parametric kernel-based estimation of the probability density function [18]. Each neuron in the input layer represents a predictor variable. In categorical variables, $(N - 1)$ neurons are used when there is N number of categories. It standardizes the range of the values by subtracting the median and dividing by the interquartile range. Then, the input neurons feed the importance to each neuron in the hidden layer.

The pattern layer contains one neuron for each case in the training dataset. It calculates the probability density function of the input vector using the parent window. It stores the values of the predictor variables for the case along with the target value. A hidden neuron computes the Euclidean distance of the test case from the neuron's center point and then applies the radial basis function kernel function using the sigma values. The output from the pattern layer is defined using Eq. 1, where $g_j(y)$ refers to the output of the jth pattern neuron, P refers to probability density function, z_{jk} refers to Euclidean distance of the jth vector to the kth vector, n_j refers to feature vectors in the input space, L refers to the particular hidden neuron vector, and μ refers to mean of the class of vector:

$$g_j(y) = \frac{1}{\mu P(2\pi)^{P/2}} \sum_{K=L}^{n_j} e^{\frac{(z_{jk}-1)}{\mu^2}} \tag{1}$$

For PNN, there is one pattern neuron for each category of the target variable. The actual target category of each training case is stored with each hidden neuron; the weighted value coming out of a hidden neuron is fed only to the pattern neuron that corresponds to the hidden neuron's category. The pattern neurons add the values for the class they represent. The weighted output corresponding to the neurons of a similar class is averaged using Eq. 2 [19], where w_{jk} represents the proportion between the between-class variance and within-class variance. The notion M_j represents the total class. It is to be noted that the probability density function of each class is derived using a Gaussian kernel, which is used in the Bayesian rule to perform the classification. The averaging of the weighted output corresponding to the neurons of a similar class is done using Eq. 2, where $P_j(y)$ refers to probability density function of the jth neuron, and z_j refers to the Euclidean distance of the jth vector to the central neuron point:

$$P_j(y) = \sum_{k=l}^{M_j} e^{\frac{z_j-1}{\mu^2}} w_{jk} \tag{2}$$

The output layer compares the weighted votes for each target category accumulated in the pattern layer and uses the largest vote to predict the target category. The architecture of a typical PNN is depicted in Fig. 7.

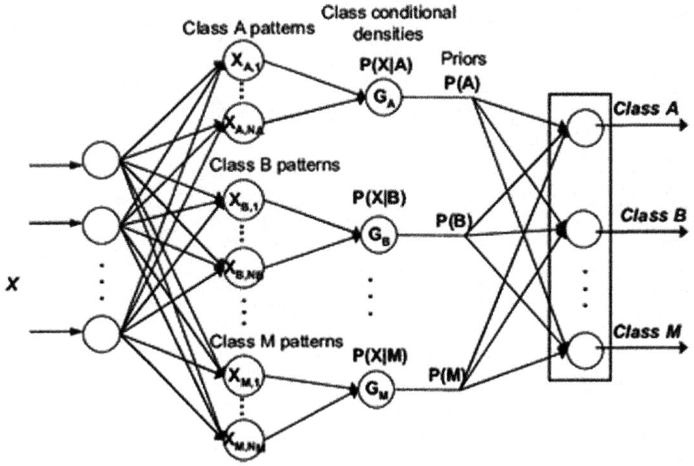

Fig. 7 The typical architecture of a PNN with four layers

The overall algorithm of a PNN works by inputting the file containing the exemplar vectors and class numbers. Sort these into k-sets, each containing one class of vectors. Create a Gaussian function centered on each exemplar vector in set k, then define the cumulative Gaussian output function for each k. Read the input vector and assign the Gaussian function according to their performance in each category. For each cluster of hidden nodes, compute all Gaussian functional, useful values at the remote nodes. Feed all these values from the hidden node cluster to the cluster's single output node. Add all the inputs for each category output node and multiply them by a constant. Determine the sum of all the valuable values added to the output nodes. For the purpose of AD detection, PNN is better compared to other classifiers like SVM and k-nearest neighbor algorithms [20]. It was found that the accuracy of PNN was 85% as compared to the 70% achieved by SVM and 77% performed by KNN.

3.3 Convolutional Neural Network

The CNN is a deep learning algorithm analogous to the traditional Artificial Neural Network (ANN). It is composed of neurons that self-optimize through learning. The CNNs are usually employed in image classification and computer vision. Figure 8 presents the typical CNN architecture.

The CNN architecture consists of a convolutional layer, a pooling layer, and a fully connected layer. The convolutional layer is the core building block of a CNN, which performs the maximum amount of computation. It comprises input data, a filter, and a feature map. The filters are of smaller spatial dimensionality compared to the input. The filter convolves (dot product) with the input data to produce a 2-D activation map,

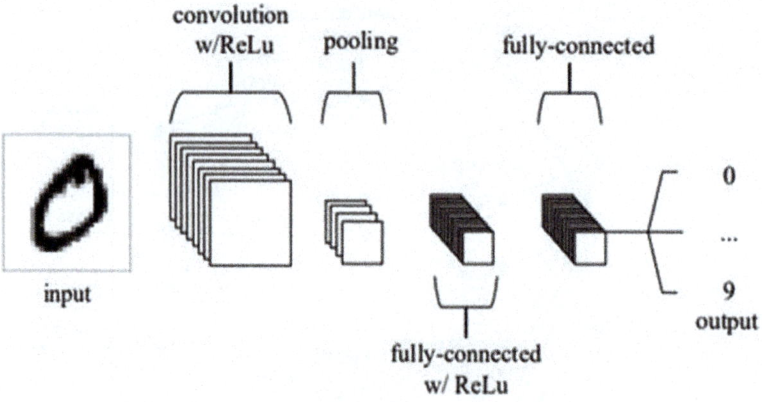

Fig. 8 The typical architecture of a CNN with three layers

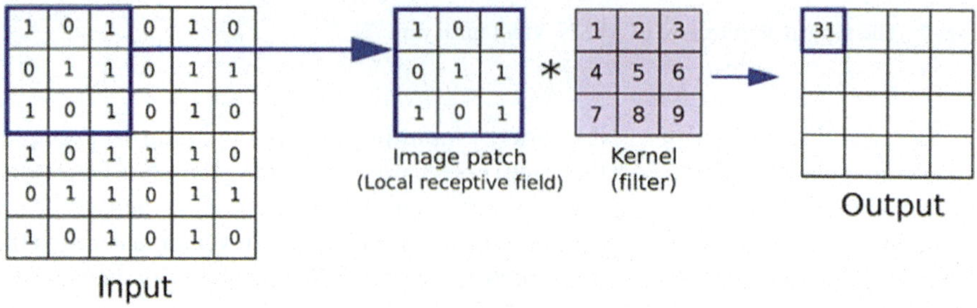

Fig. 9 The performance illustration of a convolution operation on an image

as shown in Fig. 9. Each kernel results in an equivalent activation map, which will be layered along the depth dimension to form the total output volume. For example, an image of size (64 × 64 × 3) and the receptive field size (64 × 64) would result in 108 weights on each neuron within the convolutional layer. It is much less as compared to ANN [21]. Convolutional layers can reduce complexity by optimizing their output through three hyperparameters: depth, stride, and setting zero-padding.

The depth of the output volume can be customized by the quantum of neurons in the layer corresponding to the equivalent area of the input. This is visible in the ANNs, too, where all of the neurons in the hidden layer are directly connected to every single neuron in the preceding layer, and so on. Depleting this hyperparameter can reduce the quantum of neurons but, on the other hand, will reduce the model's performance overall. The stride can be configured to create depth around the spatial dimensionality of the input to place the receptive field. If stride is 1, it will result in an extensively overlapped receptive field and substantial activations. On the other hand, if the stride is set to a sizeable number, it will diminish the extent of overlapping and result in poor resolution output. Zero-padding is the process of padding the input border to decide the dimensionality of the output volumes. All these methods alter the spatial dimensionality of the output as defined in Eq. 3, where V gives the input volume size, R is the receptive field size, Z is the amount of zero-padding, S is the stride used in the process, and O refers to the output shape:

$$O = (V - R) + 2\frac{Z}{S} + 1 \quad (3)$$

A non-integer result indicates that the stride chosen is inappropriate as the neurons cannot configure precisely as per the input. As the real-world images are of exceptionally high dimensionality, the model will be substantially large despite the efforts. Parameter sharing could be used to reduce the parameters. It assumes that if one region feature is useful to compute at a set spatial region, it will likely be helpful in another region. The restraint of having each activation map within the requisite output volume of the same weights and bias will result in a marked depletion in the number of parameters resulting from the convolutional layer. Subsequently, as the backpropagation happens, each neuron in the output stage will give the overall gradient and update a single cluster of weights instead of each and every single weight.

The role of pooling layers is to reduce the representation's dimensionality, which will further reduce the number of parameters. Finally, it results in a less complex model. The pooling layer operates on each activation map in the input and scales its dimensionality using the 'MAX' function. The popular pooling methods are average and max-pooling, as shown in Fig. 10. The most popular form of pooling is the max-pool with kernels of a dimensionality of (2 × 2). Further, a stride of 2 along the spatial dimensions of the input is applied. This reduces the size of the activation map to 25% of the original size. Usually, the stride and filters of the pooling layers are set to (2 × 2), which will allow the layer to extend through the entirety of the spatial dimensionality of the input. Moreover, overlapping pooling may be used, where the stride is set to 2 with a kernel size set to 3. Because of the destructive nature of pooling, having a kernel size above 3 will reduce the model's efficiency.

The fully connected layer, as shown in Fig. 11, is analogous to how neurons are configured in a CNN. Therefore, each node in a fully connected layer is directly related to every node in both the previous and the succeeding layer. Every node in the last frame of the pooling layer is observed to be connected as a vector to the first layer of the fully connected layer. The disadvantage of a fully connected

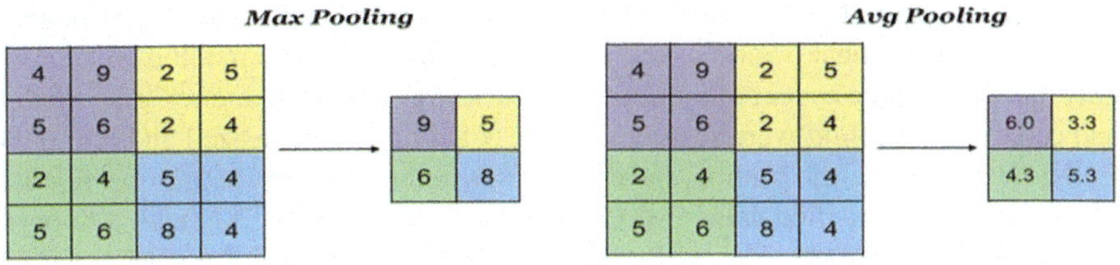

Fig. 10 Illustration of the average and max-pooling operation on the pixels

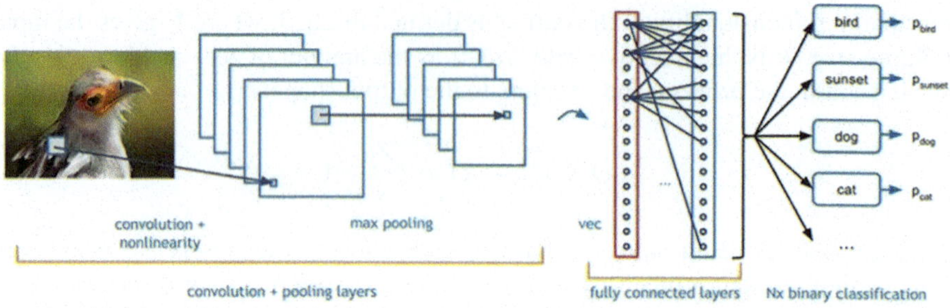

Fig. 11 Fully connected layer in CNN operation in image classification

layer is that it has many parameters requiring complex computations. Therefore, the effort is to eradicate the number of nodes and connections. The eliminated nodes and links can be performed by using the dropout technique. For example, LeNet and AlexNet designed a deep and wide network while keeping the computational complex constant. The essence of the CNN network, which is the convolution, is when the non-linearity and pooling layer are introduced.

CNN has found immense popularity and utility in the field of the detection of AD using various neuroimaging methods. In the literature, the authors have used the ADNI dataset for performing a 4-way classification as healthy, mild cognitive impairment, late cognitive impairment, and AD. An MRI scan is used in the experimental setup as a 3-D Nifti format. SPM-8 tool was used for data preprocessing like skull stripping and gray matter segmentation by bias correction and spatial normalization. This data is fed to the CNN, where various convolutional layers act on the feature maps. For this study, the authors have deployed a 22-layer deep CNN, GoogLeNet, with an inception module. It uses the Hebbian principle that results in the overall reduction of the parameters. The ResNet is marked by skip connections, which are added to the convolutional layer as a bypass. Here, the layers learn from a function $f(x)$ and function $f(x) + x$. It makes the convergence faster. Moreover, the skip connections ensure that the problem of vanishing gradient is overcome during backpropagation. The experiment results illustrated that GoogLeNet achieved an accuracy of 98.8%

In another study, the authors used 3127 MRI samples of $3 - TT - 1$ weighted from the ADNI dataset. For the study, 85:15 ratios were used for training and testing, respectively. SPSS was used to compare the results using the χ^2 test and ANOVA. The data was normalized using an SPMI-12 to ensure that each voxel corresponded to the same anatomical position. The voxel-based morphometry method created the gray matter template as part of the data preprocessing. The white matter, gray matter, and cerebral spine fluid were segmented. The 3-D images are resized and fed to the CNN. In this study, the authors have used 3-D kernels for extracting the feature. The dataset was classified using 2-D CNN, 3-D CNN, and 3-D CNN with a Support Vector Machine (SVM). The performance of ternary classification's accuracy for 2D-CNN, 3D-CNN, and 3D-CNN with SVM was $82.57 \pm 7.35\%$, $89.76 \pm 8.67\%$, and 95.74 ± 2.31%, respectively. According to accuracy, 3D-CNN-SVM yielded the best algorithm for binary and ternary classification of AD [23].

Similarly, the authors employed a novice variation of transfer learning in CNN in another study. The dataset used was ADNI, which includes 81 NC, 69 AD, and 38 MCI for training and 81 NC, 68 AD, and 37 MCI for testing. The generated off-the-shelf features have been used to get features of each single slice. The developed CNN features are dimension (609 × 512), which are reduced using principal component analysis and t-distributed stochastic neighbor embedding. Later, performance parameters are analyzed from the classifier model by applying the feature selection process. The authors selected the Naive Bayes and the KNN classifiers regarding the classifier. The study established that the optimal accuracy of an AlexNet-trained classifier with KNN was better than the scratch-trained CNN with KNN. Moreover, it is also found that transferring learning parameters from trained CNN has distinguishing characteristics for training the classifier. Secondly, the CNN feature-trained classifier performs better than CNN networks if the feature transformation, selection, and classification are done wisely [24].

3.4 Deep Belief Network

The DBN is an algorithm in deep learning. It is an effective method of solving problems from the neural network with deep layers, such as low velocity and the overfitting phenomenon in learning. Restricted Boltzmann Machine (RBM) is a stochastic neural network that can learn a distribution over its set of inputs. The network generally consists of one layer of binary-valued visible neurons and another layer of Boolean hidden units. Besides, no connections exist between neurons in the same layer, but full connections exist between neurons in different layers. It tries to learn a probability distribution from the visible layer to the hidden layer so that its configuration can exhibit desirable properties. The learning process is achieved through an energy function [25]. Given the visible units v_i, hidden units h_j, and their connection weights $W_(ij)$ of size $(n_v \times n_h)$, the offset a_i for v_i, and the bias weight b_j for h_j, and the energy function $E(v, h)$ of a certain configuration are defined using (4)

$$E(v, h) = -\sum_{i=1}^{n_v} a_i v_i - \sum_{j=1}^{n_h} b_j h_j - \sum_{i=1}^{n_v} \sum_{j=1}^{n_h} h_j w_{ij} v_i. \tag{4}$$

Then, the probability distributions $P(v, h)$ over visible and hidden layers can be defined using the energy function as in Eq. 5, where Z as defined in Eq. 6 is the partition function. Consequently, the individual activation probability of h_i given v_i, or v_i given h_i, can be deduced using Eqs. 7 and 8, respectively:

$$P(v, h) = \frac{1}{Z} e^{-E(v,h)} \tag{5}$$

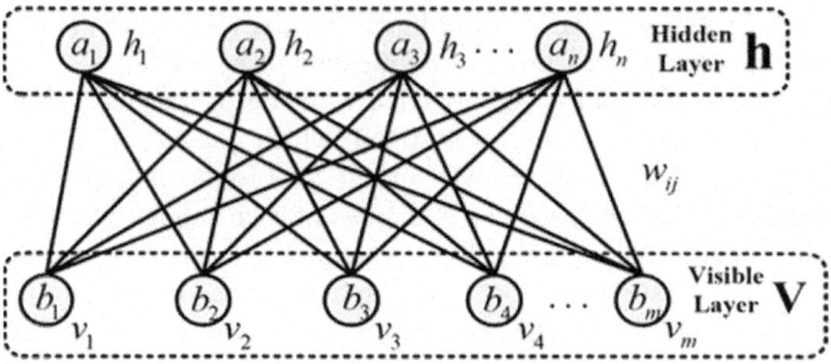

Fig. 12 The architecture of DBN implemented using RBM

$$Z = \sum_{v} \sum_{h} e^{-E(v,h)} \tag{6}$$

$$P(v_i = \frac{1}{h}) = sigm\left(a_i + \sum_{j=1}^{n_h} W_{i,j} h_j\right) \tag{7}$$

$$P(h_i = \frac{1}{v}) = sigm\left(b_i + \sum_{i=1}^{n_F} W_{i,j} v_i\right) \tag{8}$$

It is to be noted that $sigm$ denotes the logistic sigmoid function; with the aid of certain pre-training principles, the unknown RBM parameters, i.e., W, a, and b, can be determined in an unsupervised manner, stacking the binary RBM layer by layer hierarchically, and adding a logistic regression to the end of the stacks. The first layer is pre-trained as an independent RBM, with training datasets as inputs. Once the parameters of the first hidden layer are computed, the output of the hidden layer found on the layer below is taken as the input of the hidden layer above, as illustrated in Fig. 12. Then, these two hidden layers can be considered a new RBM and are thus trained similarly. The architecture of a DBN is shown in Fig. 12.

DBNs have found moderate usage in identifying AD using the MRI modality. In the literature, the authors proposed a classification method based on structural modalities using a DBN, which was compared with SVM for performance. The data was preprocessed using voxel-based morphometry and feature extraction using a 3-D mask to extract voxel values. The OASIS dataset is used for this experimental setup. This study uses two feature vectors from the voxel's mean and standard deviation (MSD) value. The voxel value is calculated from gray matter segmentation. As the epoch increases, the DBN accuracy increases. Classification based on a voxel value-based feature vector in the case of a DBN has an accuracy of 91%, which is better than the 62% accuracy of SVM. However, the MSD-based classification has an SVM accuracy of 86%, giving better results than the DBN accuracy of 73% [26].

4 Dataset Description

MRI is an effective tool in identifying the onset of AD. It aids in visualizing the volume shrinkage in the brain's hippocampus region. The shrinkage in the hippocampal region is a characteristic feature of AD. The aggregation and creation of data repositories are critical in building successful machine learning models. The neuro datasets that are primarily created and made available to assist the research community in making advancements in the field of neurosciences are presented briefly in this section.

4.1 ADNI

ADNI is a secured data repository for sharing imaging, genomic, and biomarker data related to AD. The fundamental surmise in ADNI is that AD is onset with the accumulated aggregation of amyloid-β in the brain. It is detected in cerebral spinal fluid and Positron Emission Tomography (PET) imaging. This is followed by the rise of tau species seen in the cerebral spinal fluid and, finally, the destruction of the neuron. One of the goals of ADNI is to capture the complex relationship between the various clinical and chemical biomarkers as the sample progresses from normal to AD [27].

The ADNI data's generalization is impossible as the sample does not include patients with cancer, heart failure, and substance abuse. Collaborations and engagements with academicians doing population-based studies have been initiated to ascertain the efficacy of the ADNI data on the general population. Another limitation of the dataset is the data selectivity because ADNI captures only data from 55-90 years. There are adequate reasons to suggest that AD can onset even at a much younger age. Moreover, the ADNI data cover only a few necessary tests like lifestyle information, ECG, magnetoencephalography, and magnetic resonance spectroscopy [28].

4.2 OASIS

The Open Access Series of Imaging Studies (OASIS) is a collection of brain MRI datasets distributed freely for academic and research purposes. Initially, it started with a collection of MRI data of 400 individuals. The sample had data from individuals who have dementia and healthy individuals. The robustness of various models and algorithms can be validated as the data collected is across a broad spectrum of ages. Moreover, since the dataset provides a common reference point for images, it can be used as a benchmark for various analytic techniques in comparison. Multiple procedures and measures have been adopted to ensure data is readily usable. Strict quality control and screening have ensured that only images that meet the required standards are included. The images provide adequate documentation regarding various parameters and procedures adopted during pre-acquisition. The images are associated with mini-mental state examination scores. Post-processing data required for the users are also provided as part of the package, and raw data for ease of usage. To give data anonymity, all links between the identity of the patient and the associated data are deleted before being provided to the end user. Third-party access has also been provisioned for easy distribution, and there are hardly any restrictions [29].

OASIS-1 involved the cross-sectional collection of data from 416 individuals from the age group of 18 to 96. Among these, 100 individuals were suffering from AD rated using the clinical dementia rating scale. Cross-sectional data has facilitated and validated analytic tools for atlas-based head size normalization, image segmentation, and labeling regions of interest. In some instances, where the filter could not accurately segment the facial features and remove them, the process of atlas registering and hand labeling of the segment was done to remove it. The atlas used here combined young and old targets from 12 samples. Combining templates eliminates the bias of an atlas normalization procedure due to an atrophied brain [30]. OASIS-3 is a compilation of MRI and PET images of more than 1000 individuals. The data was collected for more than 15 years of age group from 42 to 95 years composed of data from normal adults and people in various stages of neuro degradation.

4.3 COBRE

Center for Biomedical Research Excellence (COBRE) shares MRI data from 72 patients suffering from schizophrenia and 75 healthy individuals, all in the age group of 18 to 65 years. The resting fMRI, anatomical MRI and phenotype-type data are released for each individual. In COBRE-I, the primary aim was to study the neural mechanism of the brain, with emphasis on schizophrenia. This was done by integrating various imaging methods with clinical psychological tests and diagnoses. Schizophrenia is distinguished as an abnormality created by the functional, structural connectivity disorder between the cortical and subcortical regions, creating disturbances in the flow of information across the neuron circuits. It is extended to COBRE-II with imaging modalities to study psychosis and mood disorders. It is developed to COBRE-III, focusing on interdisciplinary initiatives across neuro psychiatric and neurological illnesses. The multi-modal core comprises sMRI, dMRI, MEG, spectroscopy, and genetic data [31].

4.4 MIRIAD

Minimal Interval Resonance Imaging in Alzheimer's Disease (MIRIAD) is a dataset of volumetric MRI images of AD and healthy individuals. The primary aim of this study was to explore the possibility of finding a minimal interval over which clinical research could be conducted with the measurement of atrophy from MRI. The additional aim was to identify the optimal time points for MRI measurement that could give different value additions regarding statistical capability. Additionally, the likelihood of replicability technology in the short duration over the varying conditions of variation in the subject's hydration level and scanner fluctuation is explored [32]. Over a period, a lot of research in the field of neuro disease using the MRIAD dataset has been conducted. Besides, CNN is implemented over the MIRIAD dataset for feature extraction.

4.5 ANMerge

AddNeuroMed is an initiative funded by InnoMed. The primary aim was to identify progression biomarkers for AD. Various variables were included: demographics, neuropsychological assessment, MRI, blood plasma, gene expression data, and proteomics. Despite the massive collection of data across 1700 individuals, the dataset lacked the popularity enjoyed by ADNI. On critical analysis, the probable reasons could be that the data was not officially published and the data made available was fraught with errors. To overcome these problems, AddNeuroMed underwent a few modifications and was republished as ANMerge. ANMerge is a preprocessed data cohort with a high degree of usability among the various modalities, and rigorous quality control ensured to reduce the errors [33].

4.6 AIBL

Australian Imaging Biomarkers and Lifestyle (AIBL) study is carried out to discover an analysis that can help detect AD before the visibility of symptoms and to identify lifestyle modification to prevent the onset of the disease. AIBL has more than 1100 individual participants' longitudinal data. According to the literature, the authors have devised a unique multi-modal multi-instance distillation scheme for prediction from the MRI data for progression from mild cognitive impairment to AD. The multi-modal data is composed of MRI and clinical non-imaging data. Instead of using the classical knowledge distillation method of knowledge transfer from the teacher to student mode, KL divergence decreases the difference between the probabilistic estimation. The image is partitioned into 3-D patches. These patches are known as instances and are fed into the network to get the probability estimates used for the student network. The accuracy achieved was 0.735 for AIBL [34].

Further, the authors have used the method of deep CNN to extract the features and classify the patients for AD using the ADNI dataset. The transfer learning techniques are used herein to accommodate the need for many images and handle higher dimensional data. For this experimental setup, a pre-trained CNN model (ResNext101) was used for transfer learning. The images from the AIBL dataset were used as the validation set [35].

5 Software for Neuroimaging

Neuroimaging is primarily a medical field, and many software are available to model brain images to assist doctors. Neuroimaging data are becoming of more interest among the circle of neuroimaging experts. This section talks about the popular software in this field. This information can help developers, experts, and users gain insight and better decision-making in diagnostic and mental health care.

5.1 Statistical Parametric Mapping

Statistical Parametric Mapping (SPM) is a free and open-source software written in MATLAB. It requires a prior installation of MATLAB to function on any Windows, Mac, or Linux 32- or 64-bit platform. It was developed initially for the statistical analysis of positron emission tomography and further improved for general linear models and random field theory. SPM also evolved to support the fMRI, a newly introduced imaging modality with the concept of hemodynamic response and convolution models for serially correlated time series. Over a while, new tools like registration, normalization, and segmentation were developed, which made it possible to apply on structural MRI and paved the way for voxel-based morphometry [36]. Later, SPM was utilized for the study of EEG and MEG data. This resulted in three significant advancements: the analysis of the MEG and EEG analysis of skull maps, electromagnetic source reconstruction based on Bayesian inversion of hierarchical Gaussian process models, and the approach of dynamic causal modeling.

The SPM8 is a variant of SPM that has a library of MATLAB M-files and uses MATLAB MEX for performing computationally heavy performance. The user-friendly software has various GUI functions that help switch between imaging modalities. The software also provides a provision for creating process pipelines to be applied on multiple datasets using the GUI or scripts. SPM is extensively used in the sensor and source analysis of data. In sensor analysis, since the location of the induced response cannot be predicted a priori, the topological inference is used to search over the space for critical response based on Random Field Theory (RFT). RFT gives a method of adjusting the p-values and taking the fact that the sensors are interdependent due to the smoothness of data. The statistical analysis of EEG data involves transforming it into NIfTI image format, and then the SPM analysis is similar to that performed on PET and fMRI data.

The SPM prepares the model using non-linear transformation as the MRI is processed. The individual head meshes are created using the inverse of the spatial deformation field to overcome the poor image quality. For source reconstruction, the SPM should be able to map from the sensor coordinate system to the MRI coordinate system using either a landmark-based co-registration or the surface matching with an iterative closest point algorithm.

In statistical comparison, all the datasets need to be inverted using standard or custom inversion modes. Further, the SPM has a provision for creating summary statistic images, which are a response to the subject in the form of 3-D NIfTI format. Additionally, it also contains several utility functions. The Merge option allows the concatenation of multiple datasets to analyze a subject over numerous sessions. Multi-modal fusion provides the user with the possibility of a fusion of data from various sources like EEG and MEG data simultaneously. To make the SPM generic, the makers have given the facility of developers to add toolboxes. The two main toolboxes available now are MEEGtools and Beamforming Toolbox.

5.2 *Analysis of Functional Neuro Image*

Analysis of Functional Neuro Image (AFNI) is free software made of C programs developed using a Motif 1.2 tool kit running on a Unix to load, store, analyze, and display brain MRI data. The data is stored as 3-D voxels. An open-source software developer can easily add batch functions to the software. The idea behind the development of this software was a utility that could help the researchers explore the datasets in 3-D and superimpose the functional activation maps on data slices in various orientations. The software also gave an added visualization experience while providing user-defined statistical operations [37]. Since it is modular and user-defined, there are plenty of options to control the analysis of data, visualize even the intermediate results of a process, and convert the results into the Talairach Tornoux coordinate system as given in Fig. 13. The brain atlas by Talairach Tornoux is designed to provide a standardized coordinate system to uniquely identify any location within the brain [38].

The second version of the software, which could process data in 4-D (3-D with time), is also released, as shown in Fig. 14. It also incorporated volume registration and activation analysis [39]. Further, the AFNI evolved from regular 3-D processing to triangular cortical surface models. The auxiliary data in AFNI is stored separately from the image values. However, this format makes future expansion easy but creates issues in interoperability. Presently, the software can support multiple formats like DICOM, NIfTI, and AFNI format, which the user can select. AFNI has a collection of batch programs that are useful for visualization.

Various statistical operations can be performed in the software, from simple t-tests to complex ones that give voxel-wise estimates. At the group level, the software can perform various correlation operations on the data parameters, incorporating subject-wise covariates by uploading a table with these data. There is also scope to perform

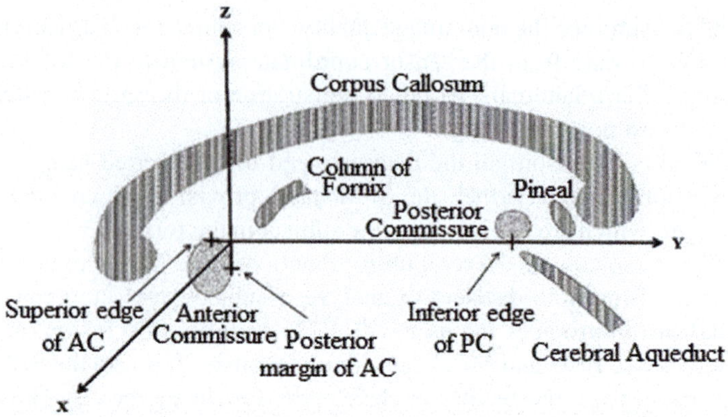

Fig. 13 A representation of Talairach Tornoux coordinate system

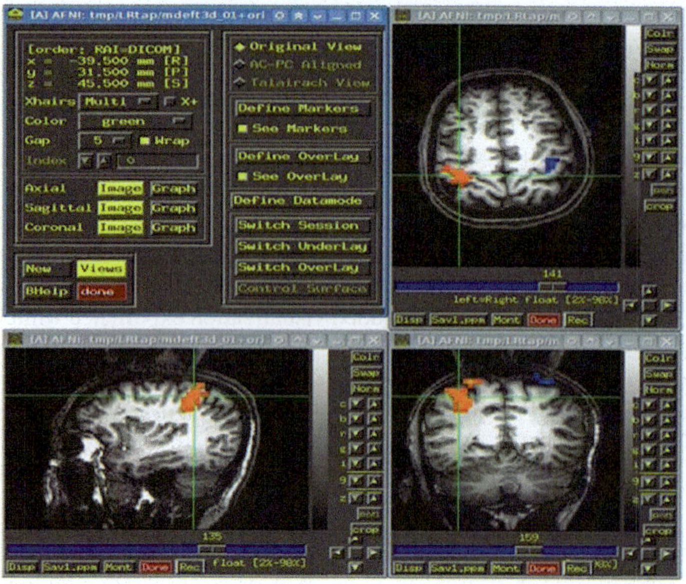

Fig. 14 MRI analysis using the AFNI software

connectivity and network analysis, including provision for rapid data analysis using the InstaCorr module that lets the user load data and preprocesses it, followed by displaying the correlation map of that voxel in respect to other voxels.

5.3 BrainVoyager

The motivation behind BrainVoyager's software development was the development of a Graphical User Interface (GUI) for the Windows operating system for processing data from a 1.5 T Siemens scanner in the field of cortex segmentation, inflation, and flattening. The software worked on the Bonferroni correction instead of the Gaussian random field approach for the statistical thresholding. Over a while, with the improvement, the user could see any point on the reconstructed cortex view getting correlated on the time course and the anatomical slice view. Later, the development of the cross-platform version was initiated by BrainVoyager QX. Further, platform-specific tweaks and library were implemented for any specific data analysis. The software was written in C++, which could be implemented in Mac and Linux.

In the subsequent development phase, the developers focused on topologically accurate segmentation using a robust pipeline for statistical analysis. Another significant development was the analysis due to independent component analysis. Further, the sliding window was implemented. It is because of the combination of the hypothesis and data-driven volume and surface space analysis. Another significant advancement was the processing of fMRI data. The turbo-BrainVoyager was intended to implement fast visualization and statistical tools like general linear modeling, in which the output variable can have an error distribution other than the normal distribution. The real-time mapping was displayed on all the aspects of the cortex, including the folded, inflated, and flattened portions. The further impetus in this field was received in the application of this software in the neurofeedback applications, primarily in Parkinson's and depression-related ailments. Another application was the development of the brain-computer interface for patients suffering from an acute motor disability [40]. The software has acquired the ability to solve the computational analysis of EEG and fMRI combined data and artifact removal. The graphical view of the software is given in Fig. 15.

The 3-D graphics of the software has been implemented using OpenGL. The software also provides analysis of covariance (ANCOVA) analysis for advanced multi-factorial designs and correlation. The ANCOVA analysis is usually carried out to compare cortical thickness maps and fractional anisotropy maps between subjects. The false discovery rate is implemented using dynamic thresholding for error correction in multiple comparisons. The software provides a cluster-threshold plugin that allows modification in various comparisons. This is additional to the FDR in the software and functions on voxel-based thresholding.

5.4 FMRIB Software Library

Functional Magnetic Resonance Imaging of the Brain (FMRIB) Software Library (FSL) is a package primarily for analyzing and statistical operation of the MRI data. The software is available for functional, structural, and diffusion MRI. The

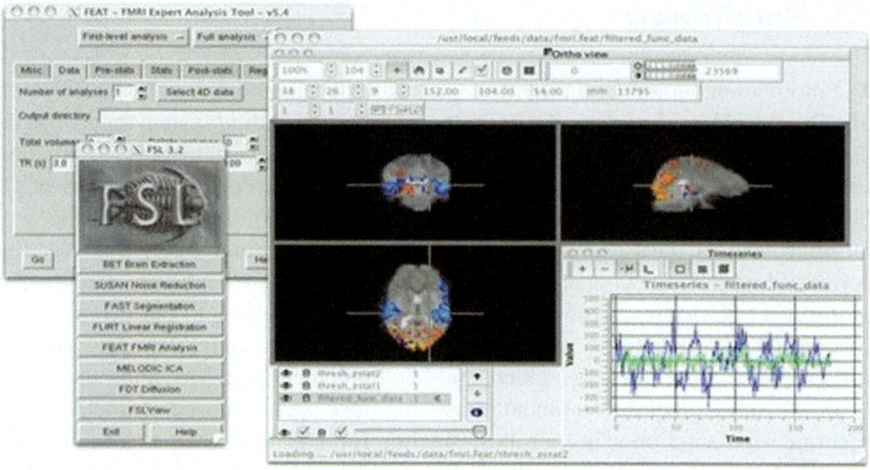

Fig. 15 Graphical view of BrainVoyager displaying the standard plugins [38]

MEDx software was the precursor to the FSL software. It began as plugins for the MEDx software written in TCL. These plugins were used extensively in the lab to understand shortcomings and underwent multiple iterations for improvement. The Unix platform was chosen for its string and file manipulation capabilities. For running on the Windows platform, the users recommend running a virtual Linux machine on Windows.

The software provides adequate GUI functionality so that the middle-level user can process the pipelines. Initially, the FSL software followed the linear registration process during the low-quality scanner images. Besides, it was less affected by artifacts, but with the enhancement in the image acquisition, the FSL shifted toward non-linear registration. The voxel-based morphometry was incorporated with FSL for analysis of change in the gray matter as it has a high sensitivity to change. The fractional anisotropic diffusion studies of white matter use tract-based spatial statistics. The snapshot of the software is given in Fig. 16.

An essential tool in the FSL is the FMRIB utility for geometrically unwarping Echo Planar Image (EPI). The input to the above utility is the EPI, which is distorted due to primarily magnetic inhomogeneity. The inhomogeneity is caused due to varying magnetic susceptibility among the tissues. The inhomogeneity will cause geometric distortion. Field inhomogeneity is measured using a field map, and this information is used to compensate for the distortion. These artifacts are removed by geometrically unwarping the images and applying the cost function masks to ignore the area where signal loss has occurred.

The other essential tool in the FSL is the FMRIB Expert Analysis Tool (FEAT). This is used for preprocessing and statistical analysis of FMRIB data and also helps automate the research to the maximum extent. The entire process of FEAT analysis can take from 5 to 20 min. The output includes a report generated, activation maps,

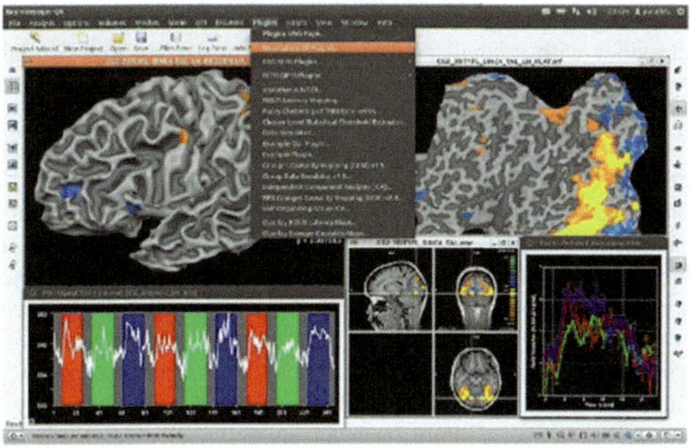

Fig. 16 Graphical view of FSL displaying the overlaid activation [39]

and plots. General linear modeling, or multiple regression, performs the FEAT. An appropriate model suited to the data can tell the user exactly which part of the brain has reacted to the stimuli. The FEAT process stores various images in a directory that can help save time in case of a statistical rerun from repeating the preprocessing steps.

FAST is a tool to segment 3-D into brain tissue types such as gray, white, CSF, and correct field inhomogeneity. This tool is implemented using a Markov random field and an expectation-maximization algorithm. The output is in the form of probabilistic and volume tissue segmentation. In FEAT, the GLM method is used in the first-level analysis. It is known as FMRIB Linear Model (FILM). FILM is a non-parametric method used to improve estimation efficiency. Here, in the data, as there is a problem in accounting for problems in temporal autocorrelation, this results in errors in the statistical analysis and inefficiency. The process of removing the autocorrelation is known as whitening. In FILM, each voxel's time series is whitened to improve efficiency. Another significant tool is the Brain Extraction Tool (BET), which is used to delete non-brain tissue from the head and results in estimating skull surface [41].

6 Conclusion

This chapter attempted to capture the various factors of neurological disease with a particular emphasis on AD. The survey highlights the fundamentals of MRI, machine learning techniques with specific reference to deep neural networks, and various repositories of AD databases. It also stated the results of the latest developments in deep learning for the neuroimaging of AD. Despite the promises made by many studies, reliable application of deep learning for neuroimaging still remains in its infancy, and many challenges remain. The various limitations include training a complex classifier with a small dataset that leads to overfitting. Similarly, larger datasets from different centers are typically acquired differently using other scanners and protocols with varying image features, leading to poor performance. Another challenge is associated with memory and compute consumption when using CNNs with higher dimensional image data.

Additionally, data is a crucially important obstacle for deep neural networks, especially in medical data analysis. When deploying deep neural networks, one is instantly faced with challenges related to data access, privacy issues, and data protection. As privacy and data protection are often required when dealing with medical data, new techniques for training models without exposing the underlying training data to the model user are necessary. Another block for the successful incorporation of deep learning methods is workflow integration. It is possible to develop clever machine learning systems for clinical use that are practically useless for actual clinicians. Attempting to augment already established procedures necessitates knowledge of the entire workflow. Involving the end user in creating and evaluating systems can make this less of an issue and increase the end user's trust in the systems. The study suggests that there should be a more concerted effort in this field in case the desired results need to be achieved. Countries should come forward to share the data and promote research. Upstream applications to image quality and value improvement are just beginning to enter the consciousness of radiologists. They will significantly impact making imaging faster, safer, and more accessible for our patients. The access to bio-sensors and edge computing on wearable devices for monitoring disease, plus an ecosystem of machine learning and other computational medicine-based technologies, will transit to a new medical paradigm that is predictive, preventive, personalized, and participatory medicine.

References

1. Ol, O.: Canadian study of health and aging: study methods and prevalence of dementia. Can. Med. Assoc. J. **150**(6), 899–913 (1994)
2. Braak, H., Braak, E.: Neuropathological stageing of Alzheimer-related changes. Acta Neuropathol. **82**(4), 239–259 (1991)
3. Letham, B., Rudin, C., McCormick, T.H., Madigan, D.: Interpretable classifiers using rules and bayesian analysis: building a better stroke prediction model. Ann. Appl. Stat. **9**(3), 1350–1371 (2015)
4. Guillozet, A.L., Mesulam, M.M., Smiley, J.F., Mash, D.C.: Butyrylcholinesterase in the life cycle of amyloid plaques. Ann. Neurol.: Official J. Am. Neurol. Assoc. Child Neurol. Soc. **42**(6), 909–918 (1997)
5. Tanzi, R.E.: The genetics of Alzheimer disease. Cold Spring Harb. Perspect. Med. **2**(10), 1–11 (2012)
6. Munoz, D.G., Feldman, H.: Causes of Alzheimer's disease. Can. Med. Assoc. J. **162**(1), 65–72 (2000)
7. Drevets, W.C.: Neuroimaging studies of mood disorders. Biol. Psychiat. **48**(8), 813–829 (2000)
8. Blamire, A.M.: The technology of MRI-the next 10 years? Br. J. Radiol. **81**(968), 601–617 (2008)

9. Heeger, D.J., Ress, D.: What does fMRI tell us about neuronal activity? Nat. Rev. Neurosci. **3**(2), 142–151 (2002)
10. Kumari, N., Acharjya, D.P.: Data classification using rough set and bioinspired computing in healthcare applications-an extensive review. Multimedia Tools Appl. **82**(9), 13479–13505 (2023)
11. Acharjya, D.P., Ahmed, P.K.: A hybridized rough set and bat-inspired algorithm for knowledge inferencing in the diagnosis of chronic liver disease. Multimedia Tools Appl. **81**(10), 13489–13512 (2022)
12. Acharjya, D.P., Anitha, A.: A comparative study of statistical and rough computing models in predictive data analysis. Int. J. Ambient Comput. Intell. **8**(2), 32–51 (2017)
13. Acharjya, D.P., Rathi, R.: An extensive study of statistical, rough, and hybridized rough computing in bankruptcy prediction. Multimedia Tools Appl. **80**(28–29), 35387–35413 (2021)
14. Acharjya, D.P., Ahmed, N.S.S.: Tracing of online assaults in 5G networks using dominance based rough set and formal concept analysis. Peer-to-Peer Netw. Appl. **14**(1), 349–374 (2021)
15. Ahmed, N.S.S., Acharjya, D.P., Sanyal, S.: A framework for phishing attack identification using rough set and formal concept analysis. Int. J. Commun. Netw. Distrib. Syst. **18**(2), 186–212 (2017)
16. Feldkamp, L.A., Puskorius, G.V.: A signal processing framework based on dynamic neural networks with application to problems in adaptation, filtering, and classification. Proc. IEEE **86**(11), 2259–2277 (1998)
17. Petrosian, A.A., Prokhorov, D.V., Lajara-Nanson, W., Schiffer, R.B.: Recurrent neural network-based approach for early recognition of Alzheimer's disease in EEG. Clin. Neurophysiol. **112**(8), 1378–1387 (2001)
18. Sankari, Z., Adeli, H.: Probabilistic neural networks for diagnosis of Alzheimer's disease using conventional and wavelet coherence. J. Neurosci. Methods **197**(1), 165–170 (2011)
19. Duraisamy, B., Shanmugam, J.V., Annamalai, J.: Alzheimer disease detection from structural MR images using FCM based weighted probabilistic neural network. Brain Imaging Behav. **13**(1), 87–110 (2019)
20. Mathew, N.A., Vivek, R.S., Anurenjan, P.R.: Early diagnosis of Alzheimer's disease from MRI images using PNN. In: Proceedings of the IEEE International CET Conference on Control, Communication, and Computing, pp. 161–164 (2018)
21. Albawi, S., Mohammed, T.A., Al-Zawi, S.: Understanding of a convolutional neural network. In: Proceedings of the IEEE International Conference on Engineering and Technology, pp. 1–6 (2017)
22. Farooq, A., Anwar, S., Awais, M., & Rehman, S.: A deep CNN based multi-class classification of Alzheimer's disease using MRI. In: Proceedings of the IEEE International Conference on Imaging Systems and Techniques, pp. 1–6 (2017)
23. Feng, W., Halm-Lutterodt, N.V., Tang, H., Mecum, A., Mesregah, M.K., Ma, Y., Li, H., Zhang, F., Wu, Z., Yao, E., Guo, X.: Automated MRI-based deep learning model for detection of Alzheimer's disease process. Int. J. Neural Syst. **30**(06), 2050032 (2020)
24. Khagi, B., Lee, C.G., Kwon, G.R.: Alzheimer's disease classification from brain MRI based on transfer learning from CNN. In: Proceedings of the 11th IEEE Biomedical Engineering International Conference, pp. 1–4 (2018)
25. Wang, H.Z., Wang, G.B., Li, G.Q., Peng, J.C., Liu, Y.T.: Deep belief network based deterministic and probabilistic wind speed forecasting approach. Appl. Energy **182**, 80–93 (2016)
26. Faturrahman, M., Wasito, I., Hanifah, N., Mufidah, R.: Structural MRI classification for Alzheimer's disease detection using deep belief network. In: Proceedings of the 11th IEEE International Conference on Information and Communication Technology and System, pp. 37–42 (2017)
27. McKhann, G.M., Knopman, D.S., Chertkow, H., Hyman, B.T., Jack, C.R., Jr., Kawas, C.H., Klunk, W.E., Koroshetz, W.J., Manly, J.J., Mayeux, R., Mohs, R.C., Morris, J.C., Rossor, M.N., Scheltens, P., Carrillo, M.C., Thies, B., Weintraub, S., Phelps, C.H.: The diagnosis of dementia due to Alzheimer's disease: recommendations from the National Institute on Aging-Alzheimer's Association workgroups on diagnostic guidelines for Alzheimer's disease. Alzheimer's Dementia **7**(3), 263–269 (2011)

28. Dubois, B., Feldman, H.H., Jacova, C., Hampel, H., Molinuevo, J.L., Blennow, K., DeKosky, S.T., Gauthier, S., Selkoe, D., Bateman, R., Cappa, S., Crutch, S., Engelborghs, S., Frisoni, G.B., Fox, N.C., Galasko, D., Habert, M.O., Jicha, G.A., Nordberg, A., Pasquier, F., Rabinovici, G., Robert, P., Rowe, C., Salloway, S., Sarazin, M., Epelbaum, S., De Souza, L.C., Vellas, B., Visser, P.J., Schneider, L., Stern, Y., Scheltens, P., Cummings, J.L.: Advancing research diagnostic criteria for Alzheimer's disease: the IWG-2 criteria. Lancet Neurol. **13**(6), 614–629 (2014)
29. Noor, M.B.T., Zenia, N.Z., Kaiser, M.S., Mamun, S.A., Mahmud, M.: Application of deep learning in detecting neurological disorders from magnetic resonance images: a survey on the detection of Alzheimer's disease. Parkinson's disease and schizophrenia. Brain Inf. **7**(1), 1–21 (2020)
30. Petrosian, A.A., Prokhorov, D.V., Lajara-Nanson, W., Schiffer, R.B.: Recurrent neural network-based approach for early recognition of Alzheimer's disease in EEG. Clin. Neurophysiol. **112**(8), 1378–1387 (2001)
31. Malone, I.B., Cash, D., Ridgway, G.R., MacManus, D.G., Ourselin, S., Fox, N.C., Schott, J.M.: MIRIAD-Public release of a multiple time point Alzheimer's MR imaging dataset. Neuroimage **70**, 33–36 (2013)
32. Silva, I.R., Silva, G.S., de Souza, R.G., dos Santos, W.P., Fagundes, R.A.D.A.: Model based on deep feature extraction for diagnosis of Alzheimer's disease. In: Proceedings of the IEEE International Joint Conference on Neural Networks, pp. 1–7 (2019)
33. Imperatori, C., Fabbricatore, M., Innamorati, M., Farina, B., Quintiliani, M.I., Lamis, D.A., Mazzucchi, E., Contardi, A., Vollono, C., Marca, G.D.: Evaluation of EEG activity and EEG power spectra in the general and population of patients with eating disorders: an eLORETA study. Brain Behav. **9**(4), 703–716 (2015)
34. Guan, H., Wang, C., Tao, D.: MRI-based Alzheimer's disease prediction via distilling the knowledge in multi-modal data. Neuroimage **244**, 118586 (2021)
35. Li, Y., Haber, A., Preuss, C., John, C., Uyar, A., Yang, H.S., Logsdon, B.A., Philip, V., Karuturi, R.K.M., Carter, G.W.: Transfer learning trained convolutional neural networks identify novel MRI biomarkers of Alzheimer's disease progression. Alzheimer's Dementia: Diagn., Assess. Disease Monitor. **13**(1), e12140 (2021)
36. Litvak, V., Mattout, J., Kiebel, S., Phillips, C., Henson, R., Kilner, J., Barnes, G., Oostenve, R., Daunizeau, J., Flandin, G., Penny, W., Friston, K.: EEG and MEG data analysis in SPM8. Comput. Intell. Neurosci. **2011**, 852961 (2011)
37. Cox, R.W., Jesmanowicz, A.: Realtime 3D image registration for functional MRI. Magnet. Reson. Med.: Official J. Int. Soc. Magnet. Reson. Med. **42**(6), 1014–1018 (1999)
38. Tang, Y., Hojatkashani, C., Dinov, I.D., Sun, B., Fan, L., Lin, X., Qi, H., Hua, X., Liu, S., Toga, A.W.: The construction of a Chinese MRI brain atlas: a morphometric comparison study between Chinese and Caucasian cohorts. Neuroimage **51**(1), 33–41 (2010)
39. Goebel, R.: BrainVoyager-past, present, future. Neuroimage **62**(2), 748–756 (2012)
40. Subramanian, L., Hindle, J.V., Johnston, S., Roberts, M.V., Husain, M., Goebel, R., Linden, D.: Real-time functional magnetic resonance imaging neurofeedback for treatment of Parkinson's disease. J. Neurosci. **31**(45), 16309–16317 (2011)
41. Smith, S.M.: Fast robust automated brain extraction. Hum. Brain Mapp. **17**(3), 143–155 (2002)